Religious Environmental Activism

This volume explores how religious and spiritual actors engage for environmental protection. Climate change and sustainability are increasingly prominent topics among religious and spiritual groups. Different faith traditions have developed "green" theologies, launched environmental protection projects, and issued public statements on climate change. Against this background, academic scholarship has raised optimistic claims about the strong potentials of religions to address environmental challenges. Taking a critical stance with regard to these claims, in this volume shows that religious environmentalism is an embattled terrain. Tensions are an inherent part of religious environmentalism. These do not necessarily manifest themselves in open clashes between different parties but in different actions, views, theologies, ambivalences, misunderstandings, and sometimes mistrust. Keeping below the surface, these tensions can create effective barriers for religious environmentalism. The chapters examine how tensions are manifested and dealt with through a range of empirical case studies in various world regions. Covering different religious and spiritual traditions, they reflect on intradenominational, interdenominational, interreligious, and religious-societal tensions. Thereby, this volume sheds new light on the problems that religions face when they seek to take an active role in today's societal challenges.

Jens Koehrsen is associate professor at the University of Oslo and a senior researcher at the University of Basel.

Julia Blanc is a theologian working at the University of Passau, Germany, and wrote her dissertation on "Ökokatholizismus" (2017).

Fabian Huber studied sociology, religious studies, and political science. He conducts research using quantitative and qualitative methods on the topics of religion and sustainability as well as religion and the media.

Routledge Studies in Religion and Environment
Series editor: Evan Berry

The Routledge Studies in Religion and Environment book series explores religious encounters with environmental challenges and strives to capture the ecological dimensions of religious life with empirical and theoretical sophistication. Resisting the urge to concentrate exclusively on religious traditions, this series conceives the term "religion" broadly, seeking to include not only religious actors, institutions and theological traditions, but also lived spiritualities, indigenous cosmovisions, para-religious organizations, and socially enacted notions of the sacred. Environmental challenges are manifest in every part of the world, but the bearing of religious actors, ideas, and institutions on these challenges is variable. Accordingly, this series is ambivalent about whether and how religion matters with respect to environmental issues. We welcome scholarly contributions that chart the dynamic relationships between systems of human meaning-making and environmental processes at all scales, from the planetary to the parochial.

Church, Cosmovision and the Environment
Religion and Social Conflict in Contemporary Latin America
Edited by Evan Berry and Robert Albro

Religious Environmental Activism
Emerging Conflicts and Tensions in Earth Stewardship
Jens Koehrsen, Julia Blanc and Fabian Huber

For more information about this series, please visit: www.routledge.com/Routledge-Studies-in-Religion-and-Environment/book-series/RSRE

Religious Environmental Activism

Emerging Conflicts and Tensions in Earth Stewardship

Edited by Jens Koehrsen, Julia Blanc, and Fabian Huber

Routledge
Taylor & Francis Group
LONDON AND NEW YORK

First published 2023
by Routledge
4 Park Square, Milton Park, Abingdon, Oxon OX14 4RN

and by Routledge
605 Third Avenue, New York, NY 10158

Routledge is an imprint of the Taylor & Francis Group, an informa business

Published with the support of the Swiss National Science Foundation (SNSF).

British Library Cataloguing-in-Publication Data
A catalogue record for this book is available from the British Library

ISBN: 978-0-367-86253-4 (hbk)
ISBN: 978-1-032-39687-3 (pbk)
ISBN: 978-1-003-01796-7 (ebk)

DOI: 10.4324/9781003017967

Typeset in Sabon
by Apex CoVantage, LLC

Contents

Tables

Foreword

The starting point for this book project was a workshop held on 15–16 November 2018 and a conference that took place on 22–24 May 2019, both organized by the Center for Religion, Economy, and Politics at the University of Basel. These events brought scholars from different disciplinary backgrounds together who undertake empirical research on religious environmentalism. Though studying different faith traditions in unlike geographical settings and through the lenses of different academic perspectives, we could identify one common denominator in our case studies: tensions. This finding then became the focus for this book project. Since our initial meetings, the participants continued to develop their ideas for individual contributions, keeping our focus on tensions in mind while new authors joined our ranks.

The initial meetings and the publishing of this volume were facilitated by the support from various institutions and colleagues.

First of all, only the steady commitment of the authors present in this volume made this academic endeavor possible. Moreover, the input from colleagues and anonymous peer reviewers helped us to further elaborate our ideas and revise the individual contributions.

The organization of the meetings was supported by Anabel Da Pra, Vanessa Heiniger, Vera Schaffer, while the latter as well as Ann-Lea Buzzi, Julius Malin, and Marc Weingart facilitated the preparation of this volume.

We would also like to thank our colleagues from the Center for Religion, Economy, and Politics and the Faculty of Theology at the University of Basel for their cooperation in undertaking this project.

The initial workshop and conference as well as this book were part of the research project "Urban Green Religions" at the University of Basel. The Swiss National Science Foundation financed this project, the meetings as well as the open-access publication of this volume. Finally, several other institutions sponsored the meetings: University of Basel's Faculty of Theology, the Swiss Association for the Academic Study of Religion (Schweizerische Gesellschaft für Religionswissenschaft), and the Voluntary Academic Society Basel (Freiwillige Akademische Gesellschaft Basel).

We are grateful for this support and believe that the outcome of this project makes a substantial contribution to our understanding of religious environmentalism.

Jens Koehrsen, Fabian Huber, Julia Blanc
Basel, February 2022

Contributors

Julia Blanc is a theologian working at the University of Passau, Germany, and wrote her dissertation on "Ökokatholizismus" (2017).

Carrie B. Dohe is a postdoctoral fellow at the University of Toronto School of the Environment and author of Jung's *Wandering Archetype: Race and Religion in Analytical Psychology* (Routledge, 2016).

Katharina Glaab is an associate professor at the Norwegian University of Life Sciences and co-author of *The Role of Religion in Struggles for Global Justice* (2018).

Jiska Gojowczyk is a researcher at SÜDWIND – Institute for Economics and Ecumenism in Bonn, Germany. As a doctoral and postdoctoral researcher at the Max Planck Institute for the Study of Societies in Cologne, Germany, she investigated environmentalism in religious orders.

Alexandre Grandjean completed his PhD at the Institute of Social Sciences of Religion at the University of Lausanne, Switzerland. He specialized in the study of the religion/spirituality and sustainability nexus in rural regions.

Derk Harmannij received his PhD from the School of Geography at the University of Exeter (UK) in 2019. He currently works outside academia.

Lior Herman is an associate professor at the Department of International Relations at the Hebrew University of Jerusalem. He is the author of *Israel's Path to Europe* (Routledge, 2019).

Fabian Huber studied sociology, religious studies, and political science. He conducts research using quantitative and qualitative methods on the topics of religion and sustainability as well as religion and the media. A recent publication, together with Jens Koehrsen (2021): "A field perspective on sustainability transitions: The case of religious organizations." In: *Environmental Innovation and Societal Transitions* 40, 408–420.

Sofiah Jamil is a PhD candidate at the Australian National University. Prior to this, she was a researcher at the Centre for Non-Traditional

Security Studies, S. Rajaratnam School of International Studies, Nanyang Technological University, Singapore.

Jens Koehrsen is an associate professor at the University of Oslo and a senior researcher at the University of Basel.

David Krantz is a doctoral candidate and National Science Foundation IGERT-SUN fellow at the School of Sustainability at Arizona State University. He is the author of "COP and the Cloth: Quantitatively and Normatively Assessing Religious NGO Participation at the Conference of Parties to the United Nations Framework Convention on Climate Change."

Stéphanie Majerus is currently affiliated with psychiatric research at the University of Fribourg and working as a journalist in Luxembourg. She was employed as a research assistant at the University of Fribourg, where she also did her PhD, which will be published under the title "Ackerbau des Lebendigen; Tier-, Landbau- und Wissenschaftsverständnis der biodynamischen Landwirtschaft."

Christophe Monnot is an associate professor in Sociology of Religion at the University of Strasbourg. He is also a researcher at the University of Lausanne where he works on projects on the relationship between churches and ecology.

George C. Nche is a postdoctoral research fellow in the Department of Religion Studies, University of Johannesburg, South Africa. He is the author of *The religion-environment (climate change) connection: Evidence from Nigeria* (2020).

Juliane Stork is John S. Mbiti Research Fellow at the Research Programme on Religious Communities and Sustainable Development at Humboldt-Universität zu Berlin and is affiliated with University of Pretoria as visiting scholar.

Charel du Toit is a research associate of the Department of New Testament and Related Literature, Faculty of Theology and Religion, University of Pretoria and is partaking in a research project titled 'Jesus and the parables.'

1 Tensions in Religious Environmentalism[1]

Jens Koehrsen, Julia Blanc, and Fabian Huber

Religious environmentalism is becoming increasingly prominent. The world's major faith traditions have developed "green" theologies, launched environmental protection projects, issued public statements on climate change, and tried to sensitize their members for more environmentally friendly lifestyles. Examples of such environmental activities range from Buddhist recycling initiatives (Lee and Han 2015; Mohamad et al. 2012), green hajj guides for Muslim pilgrims to Mecca (Mangunjaya et al. 2015; Koehrsen 2021), the ecological vision of the "green" Patriarch Bartholomew of Constantinople (Bartholomew and Chryssavgis 2012; Theokritoff 2017), and inter-faith events on climate change (Interfaith Climate Summit 2008).

Those who suggest that religious environmentalism is becoming more prominent might also point to Pope Francis' encyclical *Laudato Si'* (2015). Urging the world to tackle the looming ecological crisis, the encyclical received a glowing media echo, was applauded by climate scientists, and interpreted by Christian environmentalists as a great sign of support within the Catholic Church. Researchers used this opportunity to point toward the potentials of religions to address environmental challenges (McKim 2020; Tucker and Grim 2016): Religions could make a difference given their massive number of followers, their public visibility and credibility, and their impact on the worldviews, values, and lifestyles of their adherents. Moreover, the encyclical supported those who had long before its publication argued that religions would become "greener" over time. Some researchers even suggested a "Francis effect," arguing that the encyclical would substantially increase the public concern about climate change (Maibach et al. 2015). However, the global impact of *Laudato Si'* on the climate activism of the Church and its followers has remained unclear (Woodworth 2020). Although the influential head of the Catholic Church assumed an explicit environmental position with this encyclical, many of its segments (e.g., national churches, Bishops, members) disagreed with its contents, questioned the relevance of this call, and only backed it to varying extent (Li et al. 2016; Landrum et al. 2017). To their disappointment, many Catholic environmental groups and organization units received neither more financial support nor a stronger

DOI: 10.4324/9781003017967-1

voice in their national churches. Rather than aligning the Catholic Church under a joint environmental program and generating public support for progressive climate policies, *Laudato Si'* has rendered frictions within the institution visible.

This example illustrates the tensions of religious environmentalism. It stands for the ongoing disagreements in many religious traditions over environmental engagement. While many scholars have stressed the strong potentials of religions to address environmental challenges and suggested a "greening" of religions, the tensions show that religious environmentalism is an embattled terrain where actors with different interests, backgrounds, and understandings of their traditions and the role of religion in modern societies compete for influence. There is neither a linear direction of greening nor a univocal impact on society but multifaceted arenas with different players and visions.

This volume studies these tensions in different religious and spiritual traditions. We argue that tensions are an inherent part of religious environmentalism. They usually do not manifest themselves in open conflict and clashes between different parties but in different views and theologies, ambivalences, misunderstandings, and sometimes mistrust. Keeping below the surface, these tensions can create effective barriers for religious environmentalism. Thereby, this volume sheds new light on the problems of the supposed "greening" of religions as well as on the limitations that religions face when they seek to address the societal challenge of environmental degradation.

The remainder of this introduction is structured as follows. The first part introduces the reader to the notion of the "greening" of religions and their strong potentials to address environmental challenges. The following section contrasts this optimistic picture by accentuating empirical insights that point to the challenges of religious environmentalism. These show that empirical evidence does not support the assumed "greening" of religions and that religious practices and worldviews continue to harm the environment. Bringing both perspectives together, the next section discusses the tensions of religious environmentalism. It portrays different types of tensions and provides an overview of the individual chapters in this volume. Finally, the outlook summarizes the main insights of this volume, relates them to broader debates on religious innovations and the role of religions in societal challenges, and suggests avenues for future research.

Green Religious Optimism

The academic debate on religion and ecology has been the focal point for endeavors to study religious environmentalism (Berry 2013). An important starting point for this debate was Lynn White's (1967) seminal article in the journal *Science*. In this contribution, White argued that Western Christianity with its anthropocentrism and dominion over nature caused the environmental crisis. Interestingly, at the end of the article, he claimed that religions need to be part of the solution for the crisis: we need to develop new religions or modify the existing ones to respond to the environmental crisis.

White's contribution shaped the "greening" of Christianity (Whitney 2017). In response to his criticism, Christian theologians as well theologians from other faith traditions have sought to provide ecological reinterpretations of their traditions and to generate faith-based environmental ethics that address the environmental crisis (e.g., Tucker 2006, 2008; Gottlieb 2006; Gerten and Bergmann 2012; Blanc 2017; Veldman et al. 2014; Brockopp 2012; Foltz 2006; Saniotis 2012; Abdelzaher et al. 2019; Dessi 2013; Harris 1995; Binay and Khorchide 2019; Boff 2011; Hessel and Ruether 2000; Foltz 2003; Creegan and Shepherd 2018; Crockett and Robbins 2012). Scholars from the religion and ecology debate have interpreted these endeavors as leading to a "greening" of religions, thus suggesting that religions become more environmentally friendly over time, and, on many occasions, actively engaged in these greening efforts (Chaplin 2016; Tucker 2006, 2008). Moreover, with regard to the United States, some authors have suggested that existing environmentalism and secular endeavors to protect the environment can often be traced back to religious and, in particular, Christian values (Berry 2015; Stoll 2015).

At the same time, contributions from the religion and ecology debate have outlined the potentials of religions to address environmental problems (Gardner 2003; Gottlieb 2010; Peterson 2007). These potentials include the ability to reach broad population segments given that more than 80% of the global population are part of a religious tradition (Pew Forum 2015). Proponents of this view have suggested that the worldviews and ethical teaching of religions shape the lifestyles of their adherents and their relationships to the natural environment (Mangunjaya and McKay 2012; Sheikh 2006; Jenkins 2009; Watson and Kochore 2012). Moreover, religious leaders and umbrella organizations often enjoy a high public credibility and sometimes have close relationships with societal decision-makers (e.g., politicians, business elites) (Casanova 1994; Davie 2010). They can arguably draw on their networks and prominence to impact public debates, create awareness for environmental problems, and influence decision-making processes (Reder 2012; Schaefer 2016; Wardekker et al. 2009). Furthermore, authors underlining the potentials of religions have pointed to vast financial resources and infrastructures of religious organizations (e.g., buildings, teaching facilities), arguing that they can employ these to support societal transformations toward environmental sustainability (Gardner 2002, 2003; Palmer 2013; Blanc and Ostheimer 2019; Ostheimer and Blanc 2021).

Finally, apart from traditional forms of religion, scholarship has suggested the evolution of new spiritualities that go hand in hand with environmental protection. Though these tend to be less visible than established faith traditions (Luckmann 1967), they may be equally or even more effective in promoting care for the environment (Taylor 2004; Naess 1990; Hedlund-de Witt 2012, 2013; Sponsel 2012). For instance, Bron Taylor (2010) has suggested the emergence of a new type of religion that he calls "Dark Green Religion." It considers the natural environment as sacred, involves feelings of connectedness to nature, and promotes an ethics of reverence for nature. While Taylor assumes that this new eco-spirituality is already an essential part of the environmental milieu, others have suggested that the latter still tends to marginalize spirituality (Koehrsen 2018a) but increasingly becomes "spiritualized" through light forms of popular spirituality (Becci and Monnot 2016; Stacey 2021; Becci et al. 2021, Nita 2014).

While the relationship between ecology and religion has become an increasingly prominent topic in religious studies, environmental studies and climate change research have more recently started to regard religions increasingly as an asset for addressing environmental challenges (Hulme 2017; Haluza-DeLay 2014; Jenkins et al. 2018; Clingerman and O'Brien 2017; Edenhofer et al. 2015; Allison 2015; Smith and Leiserowitz 2013; Kilburn 2014; Murphy et al. 2016; Koehrsen 2018b, 2021). Both – environmental studies/climate change research as well as the religion and ecology debate – agree that religions can make a significant difference in addressing environmental challenges such as climate change.

At the same time, we can witness an increase of environmental activism in different faith traditions. This becomes prominently illustrated by the aforementioned encyclical *Laudato Si'* but also by the *Islamic Declaration on Global Climate Change* or the *Interfaith Climate Change Statement to World Leaders*, and many other public statements issued by religious leaders, umbrella organizations, and international religious networks. These statements appear as evidence for the "greening of religions." They indicate a rising interest in environmental issues among religious leaders, umbrella organizations, and networks. However, it remains unclear what actions follow from such statements and whether local faith communities and their adherents become "greener" to the same extent. Moreover, the dynamics behind issuing pro-environmental declarations often remain uncertain. Rather than expressing the leadership's environmental concern, on some occasions, they may be the outcome of short-term adaptation strategies to public environmental discourses or to the interests of specific religious or political constituencies (Koehrsen and Huber 2021). As such, there is a need to empirically address the local dynamics of the supposed "greening."

Therefore, this edited volume centers on empirical research and presents a selection of in-depth case studies on religious environmentalism in different faith traditions and spiritualities. Drawing on empirical research, the contributions do not necessarily share the optimism of many theoretical contributions to the religion and ecology debate. Despite increasing environmental activities among religious leaders and networks, there appears to be no linear "greening" process.

Challenges for the Religious Greening

Contrasting the optimistic pictures painted by sections of the religion and ecology debate, social science research has pointed to different challenges of religious environmentalism. These relate (a) to the absence of the supposed "greening of religions" and (b) to religious practices and worldviews that cause environmental harm.

Multiple studies have explored the assumed "greening of religions." These mostly quantitative analyses about environmental attitudes among adherents of specific faith communities show no clear evidence for a "greening" (Carlisle and Clark 2018; Clements et al. 2014; Konisky 2018; Dilmaghani 2018). Accordingly, in a review of existing research on the greening of Christianity in the United States, Bron Taylor et al. (Taylor et al. 2016a, 2016b) conclude that empirical research does not support the "greening" hypothesis. Even if green theologies have emerged and spread in different faith traditions, religious organizations and actors respond in different ways to these (Jenkins et al. 2018). Consequently, there is no encompassing and straight development path toward a "greening" in religions (Hulme 2016; Haluza-DeLay 2014).

Additionally, within single religions and religious organizations, we can find different tendencies. Qualitative research shows that even when religious head offices decide on "greening" strategies and undertake programs to improve the environmental sustainability of the given organization, congregations and branches do not necessarily implement these activities (Amri 2014; Vaidyanathan et al. 2018; Torabi and Noori 2019; Koehrsen et al. 2015, 2020; Koehrsen and Huber 2021). "Greening" programs face resistance in religious organizations. For instance, in Switzerland, religious umbrella organizations have set up measures to improve the environmental sustainability of their local congregations (Koehrsen and Huber 2021). These measures include, among others, using new environmentally friendly cleaning products in church buildings and lowering the heating temperature in churches to save energy. However, in several cases, local congregations resist implementing these measures and sometimes even enter into conflict with their umbrella organizations. This volume adds to this line of research by unfolding the tensions that produce such resistance.

Apart from the aforementioned problems in the greening, studies have also found that religious practices can involve environmental harm, as specific rituals produce negative ecological impacts (Wexler 2016). Prominent examples are the carbon emissions and waste created by religious pilgrims worldwide. Moreover, religious views can discourage adherents from actively addressing environmental problems. These views can involve (a) skepticism about the existence of environmental problems, (b) the perception of the problems as a welcomed end-of-times, or (c) as divine punishments for human sins.

In terms of skepticism about environmental problems, a prominent topic is climate change denial, reported to be prevalent among sections of evangelicals (Zaleha and Szasz 2015; Veldman 2019; Carr et al. 2012; Hayhoe et al. 2019). We can, however, sometimes find similar views among Hindus, Jews, and Muslims. For instance, some Muslim groups regard "climate change" as "poisonous knowledge from the West" (Khan 2014) or as a Western conspiracy to weaken Muslim-majority countries (Yildirim 2016). Such skeptical views may often be coupled with critical attitudes toward science and a prioritization of human needs over those of nature (Veldman et al. 2021). Economic arguments may add to this perspective, stipulating that pro-environmental policies (e.g., policies directed against fossil fuels) negatively limit economic development and affect the poor (Kearns 2014). Specific demographic factors, such as race or class, appear to facilitate religious skepticism about environmental problems. For instance, in the United States, white evangelicals tend to show a stronger inclination toward skepticism than evangelicals from other racial backgrounds (Veldman et al. 2021).

Apart from skepticism about climate change, its acceptance as a sign of the looming end-of-times can constitute a hindrance for climate engagement. While the academic debate may have exaggerated this argument in the United States, as Veldman shows (2019), in other world regions such as Sub-Saharan Africa, the acceptance of the coming end-of-times may play a role for local populations (Nche 2020; Jooste et al. 2018; Makame and Shackleton 2019; Barker and Bearce 2013; Artur and Hilhorst 2012). In this view, humans are not responsible for climate change given that it is the fulfilment of an end-of-time prophecy. From this perspective, believers may welcome and embrace climate change.

Finally, a third interpretation that does not encourage direct environmental action perceives ecological problems as a divine punishment for human sins. In this perspective, God responds with different forms of environmental degradation or disasters to the sinful behavior of political leaders (e.g., corruption) or local populations (e.g., stealing, lying). Research has reported this view, for instance, for different regions of Sub-Saharan Africa (Bell 2014; Shehu and Molyneux-Hodgson 2014; Abegunde 2017; Haron 2017; Watson and Kochore 2012).

In total, apart from the optimistic outlooks, empirical research on religion and ecology has pointed to the challenges of religious environmentalism. These challenges relate to the tensions that this volume focuses on. By drawing on qualitative empirical research, the contributions to this volume explore the tensions inherent in religious environmentalism and thereby contrast the sometimes romantic views on religious earth stewardship.

Tensions in Religious Environmentalism: Toward a New Research Perspective

Research on religious environmentalism has tended to explore how religion matches or mismatches environmental protection. In so doing, it often suggests a forthright relationship between religion and the environment, leading to environmental protection (Gardner 2003; Tucker 2006, 2008), to a neglect of the environment (Taylor et al. 2016b), or even environmental degradation (White, Lynn, Jr. 1967; Barker and Bearce 2013; Amery 1974). However, to understand the multifaceted ways in which religions position themselves vis-a-vis the environment and the ambivalences underlying their positions, there is a need to study the tensions inherent in religious environmentalism.

This volume suggests that tensions are part of religious environmentalism. Religious environmentalism implies innovation processes within religions as well as in the broader societies in which religions are embedded. These innovation processes entail different visions of the envisioned change process and, therefore, lead to tensions. As such, the tensions are not a product of the given religious traditions but of opposing actors, interests, visions, and structures in these traditions, or of oppositions between different traditions and their societal environments. By undertaking empirical research on tensions, we can grasp these oppositions and the negotiation processes that take place around the supposed "greening" of religions. In this way, we gain insights into the ongoing "(non)greening" processes in different types of religious communities and at different organizational levels (e.g., national umbrella organizations, local congregations).

By pointing to the tensions, this volume aims to generate a more nuanced picture of religious environmentalism. Religious environmentalism is neither a linear nor a smooth process that finally leads to the "greening" of a given tradition and its social environment. Paralleling other types of environmental engagement, religious environmentalism faces resistance, limitations, and conflicts while seeking to change the existing order. Sometimes, said tensions can be located in the fabric of religious organizations, in the form of countermovements, rivaling theological schools, or simply a lack of financial resources, obstructing internal transformations and tearing to pieces the prospects of a "greening" process. In other cases, rivalry between different religious traditions, resistance of nonreligious actors, the obduracy of secular institutions, or the marginalization of religion can thwart the impact of religious environmentalism.

This volume focuses on four types of tensions: (1) intradenominational tensions, (2) interdenominational tensions, (3) interreligious tensions, and (4) religious-societal tensions.

Intradenominational Tensions: Struggles and Institutional Barriers Within Communities

Intradenominational tensions occur within a specific religious denomination or branch/group. These can become, for instance, visible as frictions between ethical teachings and deeds, between different understandings of the teachings and the environmental problems within the same religious community, or between the environmental ambitions of some members and the hierarchal structures of the given religious organization. Often, such oppositions are overseen in presenting, for example, "the" Christian or "the" Muslim view on ecology. Small-scale findings on internal oppositions are precious for understanding the broader dynamics of religious environmentalism. Not only do they show that frictions emerge even within small and theoretically homogeneous groups, but that they are also a fundamental part of religious environmentalism.

Jiska Gojowczyk's contribution in this volume provides a first example of such intradenominational tensions. Her study of Roman Catholic orders reveals different positions regarding environmental commitment. Actors within the orders agree on the need to preserve God's creation. Yet, they disagree in their interpretation of this goal and the strategies needed to achieve this. Some prioritize a scientific approach; others emphasize social work or forward a spiritual approach. In particular, they disagree on whether they should direct environmental actions inward or outward. Studying Friars Minor and Jesuits in Germany as well as in the Philippines shows that context matters. However, in this case, the positions the different actors take with regard to Catholic environmentalism do not necessarily depend on the given order or on its geographical place but on the specific professional roles the actors occupy and assume in the orders and their background. Unity as a Roman Catholic order does not imply holding a unified environmental perspective. Environmental engagement as well as the perception of nature is never independent from the context in which actors move. In this respect, different contexts are likely to lead to different views and thus create tensions within a denomination.

Nevertheless, also within the same context, tensions within a denomination or spiritual tradition can be observed, such as those between conservative and progressive currents. In her contribution, Stéphanie Majerus examines biodynamic agriculture and its roots in anthroposophy. Based on ethnographic field research and interviews, she shows how tensions arise between two types of biodynamic farmers. In the tradition of Rudolph Steiner, the older, strongly anthroposophical farmers hold anthropocentric views, placing human beings at the center of their biodynamic activities. By contrast, novices question this anthropocentrism and orient themselves toward other eco-spiritual movements that share elements of deep ecology or Dark Green Religion. This contribution thus outlines tensions that are frequent in religions: those between tradition and innovation. In this specific case, newcomers question the tradition and introduce innovations regarded as critical by the established biodynamic farmers.

Finally, intradenominational tensions can also occur when individual ambitions face organizational barriers. This is the case when members of a religious organization want to push environmentalism forward but face important institutional obstacles within their organization. In his article, Christophe Monnot describes cases of two congregations strongly committed to environmental activism: a Catholic parish from the German-speaking part of Switzerland and a reformed congregation from the French-speaking part. His contribution explores the organizational challenges that the congregations face in their endeavor to "green" their denominations. Drawing upon a neo-institutional perspective as applied by Chaves, Monnot identifies three levels of pressure needed for a "greening": internal pressure, institutional pressure, and pressure from society. A successful "greening" process requires committed members, organizational structures that allow for changes, and a society that promotes environmentalism. If these conditions are not given, the religious greening processes will face difficulties and only slowly take shape, as in the case of Switzerland.

The three contributions in this first part address different forms of intra-denominational tensions. These show that religious environmentalism is always embedded in a particular context that includes the region, the culture, and the specific professional background of the given actor. Rather than the particular religious tradition, the context appears to inform the environmental views of actors and their environmental engagement. Moreover, tensions emerge between older traditional and younger innovation-oriented members as well as between individual greening efforts and the institutional structures of the given organization. As such, the power positions of traditional members combined with the hierarchical structures of many religious organizations can generate massive barriers for environmental transformations.

Interdenominational Tensions: Dissonance Between Communities From the Same Faith Background

Interdenominational tensions are those that unfold between different denominations or branches of the same religion. Different branches and communities that belong to the same religious tradition may pursue different ecological agendas and disagree on their understandings of environmental challenges such as climate change and promote divergent responses to it. A prominent example is the smoldering conflict between different camps of US evangelicals with regard to the government's climate policy (Carr et al. 2012; Wardekker et al. 2009; Nagle 2008; Veldman 2019; McCammack 2007; Kearns 1996, 1997, 2014).

However, as Sofiah Jamil's contribution to this volume shows, such tensions may also become apparent in other religious traditions such as Islam. Studying the example of halal wastewater recycling in Indonesia, Jamil unfolds the different positions that the three biggest Muslim organizations in this country assume with regard to the re-use of wastewater. Thereby, these organizations address the question whether and under what conditions water recycling is halal. Whereas the *Majlis Ulama Indonesia* (a state-supported umbrella organization of Islamic scholars and clerics) and *Muhammadiyah* (the representatives of a "modernist" position) position themselves in favor of re-use, *Nahdahtul Ulama* (the "traditionalist" camp with strong influence in rural areas) opposes the recycling of wastewater and its re-use for ritual ablution or consumption. The contribution illustrates that there is not "one Islamic" view on water recycling. Different currents within Islam will favor distinct visions on environmental issues.

Disparities between theology and practice or between ideas provided by a religious tradition and their implementation are another form of interdenominational tension. In her contribution, Julia Blanc addresses Christian eco-theology and its practical implementation in local churches. Drawing on interviews with representatives of Catholic and Protestant congregations in Germany and Switzerland, Blanc shows that there is a tension between Christian ethical requirements and their implementation, as the environmental commitment at the local level is rather low. Interestingly, the interviewees only rarely refer to eco-theological reasoning when discussing religious environmental commitment. As such, the article points to a major gap between Christian eco-theology and the reasoning in local congregations, producing problems in the Christian greening.

Finally, tensions arise when religious groups of a given tradition hold divergent environmental views. The contribution by George C. Nche presents cases of dissenting voices on climate change among Christians in Nigeria. Nche focuses on the voices that disagree with views on climate change prevalent among Christian leaders in Nigeria. These dissenting voices question the anthropogenic dimensions of climate change and become manifest in two different views: (a) humans as not influencing climate change and (b) a Christian focus on soul winning and Parousia. Both views lead to limitations in addressing environmental issues. While the first view limits the human potential to resolve the problems, the second one restricts the role of religions in society.

These three contributions show how tensions within a tradition can become manifest. The findings indicate a broad field for intra-religious tensions involving a vast range of potential positions. Religious environmentalism often depends on the interpretation of the given tradition. As religious organizations and leaders from different denominations are likely to interpret the tradition in different ways, they may also hold different views on the environmental implications of the given tradition. Moreover, the tensions relate to general questions around the role of religion in society: should religions only focus on their core religious functions (e.g., providing salvation "goods") or also address societal challenges?

Interreligious Tensions: Competition and Boundary-Drawing Between Different Faith Traditions

Religious environmentalism has been regarded as a suitable topic for interreligious dialogue and, therefore, for bringing actors from different traditions together to establish joint grounds for enduring interfaith partnerships. Nevertheless, this can also bring prejudices against each other to the surface and, therefore, involve tensions. Representatives of different religious traditions may hold different views on environmental problems and the role of religions in addressing them, thereby potentially creating *interreligious tensions*. If they agree on the general importance of the subject, they may compete over the authenticity of their environmental engagement. They aim to become pioneering "green" religions in their given societies, as this can lead to an advantage in their public perception and social legitimacy.

Drawing on the sociological theories of Max Weber and Pierre Bourdieu, Fabian Huber conceives environmental engagement as a battlefield between established religious actors (Reformed and Catholic Christians) and non-established ones (other religious groups) in Switzerland. The established actors show a much closer link to the political mainstream, enjoy higher societal legitimacy due to their legal recognition, and have access to more financial resources than the others do. While representatives from the first group see environmental engagement as a part of their engagement for society, newcomers find it more difficult to engage in environmental issues. Religious environmental commitment can generate competitive advantages. Tensions between the religious actors become manifest as competition for such advantages within the religious field.

While the tensions between religions remain latent in the aforementioned example, they might also manifest themselves in settings where religious actors directly collaborate, namely, in interreligious cooperation. This is demonstrated by Carrie B. Dohe's investigation of the "Religious Week of Nature Conservation 2017." As part of an interreligious consultation, religious prejudices became apparent in the process. These tensions were foremost interreligious, as many participants were reserved vis-à-vis the Muslim partners, questioning the credibility of their environmental engagement. Moreover, religious-societal tensions emerged because atheistic participants were generally critical of religions and struggled to overcome these "reservations" in the light of nature protection.

Interestingly, interreligious tensions can even become manifest within a single religious community. This is the case when the given community draws its concepts and practices from different religious traditions. Juliane Stork and Charel du Toit illustrate this by studying the emerging "greening" of the biggest African Initiated Church in South Africa: the Zion Christian Church (ZCC). They analyze a speech on ecological sustainability given by Bishop Lekganyane, the leader of this church. Their work reveals that the speech negotiates the tension between the African Initiated Churches and African Traditional Religions. On the one hand, it draws on concepts from African Traditional Religions; on the other hand, it aims to distance itself from these. The ZCC seems entangled in this tension between African Traditional Religions and Christian teachings. In this way, environmental commitment ultimately reveals itself as a negotiation process that involves balancing between different concepts and the drawing of symbolic boundaries to demarcate the unique identity of the own community.

These three examples reveal that environmentalism can involve various forms of tension between religions. Interestingly, these do not arise from environmental commitment itself but often relate to the competition between religious organizations over societal legitimacy, religious authenticity, and access to resources.

Religious-Societal Tensions: Oppositions and Negotiations Between Religion and Society

Historically, societal differentiation has led to the evolution of relatively autonomous social spheres in modern societies, involving also a partial separation between religion and other social domains (e.g., politics, science). Inhabiting different social worlds, religions and their societal environment can mutually influence each other. But it depends on the boundaries between both whether and to what extent religions can participate in addressing societal challenges such as environmental degradation. *Religious-societal tensions* refer to frictions between religion(s) and their given societal environment. Religious groups and societal actors (e.g., secular environmental groups) may struggle with each other over the shaping of climate policies. At the same time, societies may create barriers that inhibit the engagement of religious actors in environmental affairs (e.g., exclusion of religious actors or religious reasoning from specific institutions), while secular environmental groups may stigmatize specific religious groups or religion in general.

A first example of this form of tension relates to the Jewish concept of "kosher," which is known to go beyond dietary regulations. However, it may surprise some that it also touches electricity. In his contribution to this volume, Lior Herman addresses kosher electricity by exploring the use of different electricity sources on Shabbat and other Jewish holidays in a rising number of Israeli ultra-orthodox communities. Based on their religious interpretations, these communities regard the work of Jewish people during Shabbat and other Jewish holidays as unkosher (forbidden), including also work in the production of electricity. For this reason, traditionally, they tend to use diesel generators and batteries during these days. However, this use causes several tensions. The majority of the population, consisting of secular and non-orthodox Jews, does not consider the production of electricity during Shabbat and Jewish as unkosher. Moreover, the state has prohibited the use of generators given that it is dangerous and that they pollute the environment. Parts of the civil society (including ultra-orthodox environmentalists) engage against environmentally unsustainable solutions for the generation of kosher electricity (e.g., legalization of diesel generators). However, most ultra-orthodox communities are against renewable energies because they are expensive and difficult to implement. The case of kosher electricity shows that religious interpretations and lifestyles can have direct implications for societal transformation efforts such as the energy transition and involve multiple tensions between different types of actors (e.g., state, environmental groups, ultra-orthodox communities).

The weighting of spirituality and sustainability also leads to tensions in the chapter by Alexandre Grandjean where they manifest themselves among biodynamic wine-crafters that seek to distinguish themselves from broader society. Grandjean investigates the spiritual dimension of biodynamic wine crafting. Interviewees from this field underline the importance and significance of individual experience and relation to their soil and plants. Nevertheless, there are quarrels around two subjects: first, the distinction from "normal" sustainable agriculture and, second, the question what else should be taken into account in biodynamic wine crafting beyond Rudolf Steiner's cosmological legacy. This case sheds light on the importance of individual experiences and feelings (as lived and promulgated by the interviewees) in religious environmentalism.

Similarly, Katharina Glaab's contribution addresses the question of how to stick to personal convictions in a context that does not share them. Contrasting the aforementioned study, she tackles this question for a different setting: the United Nations Framework Convention of Climate Change (UNFCCC) where she examines the role of Faith-Based-Actors (FBAs). Drawing on Bourdieu, she conceives the UNFCCC as a social field in which a secular logic prevails. Based on participatory observation and semi-structured interviews with FBAs, the chapter explores the question to what extent religious actors can become involved in this field. Engaging in a secular field generates tensions for religious actors. These employ two different strategies to deal with these tensions: (a) adaptation to the secular field by concealing the religious identity or (b) demarcation of the religious background. However, the choice of the strategy ultimately also depends on the concrete context: as such, FBAs tend to downplay the religious background in meetings with secular actors and emphasize their religiosity in interfaith meetings. Concrete tensions do not occur between the distinction of the religious and the secular as such but arise in the interrelation between their respective understandings and can also arise between FBAs. Religious environmentalism can manifest itself in different ways, but the question is to what extent it can maintain its religious identity.

Similarly, David Krantz's contribution deals with tensions at the UNFCCC. Using literature, participant observation, and qualitative interviews at COP 23 (2017 in Bonn), Krantz distinguishes various functions and goals that faith actors pursue. These include, for example, networking, lobbying, and education. He observes tensions in most of the functions. A few also turn out to be innerreligious tensions, while three concern tensions between religious groups and society. At COP 23, there were different views on whether climate change should be discussed together with its consequences (e.g., migration), how to present oneself in public as religious groups (whether to communicate secularly or not), or how to deal with meat consumption at the COP. The article illustrates that different challenges arise when different religious actors and secular actors come together in an international public setting. Although they may have similar environmental goals, negotiating about their implementation leads to diverse forms of tension.

While the previous contributions found tensions between existing actors in specific settings, Derk Harmannij's contribution addresses the general tension between societal expectations vis-à-vis religion and the activities of religious actors. Drawing on data collected in focus groups, the contribution examines such tensions by studying how Christian churches in the United Kingdom engage with environmental issues. He starts from the premise that religious actors are important for addressing environmental problems: scientists, the media, activists, and faith leaders emphasize the importance of religion. However, the results of his empirical fieldwork show that churches are struggling to engage with the environment, and that their engagement is often limited to the promotion of individual lifestyle changes (i.e., reductions in private consumption behavior). In this respect, congregations appear unable to meet the high societal expectations. While churches can engage green theologies and encourage followers to adopt greener lifestyles, religion alone cannot render people environmentally friendly.

Tensions between religion and society can emerge in many different ways. The contributions referred to tensions between religious and state actors, the drawing of distinctions toward a secular society, the religious engagement in secular settings such as the UN, and the gap between high societal expectations and low ecological engagement of religious actors. Religious actors sometimes grapple with the strong societal expectations in terms of their engagement. However, even when they participate in the societal struggles for more environmental sustainability, they may face secular barriers and feel the need to adapt to the nonreligious working principles of their societal environment.

Outlook: The Tensions of Religious Transformation Endeavors

This volume explores the tensions inherent in religious environmentalism. The tensions indicate that there is no linear and smooth religious "greening" process. Religious environmentalism is not as straightforward as one may suppose (or hope). The contributions illustrate the messiness of religious environmentalism in action. Even if religious leaders publically voice a strong environmental commitment, putting this commitment into practice remains a messy and embattled process, as various types of actors, interests, and social structures interact with each other. We can find "greening" tendencies that struggle against "ungreen" oppositions as well as among each other. Therefore, this volume sheds critical light on optimistic perspectives in the religion and ecology debate that stress global greening processes.

Studying the tensions of religious environmentalism helps to explain why the supposed "greening" faces barriers and diffuses only slowly within religious traditions or is not taking effect at all. These can be institutional barriers in the given organization (Monnot 2022) or related to the power of traditionalist camps (Majerus 2022; Jamil 2022). However, even when members of a religious tradition agree on the importance of addressing environmental issues, tensions may become manifest in different ecological interpretations of the given tradition and lead to diverging practical approaches (Gojowczyk 2022; Blanc 2022; Nche 2022). Apart from internal ones, tensions may also evolve between different religious communities as well as between religions and their broader societal environment, as this volume illustrates. Religious communities may compete over societal influence (Huber 2022) and addressing environmental issues in the most authentic way (Dohe 2022; Stork and du Toit 2022). At the same time, the societal environment may create barriers for the successful engagement of religious communities in public debates. For instance, it may establish mostly secular negotiation arenas that tend to exclude religious reasoning (Glaab 2022; Krantz 2022). But not only "secular" society may seek to ban religion from its public arenas. Religious actors, too, may pursue distancing themselves from mainstream society (Herman 2022; Grandjean 2022). Yet even when aligning with mainline society, they may find it difficult to fulfill societal expectations with regard to their environmental engagement (Harmannij 2022).

Despite these divergences and contentions, visible conflicts over religious environmentalism barely emerge. Tensions become manifest in different positions, ambivalences, misunderstandings, and sometimes mistrust. They lead to institutional barriers and challenges in greening endeavors as well as in the ambitions of religious communities to participate in societal transitions toward sustainability, but they hardly erupt into visible clashes. However, keeping below the surface, tensions even more become effective barriers for transformations. Being maintained invisible, they can hardly be addressed in a direct manner and create frustration rather than enthusiasm for change.

From a frame alignment perspective (Snow et al. 1986, 2014), we can perceive the dynamics within religious traditions as failing alignment processes between diverging schemes of interpretation (frames). Different sections of the same tradition embrace different interpretive frames of environmental problems and measures. We can witness diverging frames, for instance, among biodynamic farmers with some supporting anthropocentric views and others favoring eco-spiritual views (Majerus 2022), or among actors working in different professional roles in Catholic orders where some embrace a scientific approach and others promote a social work approach (Gojowczyk 2022). When religious frames are unable to connect to each other (e.g., by bridging their differences), there will be no unified mobilization for specific environmental measures. For instance, an extensive eco-theological tradition has evolved within progressive Christianity. However, local congregations in Switzerland and Germany frequently find it difficult to connect to this eco-theological frame and to put it into practice (Blanc 2022). Similarly,

Muslim environmentalists struggle with linking Islamic environmentalism with the views of broader Muslim communities, as research in the United Kingdom and elsewhere shows (Hancock 2015, 2018; Gilliat-Ray and Bryant 2011; Nita 2014; Koehrsen 2021; Gade 2019). As eco-religious frames fail to link with the interpretative frames of wider sections of the given tradition, they do not succeed in generating a broad religious mobilization for environmental action.

Scholars of religion need to address these tensions to understand the challenges of religious environmentalism. This volume takes a step in this direction. Future research can add to this by generating broader quantitative insights that measure the diffusion of specific "green" and "ungreen" theologies in different traditions and communities. This also includes a closer understanding of what is meant in the given case by "green" and who engages with specific "un/green" theologies, taking into account backgrounds of class, race, and gender (Baugh 2020; Veldman et al. 2021). Moreover, there is a need to strengthen research endeavors about the role of religions in environmental sustainability and climate change in the Global South. Given the importance of religion for the lifestyles and worldviews of broad population segments in many Global South societies, it is here where religion can make a fundamental difference. To get a better understanding of effects of religions on the environment in these regions, the academic study of religion may engage in collaborations with environmental studies and climate change research.

This volume illustrates how change is a constant topic of religions, be it internal or external societal change. Exploring the greening processes in religions and the role of religions in environmental challenges, it addresses two general topics in the study of religion: (a) innovation processes in religions and (b) the role of religions in societal transformation processes.

The contributions in this volume point to the challenges of religious innovation processes (Koehrsen 2019; Nagel 2018; Finke 2004). Ambitions to transform religions go hand in hand with the emergence of different visions with regard to the intended change. The diverging visions imply tensions that will affect the innovation process. The insights from this volume may also be applicable to other religious innovation processes – such as the evolution of European Islam, the rising digitalization of religions, or the diffusion of prosperity gospel within Pentecostalism and, more broadly, in Evangelicalism – by pointing to the tensions inherent in these transformation activities. Religious innovations face barriers, become negotiated and modified, and finally divert from the originally intended pathway.

Furthermore, the volume also examines how the relationship between society and religion becomes negotiated in the face of societal challenges, becoming visible in debates around "post-secularity" (Habermas 2008; Baker and Beaumont 2011; Casanova 2006; Berger 1999). It opens up the question of what position religious groups should assume in their societies: Should they address societal problems and contribute to societal transformation endeavors? Should they have a voice in debates around sustainability? These questions need to be addressed within the very religious communities as well as by their surrounding societies. On both sides, we can find different opinions as well as negotiation processes. Global North societies sometimes generate considerable hindrances for the participation of religions in societal negotiations about the future development of their societies. By contrast, in other contexts, religious groups may become key stakeholders for societal transformations that can potentially block or promote the intended transition process. Religious and societal actors can contest existing secular-religious boundaries, pursuing to shift, dissolve, or strengthen them (Berry 2014). These processes can become visible when religious actors seek to engage in societal debates on sustainability or when societal actors strive to exploit the potentials of religions for sustainability transitions. In both cases, existing secular-religious boundaries are likely to structure the engagement of religions in these societal transformation efforts. However, their involvement may also lead to negotiations about the role of religions in their societies that can affect these boundaries.

Note

1 A revised version of this introduction has been published under the title "How "green" can religions be? Tensions in Religious Environmentalism" in the Journal for Religion, Society, and Politics Koehrsen *et al.* (2022).

References

Abdelzaher, D., Kotb, A., and Helfaya, A., 2019. Eco-Islam. Beyond the Principles of Why and What, and into the Principles of How. *Journal of Business Ethics* (155), 623–643.

Abegunde, A.A., 2017. Local Communities' Belief in Climate Change in a Rural Region of Sub-Saharan Africa. *Environment, Development and Sustainability*, 19 (4), 1489–1522.

Allison, E.A., 2015. The Spiritual Significance of Glaciers in an Age of Climate Change. *Wiley Interdisciplinary Reviews: Climate Change*, 6 (5), 493–508.

Amery, C., 1974. *Das Ende der Vorsehung: die gnadenlosen Folgen des Christentums*. Reinbeck: Rowohlt.

Amri, U., 2014. From Theology to a Praxis of "Eco-Jihad". The Role of Religious Civil Society Organizations in Combating Climate Change in Indonesia. *In:* R.G. Veldman, A. Szasz, and R. Haluza-DeLay, eds. *How the World's Religions Are Responding to Climate Change. Social Scientific Investigations*. London: Routledge, 75–93.

Artur, L., and Hilhorst, D., 2012. Everyday Realities of Climate Change Adaptation in Mozambique. *Global Environmental Change*, 22 (2), 529–536.

Baker, C.R., and Beaumont, J., eds., 2011. *Postsecular Cities. Space, Theory and Practice/Edited by Christopher Baker and Justin Beaumont*. London: Continuum.

Barker, D.C., and Bearce, D.H., 2013. End-Times Theology, the Shadow of the Future, and Public Resistance to Addressing Global Climate Change. *Political Research Quarterly*, 66 (2), 267–279.

Bartholomew, and Chryssavgis, J., 2012. *On Earth as in Heaven. Ecological Vision and Initiatives of Ecumenical Patriarch Bartholomew*. 1st ed. New York: Fordham University Press.

Baugh, A.J., 2020. Nepantla Environmentalism: Challenging Dominant Frameworks for Green Religion. *Journal of the American Academy of Religion*, 88 (3), 832–858.

Becci, I., and Monnot, C., 2016. Spiritualité et religion. Nouveaux carburants vers la transition énergétique? *Histoire, monde et cultures religieuses*, 40 (4), 93–109.

Becci, I., Monnot, C., and Wernli, B., 2021. Sensing 'Subtle Spirituality' Among Environmentalists. *Journal for the Study of Religion, Nature & Culture*, 15 (3), 344–367.

Bell, D., 2014. Understanding a 'Broken World'. Islam, Ritual, and Climate Change in Mali, West Africa. *Journal for the Study of Religion, Nature & Culture*, 8 (3), 287–306.

Berger, P.L., ed., 1999. *The Desecularization of the World. Resurgent Religion and World Politics*. Washington DC: Wm. B. Eerdmans Publishing Company.

Berry, E., 2013. Religious Environmentalism and Environmental Religion in America. *Religion Compass*, 7 (10), 454–466.

Berry, E., 2014. Religion and Sustainability in Global Civil Society. Some Basic Findings from Rio+20. *Worldviews: Global Religions, Culture, and Ecology*, 18 (3), 269–288.

Berry, E., 2015. *Devoted to Nature. The Religious Roots of American Environmentalism*. Oakland, CA: University of California Press.

Binay, S., and Khorchide, M., eds., 2019. *Islamische Umwelttheologie. Ethik, Norm und Praxis*. Freiburg: Herder.

Blanc, J., 2017. *Ökokatholizismus. Sozialethische Untersuchungen zu ausgewählten Ländern und Institutionen in Europa*. Marburg: Metropolis-Verl.

Blanc, J., 2022. From "Why Should?" to "Why Do?". Tensions in the Christian Context While Acting for the Environment. *In:* J. Koehrsen, J. Blanc, and F. Huber, eds. *Religious Environmental Activism. Emerging Conflicts and Tensions in Earth Stewardship*. London: Routledge.

Blanc, J., and Ostheimer, J., 2019. Religion as Transmission Belt for Promoting the Sustainable Development Goals. *In:* W. Leal Filho et al., eds. *Partnerships for the Goals. Encyclopedia of the UN Sustainable Development Goals*. Cham: Springer, 1–10.

Boff, L., 2011. *Ecología. Grito de la tierra, grito de los pobres*. 5th ed. Madrid: Trotta.

Brockopp, J., 2012. Islam and Ecology. Theology, Law and Practice of Muslim Environmentalism. *Worldviews*, 16 (3), 213–217.

Carlisle, J.E., and Clark, A.K., 2018. Green for God: Religion and Environmentalism by Cohort and Time. *Environment and Behavior*, 50 (2), 213–241.

Carr, W.A. et al., 2012. The Faithful Skeptics: Evangelical Religious Beliefs and Perceptions of Climate Change. *Journal for the Study of Religion, Nature and Culture*, 6 (3), 276–299.

Casanova, J., 1994. *Public Religions in the Modern World*. Chicago: University of Chicago Press.

Casanova, J., 2006. Rethinking Secularization: A Global Comparative Perspective. *The Hedgehog Review*, 8 (1–2), 7–22.

Chaplin, J., 2016. The Global Greening of Religion. *Palgrave Communications*, 2. Available from: https://doi.org/10.1057/palcomms.2016.47.

Clements, J.M., Xiao, C., and McCright, A.M., 2014. An Examination of the "Greening of Christianity" Thesis Among Americans, 1993–2010. *Journal for the Scientific Study of Religion*, 53 (2), 373–391.

Clingerman, F., and O'Brien, K.J., 2017. Is Climate Change a New Kind of Problem? The Role of Theology and Imagination in Climate Ethics. *Wiley Interdisciplinary Reviews: Climate Change*, 8 (5), e480.

Creegan, N.H., and Shepherd, A., 2018. *Creation and Hope: Reflections on Ecological Anticipation and Action from Aotearoa New Zealand*. Eugene: Wipf and Stock Publishers.

Crockett, C., and Robbins, J., 2012. *Religion, Politics, and the Earth: The New Materialism*. Berlin: Springer.

Davie, G., 2010. Resacralization. *In:* B.S. Turner, ed. *The New Blackwell Companion to the Sociology of Religion*. Malden, MA: Wiley-Blackwell, 160–177.

Dessi, U., 2013. 'Greening Dharma': Contemporary Japanese Buddhism and Ecology. *Journal for the Study of Religion, Nature and Culture*, 7 (3), 334–355.

Dilmaghani, M., 2018. Which Is Greener: Secularity or Religiosity? Environmental Philanthropy Along Religiosity Spectrum. *Environmental Economics and Policy Studies*, 20 (2), 477–502.

Dohe, C.B., 2022. "What Does Religion Have to Do with Nature Conservation?" Investigating the Tensions in an Interreligious Nature Conservation Project in Germany. *In:* J. Koehrsen, J. Blanc, and F. Huber, eds. *Religious Environmental Activism. Emerging Conflicts and Tensions in Earth Stewardship*. London: Routledge.

Edenhofer, O., Flachsland, C., and Knopf, B., 2015. Science and Religion in Dialogue Over the Global Commons. *Nature Climate Change*, 5 (10), 907–909.

Finke, R., 2004. Innovative Returns to Tradition. Using Core Teachings as the Foundation for Innovative Accommodation. *Journal for the Scientific Study of Religion*, 43 (1), 19–34.

Foltz, R.C., ed., 2003. *Worldviews, Religion, and the Environment*. Belmont, CA: Wadsworth.

Foltz, R.C., 2006. Islam. *In:* R.S. Gottlieb, ed. *The Oxford Handbook of Religion and Ecology*. Oxford: Oxford University Press, 207–220.

Gade, A.M., 2019. *Muslim Environmentalisms. Religious and Social Foundations*. New York: Columbia University Press.

Gardner, G.T., 2002. *Invoking the Spirit. Religion and Spirituality in the Quest for a Sustainable World* [online]. Available from: www.worldwatch.org/system/files/EWP164.pdf [Accessed 10 October 2016].

Gardner, G.T., 2003. Engaging Religion in the Quest for a Sustainable World. *In:* Worldwatch Institute, ed. *State of the World, 2003. A Worldwatch Institute Report on Progress Toward a Sustainable Society*. New York: W.W. Norton & Company, 152–175.

Gerten, D., and Bergmann, S., eds., 2012. *Religion in Environmental and Climate Change. Suffering, Values, Lifestyles*. London, New York: Continuum.

Gilliat-Ray, S., and Bryant, M., 2011. Are British Muslims 'Green'? An Overview of Environmental Activism among Muslims in Britain. *Journal for the Study of Religion, Nature & Culture*, 5 (3), 284–306.

Glaab, K., 2022. The Green, the Secular, and the Religious. The Legitimacy of Religious Environmentalism in Global Climate Politics. *In:* J. Koehrsen, J. Blanc, and F. Huber, eds. *Religious Environmental Activism. Emerging Conflicts and Tensions in Earth Stewardship*. London: Routledge.

Gojowczyk, J., 2022. From Global Goal to Local Practice. Potential Lines of Tension in Religious Environmentalism in Catholic Religious Orders. *In:* J. Koehrsen, J. Blanc, and F. Huber, eds. *Religious Environmental Activism. Emerging Conflicts and Tensions in Earth Stewardship*. London: Routledge.

Gottlieb, R.S., ed., 2006. *The Oxford Handbook of Religion and Ecology*. Oxford: Oxford University Press.

Gottlieb, R.S., 2010. *A Greener faith. Religious Environmentalism and Our Planet's Future*. New York, Oxford: Oxford University Press.

Grandjean, A., 2022. The Negotiation of Self-Identity in Swiss Biodynamic Wine-Crafting. Facets of a Sentient and Practitioner-based Sustainable Agronomy. *In:* J. Koehrsen, J. Blanc, and F. Huber, eds. *Religious Environmental Activism. Emerging Conflicts and Tensions in Earth Stewardship*. London: Routledge.

Habermas, J., 2008. Notes on Post-Secular Society. *New Perspectives Quarterly* (25), 17–29.

Haluza-DeLay, R., 2014. Religion and Climate Change. Varieties in Viewpoints and Practices. *Wiley Interdisciplinary Reviews: Climate Change*, 5 (2), 261–279.

Hancock, R., 2015. 'Islamic' Environmentalism in Great Britain. *In:* T. Peace, ed. *Muslims and Political Participation in Britain*. London, New York: Routledge, 119–139.

Hancock, R., 2018. *Islamic Environmentalism. Activism in the United States and Great Britain*. Abingdon, Oxon, New York: Routledge.

Harmannij, D., 2022. Environmental Action Within Local Faith Communities. Navigating Between High Expectations and Practical Action. *In:* J. Koehrsen, J. Blanc, and F. Huber, eds. *Religious Environmental Activism. Emerging Conflicts and Tensions in Earth Stewardship*. London: Routledge.

Haron, M., 2017. Drawing on African Muslims' Intangible Assets. Doing Jihad Against Climate Change. *The Ecumenical Review*, 69 (3), 348–361.

Harris, I., 1995. Buddhist Environmental Ethics and Detraditionalization. The Case of ecoBuddhism. *Religion*, 25 (3), 199–211.

Hayhoe, D., Bloom, M.A., and Webb, B.S., 2019. Changing Evangelical Minds on Climate Change. *Environmental Research Letters*, 14 (2), 24016.

Hedlund-de Witt, A., 2012. Exploring Worldviews and Their Relationships to Sustainable Lifestyles. Towards a New Conceptual and Methodological Approach. *Ecological Economics*, 84, 74–83.

Hedlund-de Witt, A., 2013. Pathways to Environmental Responsibility. A Qualitative Exploration of the Spiritual Dimension of Nature Experience. *Journal for the Study of Religion, Nature and Culture*, 7 (2).

Herman, L., 2022. Kosher Electricity and Sustainability. Building Block or Stumbling Stone? *In:* J. Koehrsen, J. Blanc, and F. Huber, eds. *Religious Environmental Activism. Emerging Conflicts and Tensions in Earth Stewardship*. London: Routledge.

Hessel, D.T., and Ruether, R.R., eds., 2000. *Christianity and Ecology. Seeking the Well-Being of Earth and Humans/Edited by Dieter T. Hessel and Rosemary Radford Ruether*. Cambridge, MA: Distributed by Harvard University Press for the Harvard University Center for the Study of World Religions.

Huber, F., 2022. Environmentalism in the Religious Field. The Role of the Establishment for Competition in Switzerland. *In:* J. Koehrsen, J. Blanc, and F. Huber, eds. *Religious Environmental Activism. Emerging Conflicts and Tensions in Earth Stewardship*. London: Routledge.

Hulme, M., 2016. Climate Change. Varieties of Religious Engagement. *In:* W. Jenkins, M.E. Tucker, and J. Grim, eds. *Routledge Handbook of Religion and Ecology*. London: Routledge, 239–248.

Hulme, M., 2017. Climate Change and the Significance of Religion. *Economic & Political Weekly*, 52 (28), 15.

Interfaith Climate Summit, 2008. *Hope for the Future! The Uppsala Interfaith Climate Manifesto 2008* [online]. Available from: www.svenskakyrkan.se/default.aspx?id=664984 [Accessed 7 December 2015].

Jamil, S., 2022. Halal Wastewater Recycling. Environmental Solution or Religious Complication? *In:* J. Koehrsen, J. Blanc, and F. Huber, eds. *Religious Environmental Activism. Emerging Conflicts and Tensions in Earth Stewardship*. London: Routledge.

Jenkins, W., 2009. After Lynn White: Religious Ethics and Environmental Problems. *Journal of Religious Ethics*, 37 (2), 283–309.

Jenkins, W., Berry, E., and Kreider, L.B., 2018. Religion and Climate Change. *Annual Review of Environment and Resources*, 43, 85–108.

Jooste, B.S. et al., 2018. Challenges to Belief Systems in the Context of Climate Change Adaptation. *Jamba Journal of Disaster Risk Studies*, 10 (1), 1–10.

Kearns, L., 1996. Saving the Creation: Christian Environmentalism in the United States. *Sociology of Religion*, 57 (1), 55–70.

Kearns, L., 1997. Noah's Ark Goes to Washington: A Profile of Evangelical Environmentalism. *Social Compass*, 44 (3), 349–366.

Kearns, L., 2014. Green Evangelicals. *In:* B. Steensland, ed. *The New Evangelical Social Engagement*. New York: Oxford University Press, 157–173.

Khan, N., 2014. Dogs and Humans and What Earth Can Be. *HAU: Journal of Ethnographic Theory*, 4 (3), 245–264.

Kilburn, H.W., 2014. Religion and Foundations of American Public Opinion Towards Global Climate Change. *Environmental Politics*, 23 (3), 473–489.

Koehrsen, J., 2015. Does Religion Promote Environmental Sustainability? Exploring the Role of Religion in Local Energy Transitions. *Social Compass*, 62 (3), 296–310.

Koehrsen, J., 2018a. Eco-Spirituality in Environmental Action: Studying Dark Green Religion in the German Energy Transition. *Journal for the Study of Religion, Nature and Culture*, 12 (1), 34–54.

Koehrsen, J., 2018b. Religious Agency in Sustainability Transitions: Between Experimentation, Upscaling, and Regime Support. *Environmental Innovation and Societal Transitions*, 27, 4–15.

Koehrsen, J., 2019. Innovation und Religion. *In:* B. Blättel-Mink, I. Schulz-Schaeffer, and A. Windeler, eds. *Handbuch Innovationsforschung*. Wiesbaden: Springer Fachmedien Wiesbaden, 1–17.

Koehrsen, J., 2021. Muslims and Climate Change: How Islam, Muslim Organizations, and Religious Leaders Influence Climate Change Perceptions and Mitigation Activities. How Islam, Muslim Organizations, and Religious Leaders Influence Climate Change Perceptions and Mitigation Activities. *Wiley Interdisciplinary Reviews. Climate Change*, 12 (3).

Koehrsen, J., Blanc, J., and Huber, F., 2022. How "Green" Can Religions Be? Tensions About Religious Environmentalism. *Journal for Religion, Society and Politics*, 6, 43–64.

Koehrsen, J., and Huber, F., 2021. A Field Perspective on Sustainability Transitions: The Case of Religious Organizations. *Environmental Innovation and Societal Transitions*, 40, 408–420. Available from: www.sciencedirect.com/science/article/pii/S2210422421000630.

Konisky, D., 2018. The Greening of Christianity? A Study of Environmental Attitudes Over Time. *Environmental Politics*, 27 (2), 267–291.

Krantz, D., 2022. Climate and Covenant. A Case Study of the Functions, Goals and Tensions of Faith at the 23rd Conference of Parties to the United Nations Framework Convention on Climate Change. *In:* J. Koehrsen, J. Blanc, and F. Huber, eds. *Religious Environmental Activism. Emerging Conflicts and Tensions in Earth Stewardship*. London: Routledge.

Landrum, A.R. et al., 2017. Processing the Papal Encyclical Through Perceptual Filters: Pope Francis, Identity-Protective Cognition, and Climate Change Concern. *Cognition*, 166, 1–12.

Lee, C., and Han, L., 2015. Recycling Bodhisattva: The Tzu-Chi Movement's Response to Global Climate Change. *Social Compass*, 62 (3), 311–325.

Li, N. et al., 2016. Cross-Pressuring Conservative Catholics? Effects of Pope Francis' Encyclical on the U.S. Public Opinion on Climate Change. *Climatic Change*, 139 (3–4), 367–380.

Luckmann, T., 1967. *The Invisible Religion. The Problem of Religion in Modern Society*. New York: Macmillan.

Maibach, E. et al., 2015. *The Francis Effect: How Pope Francis Changed the Conversation About Global Warming* [online]. George Mason University Center for Climate Change Communication. Available from: https://climatecommunication.yale.edu/wp-content/uploads/2015/11/The_Francis_Effect.pdf [Accessed 28 July 2019].

Majerus, S., 2022. Cosmological Tensions. Biodynamic Agriculture's Anthropocentrism and Its Contestation. *In:* J. Koehrsen, J. Blanc, and F. Huber, eds. *Religious Environmental Activism. Emerging Conflicts and Tensions in Earth Stewardship*. London: Routledge.

Makame, M.O., and Shackleton, S., 2019. Perceptions of Climate Variability and Change in Relation to Observed Data Among Two East Coast Communities in Zanzibar, East Africa. *Climate and Development*, 3 (7), 1–13.

Mangunjaya, F.M., and McKay, J.E., 2012. Reviving an Islamic Approach for Environmental Conservation in Indonesia. *Worldviews: Global Religions, Culture, and Ecology*, 16 (3), 286–305.

Mangunjaya, F.M. et al., 2015. Faiths from the Archipelago. Action on the Environment and Climate Change. *Worldviews*, 19 (2), 103–122.

McCammack, B., 2007. Hot Damned America: Evangelicalism and the Climate Change Policy Debate. *American Quarterly*, 59 (3), 645–668.

McKim, R., 2020. The Challenge and the Opportunity. Some Perspectives on Laudato Si'. *In:* R. McKim, ed. *Laudato Si' and the Environment. Pope Francis' Green Encyclical*. London: Routledge, 3–18.

Mohamad, Z.F., Idris, N., and Mamat, Z., 2012. Role of Religious Communities in Enhancing Transition Experiments. A Localised Strategy for Sustainable Solid Waste Management in Malaysia. *Sustainability Science*, 7 (2), 237–251.

Monnot, C., 2022. The Slow Greening of Established Churches in Switzerland. Tensions Between Local Parishes and Church Head Organizations. *In:* J. Koehrsen, J. Blanc, and F. Huber, eds. *Religious Environmental Activism. Emerging Conflicts and Tensions in Earth Stewardship*. London: Routledge.

Murphy, C. et al., 2016. Adapting to Climate Change in Shifting Landscapes of Belief. *Climatic Change*, 134 (1–2), 101–114.

Naess, A., 1990. *Ecology, Community and Lifestyle: Outline of an Ecosophy*. Cambridge: Cambridge University Press.

Nagel, A.-K., 2018. Bedeutung und Wandel von Religion im Migrationskontext. *In:* R. Ceylan, Uslucan, and Haci-Halil, eds. *Transformation religiöser Symbole und religiöser Kommunikation in der Diaspora. Sozialpsychologische und religionssoziologische Annäherungen an das Diskursfeld Islam in Deutschland*. Wiesbaden: Springer VS, 19–39.

Nagle, J.C., 2008. *The Evangelical Debate Over Climate Change* [online]. Available from: http://scholarship.law.nd.edu/law_faculty_scholarship/433 [Accessed 31 September 2014].

Nche, G.C., 2020. Beyond Spiritual Focus: Climate Change Awareness, Role Perception, and Action Among Church Leaders in Nigeria. *Weather, Climate, and Society*, 12 (1), 149–169.

Nche, G.C., 2022. The Dissenting Voices. Perception of Climate Change and Church's Responsibility in Nigeria. *In:* J. Koehrsen, J. Blanc, and F. Huber, eds. *Religious Environmental Activism. Emerging Conflicts and Tensions in Earth Stewardship*. London: Routledge.

Nita, M., 2014. Christian and Muslim Climate Activists Fasting and Praying for the Planet. Emotional Translation of "Dark Green" Activism and Green-Faith Identities. *In:* R.G. Veldman, A. Szasz, and R. Haluza-DeLay, eds. *How the World's Religions Are Responding to Climate Change. Social Scientific Investigations*. London: Routledge, 229–243.

Ostheimer, J., and Blanc, J., 2021. The Catholic Church as a Change Agent in the Application of the SDGs. *Sustainability*, 13.

Palmer, M., 2013. The Quiet Revolutionaries: Why the Faith-Based Environmental Movement Has Become the Largest Civil Society Movement in the Conservation World. *Biodiversity*, 14 (3), 180–182.

Peterson, A.L., 2007. Talking the Walk: A Practice-Based Environmental Ethic as Grounds for Hope. *In:* L. Kearns and C. Keller, eds. *Ecospirit. Religions and Philosophies for the Earth/Edited by Laurel Kearns and Catherine Keller*. New York: Fordham University Press, 45–62.

Pew Forum, 2015. *Global Religious Futures. Country Profiles* [online]. Available from: www.globalreligiousfutures.org/countries [Accessed 15 May 2020].

Pope Francis, 2015. *Laudato Sí. On Care for Our Common Home*. Brooklyn: Melville House.

Reder, M., 2012. Religion in the Public Sphere. The Social Function of Religion in the Context of Climate and Development Policy. *In:* D. Gerten and S. Bergmann, eds. *Religion in Environmental and Climate Change. Suffering, Values, Lifestyles*. London, New York: Continuum, 32–45.

Saniotis, A., 2012. Muslims and Ecology. Fostering Islamic Environmental Ethics. *Contemporary Islam*, 6 (2), 155–171.

Schaefer, J., 2016. Motivated for Action and Collaboration: The Abrahamic Religions and Climate Change. *Geosciences*, 6 (3), 31.

Shehu, M., and Molyneux-Hodgson, S., 2014. Faith Communities and Environmental Degradation in Northeast Nigeria. *International Journal of Environmental Sustainability*, 10 (1), 27–40.

Sheikh, K.M., 2006. Involving Religious Leaders in Conservation Education in the Western Karakorum, Pakistan. *Mountain Research and Development*, 26 (4), 319–322.

Smith, N., and Leiserowitz, A., 2013. American Evangelicals and Global Warming. *Global Environmental Change*, 23 (5), 1009–1017.

Snow, D.A. et al., 1986. Frame Alignment Processes, Micromobilization, and Movement Participation. *American Sociological Review*, 51 (4), 464–481. Available from: www.jstor.org/stable/2095581.

Snow, D.A. et al., 2014. The Emergence, Development, and Future of the Framing Perspective: 25+ Years Since "Frame Alignment". *Mobilization: An International Quarterly*, 19 (1), 23–46.

Sponsel, L.E., 2012. *Spiritual Ecology: A Quiet Revolution*. Santa Barbara: ABC-CLIO.

Stacey, T., 2021. Toying with Animism. *Nature and Culture*, 16 (3), 83–109.

Stoll, M., 2015. *Inherit the Holy Mountain: Religion and the Rise of American Environmentalism*. Oxford: Oxford University Press.

Stork, J., and du Toit, C., 2022. Finding Ubuntu in the Bible. How the Zion Christian Church in South Africa Relates to Concepts of Ecology in African Traditional Religions. *In:* J. Koehrsen, J. Blanc, and F. Huber, eds. *Religious Environmental Activism. Emerging Conflicts and Tensions in Earth Stewardship*. London: Routledge.

Taylor, B., 2004. A Green Future for Religion? *Futures*, 36 (9), 991–1008.

Taylor, B., 2010. *Dark Green Religion: Nature Spirituality and the Planetary Future*. Berkeley: University of California Press.

Taylor, B., van Wieren, G., and Zaleha, B.D., 2016a. The Greening of Religion Hypothesis (Part Two). Assessing the Data from Lynn White, Jr, to Pope Francis. *Journal for the Study of Religion, Nature & Culture*, 10 (3), 306–378.

Taylor, B., van Wieren, G., and Zaleha, B.D., 2016b. Lynn White Jr. and the Greening-of-Religion Hypothesis. *Conservation Biology*, 30 (5), 1000–1009.

Theokritoff, E., 2017. Green Patriarch, Green Patristics: Reclaiming the Deep Ecology of Christian Tradition. *Religions*, 8 (7), 116.

Torabi, M., and Noori, S.M., 2019. Religious Leaders and the Environmental Crisis. *The Ecumenical Review*, 71 (3), 344–355.

Tucker, M.E., 2006. Religion and Ecology. Survey of the Field. *In:* R.S. Gottlieb, ed. *The Oxford Handbook of Religion and Ecology*. Oxford: Oxford University Press, 398–418.

Tucker, M.E., 2008. World Religions, the Earth Charter, and Sustainability. *Worldviews: Global Religions, Culture, and Ecology*, 12 (2), 115–128.

Tucker, M.E., and Grim, J., 2016. The Movement of Religion and Ecology. Emerging Field and Dynamic Force. *In:* W. Jenkins, M.E. Tucker, and J. Grim, eds. *Routledge Handbook of Religion and Ecology*. London: Routledge, 3–12.

Vaidyanathan, B., Khalsa, S., and Ecklund, E.H., 2018. Naturally Ambivalent: Religion's Role in Shaping Environmental Action. *Sociology of Religion*, 79 (4), 472–494.

Veldman, R.G., 2019. *The Gospel of Climate Skepticism: Why Evangelical Christians Oppose Action on Climate Change*. Berkeley: University of California Press.

Veldman, R.G. et al., 2021. Who Are American Evangelical Protestants and Why Do They Matter for US Climate Policy? *Wiley Interdisciplinary Reviews: Climate Change*, 12 (2).

Veldman, R.G., Szasz, A., and Haluza-DeLay, R., eds., 2014. *How the World's Religions Are Responding to Climate Change. Social Scientific Investigations.* 1st ed. London: Routledge.

Wardekker, A., Petersen, A.C., van der Sluijs, and Jeroen, P., 2009. Ethics and Public Perception of Climate Change. Exploring the Christian Voices in the US Public Debate. *Global Environmental Change*, 19 (4), 512–521.

Watson, E.E., and Kochore, H.H., 2012. Religion and Climate Change in Northern Kenya: New Moral Frameworks for New Environmental Challenges? *Journal for the Study of Religion, Nature and Culture*, 6 (3), 319–343.

Wexler, J., 2016. *When God Isn't Green. A World-Wide Journey to Places Where Religious Practice and Environmentalism Collide.* Boston: Beacon Press.

White, Lynn, Jr., 1967. The Historical Roots of Our Ecologic Crisis. *Science*, 155 (3767), 1203–1207.

Whitney, E., 2017. Lynn White Jr.'s 'The Historical Roots of Our Ecologic Crisis' After Fifty Years. *In:* T. LeVasseur and A. Peterson, eds. *Religion and Ecological Crisis. The "Lynn White Thesis" at Fifty*. New York: Routledge, Taylor & Francis Group, 18–32.

Woodworth, P., 2020. "We Were Nowhere. We've Got Somewhere." Does Laudato Si Go Far Enough, and Is the Church on Board for the Climate Journey? *In:* R. McKim, ed. *Laudato Si' and the Environment. Pope Francis' Green Encyclical.* London: Routledge, 105–134.

Yildirim, A.K., 2016. Between Anti-Westernism and Development. Political Islam and Environmentalism. *Middle Eastern Studies*, 52 (2), 215–232.

Zaleha, B.D., and Szasz, A., 2015. Why Conservative Christians Don't Believe in Climate Change. *Bulletin of the Atomic Scientists*, 71 (5), 19–30.

Part I

Intradenominational Tensions

2 From Global Goal to Local Practice

Potential Lines of Tension in Religious Environmentalism in Catholic Religious Orders

Jiska Gojowczyk

Introduction

Are Catholic religious orders "greening"? Decisions of the highest decision-making bodies of several congregations point in this direction. In 2008, roughly 17,000 Jesuits were asked "to assist [Christ] as he sets right our relationships with God, with other human beings, and with creation." The General Congregation (GC35) decided that reconciliation with creation was one of three major aspects of the mission of the Society of Jesus (SJ), the largest religious order in the Roman Catholic Church. This decision was a simple reflection of the demands of the time for some members. For others, it was a new challenge to be addressed.

When the General Chapter of the Order of the Friars Minor (OFM) convened in 2009, of the roughly 13,500 Franciscans probably no one was surprised to see "integrity of creation" as one of community's goals for the future. "Justice, peace and integrity of creation" (JPIC) had long been recorded as an important mission area. In 2009, the Chapter had renewed former mandates with some actualizations, reflecting the observation that still much was left to be done in the area of JPIC in the world and within the community.

Such goals and similar global statements by religious leaders have received considerable attention from academics inquiring into the field of religion and ecology. Some scholars have been looking at them with hope (e.g., Hall et al. 2009; Tucker and Grim 2017; Veldman et al. 2014). Updated doctrines and respective explications of religious leaders are understood as vehicles to accelerate social change toward sustainability transitions worldwide. Together with flagship engagements and other forms of "religious environmentalism" activism, they are taken as proof for the "greening" of religious communities, assuming implicitly that the rest of the community members will follow eventually. Other scholars have argued that what is stated on paper is not necessarily what communities really do – pointing to struggles and inertia hindering *any* progressive changes in communities and

DOI: 10.4324/9781003017967-3

beyond, and to religious communities who are in fact opposing necessary environmental efforts, for example, to mitigate climate change (e.g., Ronan 2017; Taylor 2005). There is no proof, they say, that religion will help to address ecological challenges (cf. Taylor et al. 2016). Studies on religious environmentalism hence seem to draw two different pictures of trends, leaving interested readers so puzzled that one nearly forgets that religious-environmental goals actually exist and that their realization can therefore be studied empirically.

Fortunately, an increasing number of social scientists are taking up the brush. With more nuanced empirical case studies investigating religious environmentalism, it becomes clear that to answer in how far and in which ways religious communities are relevant change agents, we need to inquire into religious environmentalism in its diversity (e.g., Amri 2014; Baugh 2019; Darlington 2012; Ellingson 2016; Glaab 2017; Hancock 2018). One of the crucial areas which still remains understudied, however, is the link between global goals and community members' practices.

That is what I explore in this chapter. I ask how members of the SJ and the OFM interpret their key environmental goals to reconcile with creation (SJ) or to foster the integrity of creation (OFM) in the Philippine and German provinces. With this, I seek to problematize the existing argument which formulates hopes for sustainable transformations deduced from statements by religious authorities and to expand our knowledge on the question of how abstract religious environmental goals are in fact put into practice in everyday life.

Based on a reconstructive approach, the analysis reveals that potential lines of tension exist even in such "rigidly defined" and hierarchically organized communities as Catholic religious orders because members interpret the key environmental goals quite differently as they draw on knowledge from different social worlds which they are part of. I argue that diversity in religious environmentalism(s) is not just grounded in different religious traditions or geographical contexts but is also related to professional, political, and possibly demographic factors. Members of the religious orders understand creation ecologically or human-focused. Action based on the environmental goals may be oriented inward or outward with respect to the religious community. In detail, I reconstructed five different types of interpretations of the orders' key environmental goals. They are addressed through science and teaching, social work, internal education, (political) activism, or spirituality.

Exemplifying the ambiguity of theological concepts, the results question some latent assumptions of the existing literature on religious environmentalism. Neither does religious environmentalism necessarily express itself in classical "activist" forms nor does it conclusively follow from such activism of few that many can be mobilized for similar purposes. However, that does not mean that hope for social changes toward more sustainability in or through religious communities is groundless. But it does not come easy: It has to be negotiated and fought for, not just with leaders, but locally, taking into account the messiness of social life beyond doctrines. Given the diversity of interpretations of environmental goals, it is evident that the field of religious environmentalism is in fact also one of tensions. The results presented in this chapter suggest on which differences they may be founded.

Popular Academic Perspectives on Religious Environmentalism

For many years, academics as well as non-academics have highlighted the potential relevance of religious communities in times of human-caused environmental crises (e.g., Hall et al. 2009; Tucker and Grim 2017; Veldman et al. 2014 at the global level; at the national level Nche et al. 2017; see also the introduction of this volume). Looking at the academic roots of the most vibrant strands of research in this area helps to identify at least three pre-assumptions which any social scientist studying the relationship of religion and environmental protection is confronted with. Those assumptions have become the subject of justified criticism within the last years, but they are still helpful working hypotheses which have not been sufficiently addressed.

Theology Is a Relevant Drive for Religious Actors Protecting the Environment

In 1967, White argued that we need to shed light on the religious component of destructive human-nature relations. On the fertile grounds of his provocative argument, a strong trunk of literature has been growing in the disciplines of theology and ethics. As "attitudes toward nature have been consciously and unconsciously conditioned by . . . religious worldviews" (Tucker and Grim 2001, p. 4), "religions" are seen as "necessary partners in the current ecological movement" (ibid., p. 3). Early studies have analyzed religious texts and ideas with respect to their sensibility to ecological concerns, assuming that with respective theological motifs, "solutions" for environmental crises could be found there, especially if they are disseminated by the religious traditions' leadership (Tucker and Grim 1997). This focus on specific traditions and theologies has remained a core around which further studies grew, including the tendency to structure the discussion by religious traditions (cf. Jenkins and Chapple 2011). In much of this literature, it appears to be obvious that for religious actors, theology is a relevant basis for action and therefore, change toward more ecologically friendly practices can be initiated through changes in theological thought.

Prominently, Taylor has been criticizing this pre-assumption as "undemonstrated idealism" (Taylor 2005, p. 1376). In fact, recent empirical studies reveal reasons for skepticism. Baugh (2019) demonstrates a crucial causation problem of the hypothesis in contexts where religious belonging is also a question of personal choice. In the case of Unitarian Universalism in the United States, she argues, religion is *not* the major drive for community members to be environmentalists. Instead, environmental concern motivates members to be part of that specific religious community. Empirical studies also show that especially local and national context factors influence the forms religious environmentalism may take (e.g., Smith 2017; Witt 2016; cf. also Jamil 2022; Herman 2022; Dohe 2022 in this book).

Theology is inevitably characterized by plurality, ambiguity, and contradictions. While not many studies on religion and ecology have addressed related questions, indeed, in any situation, actors *need* to get involved in interpretative efforts to adapt the abstract to the concrete. Different ideas within the large religious traditions also offer followers several concepts about human-nature relations to choose from. Kearns (1996) was among the first to empirically investigate this within the Christian tradition in the United States. With data collected between 1987 and 1992, she identified stewardship, eco-justice, and creation spirituality as relevant ideal types of eco-theology accessible for Christians. Those types may change over time and differ geographically, but her findings support the proposition that "certain ideas [can be] mobilized by certain religious groups as a part of broader political agendas" (Proctor and Berry 2005, p. 1572), leading to questions about what constitutes such agendas and mobilizations. It is an empirical question if and how theology is relevant for religious communities in particular contexts.

Religious Environmentalism Expresses Itself in Classical "Activist" Forms

Empirical studies of religious environmentalism cover a large variety of themes, forms, strategies, and arenas, but, as Baugh also observes, they focus mainly on "mainstream constructions of explicit activism" (Baugh 2019, p. 3), such as advocacy meant to influence political decision-makers, local protests, or explicated flagship initiatives (e.g., cf. Amri 2014; Darlington 2012; Glaab 2017; Gojowczyk 2015; Moyer and Scharper 2019). Baugh shows that other forms of religious environmentalism – "embedded" rather than explicit – exist but tend to be overlooked systematically by research which majorly targets social movement activities and actors.

The tendency to understand the phenomenon as a classical social movement is a shortcoming across academic "camps."[1] To illustrate that religious communities are valuable "partners in the current ecological movement" (Tucker and Grim 2001, p. 3) and to promote them further, Tucker and Grim refer to anecdotal local and global interdenominational religious movement initiatives (Tucker and Grim 2001; cf. also Hitzhusen and Tucker 2013; O'Brien and Palmer 2007). Looking at it from a different angle, Taylor is especially interested in the religious dimension in apparently secular environmental movements (Taylor 2010). Partly based on the argument of intersectionality across diverse social movement issues, Witt and Taylor (2017) discuss a wide range of current social movement developments under the title "Religion and Eco-Resistance Movements."

The focus on specific forms of environmentalism may be due to an implicit theory of change which scholars share: While the precise mechanisms (diffusion, obedience, conversion, etc.) are not explicated, next to the pronouncements of religious leadership, it is that activist form which is expected (or questioned) to cause the *social* change to save the planet. This leads to the third working hypothesis, which is especially crucial for the "idealist" scholars' argument.

Activism of Few Hints Toward the Potential to Mobilize Many

One of the simple, but major arguments, why religious environmentalism is seen as a sign of hope for sustainable transformation is that religious communities are large, and hence, the activism of some might mobilize masses (cf. O'Brien and Palmer 2007). However, several studies show that the hypothesis of mass mobilization needs to be further investigated: While there *are* a number of (inter)religious initiatives promoting environmental protection, not every religious community is one of environmentalists (but some can actually stymie such efforts; cf. Ronan 2017) and not everyone in a religious community agrees on the relevance and the meaning of "environmental" goals (cf. Vincentnathan et al. 2016; cf. also Haluza-DeLay 2014). Activists who advocate for environmental concerns might find themselves in complicated identity conflicts between political cultures within religious communities and among secular environmentalists. In such circumstances, Ellingson (2016) shows, activism may lead into strategic dead ends, leading Taylor to argue that "religion is not coming to the environmental rescue any time soon, if ever" (Taylor 2017, p. 931).

But this conclusion might be just as over hasty as the idealistic argumentation, considering that members of religious communities who are not directly identified as activists are systematically understudied, especially if they do not live in the United States (cf. Taylor et al. 2016, p. 311). Qualitative research designs often treat them either as embellishments or as opponents in political struggles. Quantitative studies take large communities into account, but results are inconclusive and designs are only suitable for very specific research questions. Among other aspects, public opinion research has investigated how statements such as the Papal encyclical *Laudato Si'* have affected public opinion on climate change (cf. Jenkins et al. 2018 for a good summary). However, inquiring into a larger variety of local translations of global statements, quantitative studies bear the burden to predefine what counts as relevant "pro-environmental" attitudes or behaviors. While some authors problematize what counts as religious in the literature as a too "narrow focus and privileging of . . . mainstreams" (Taylor 2005, p. 1376; cf. also Haluza-DeLay 2014), just the same can be observed with regard to the "environmental" or "ecological." An important reason for this is little research on the question of how believers translate religious environmental goals into practices in different contexts.

Inductive approaches are necessary that reconstruct what the environmental goals formulated by religious authorities mean for different members of the respective communities and if and how they are translated into local practices. Those in-depth reconstructions need to move beyond the explicit flagship projects and most vocal advocates of social movement environmentalism in religious communities to reveal possible lines of tension within communities and to learn more about their transformative potential.

The Ethnographic Study in the SJ and the OFM

The basis for this contribution is a multi-level, multi-site ethnographic study which I conducted between spring 2014 and spring 2015 in the SJ and the OFM (with more than 70 interviews, several group discussions, and participatory observations).[2] The two communities selected stand for different traditions or "charisms" of religious orders within the Roman Catholic Church. The SJ, based on the teachings of Ignatius of Loyola, is known for an intellectual approach and high mobility of members. Since the mid-1970s, justice has also become an important concern. The OFM is the oldest order of the "Franciscan family" of female and male congregations inspired by Francis and Clare of Assisi. As a mendicant order, humility and efforts for the poor are central. Because Francis of Assisi was named patron of ecology by Pope John Paul II in 1979, the OFM may perceive additional pressure to become pro-environmentally active. Despite the differences, the religious orders are not independent of each other. Both are embedded in the Roman Catholic Church; members interact across congregational boundaries. As sketched in the introduction, both religious orders have written environmental concerns into their decrees and mandates.

Members can be expected to be very committed toward the collective goals and toward the community. When individuals enter after intensive formation lasting several years, they make a choice considered to be for life which results in sweeping life changes. An engaged interpretation of the environmental goals set by the orders' leadership appears more likely compared to religious communities in which costs and consequences of membership are less extensive (cf. Feldbauer-Durstmüller et al. 2012).

At the same time, both communities are internally heterogeneous. Geographically, they extend over continents. Demographically, several generations are working and living together. Organizationally, their complicated structure embraces diverse specializations and careers within and paralleling the dominant hierarchy. That is why members may pursue a common goal in very different ways, especially if it is relatively new and only weakly governed, as is the case for the environmental goals.

Regionally, I focus on the Philippine and German/Central European provinces. Collecting the data, I took formal members of other provinces into account if they took an active part in the Philippine or German communities or were experts on the topic following the internal specialization structures.

The explorative design allowed discovering diversity with regard to charism and societal, political, socioeconomic, and geographical context.

The Reconstructive Approach

The results presented in this chapter are based on the inductive analysis of 12 semi-structured narrative interviews. Two were pair interviews; one interview was conducted via video chat (skype). They took place as part of the ethnographic study previously outlined. The interviews were selected based on their feasibility for the method of analysis.

With some variation due to the flow of the conversations, the informants were asked, *inter alia*, to define the respective main term of the goal of the community ("What does integrity of creation (OFM)/reconciliation with creation (SJ) mean for you?") and to recall instances of the previous week which were related to those main terms. I usually did not explicitly ask for the leadership decisions cited previously if those were not brought up by the informants themselves. However, especially in the Society of Jesus in the Philippines the term led many to talk about the GC35, probably reflecting the strong echo which the decision had in the community.

I analyzed the data with a reconstructive approach in general and the documentary method in particular. For the approach, two theoretical assumptions are crucial: First, that practices are based on implicit knowledge gained through social experience, meaning that second, scholars can access practices which informants may otherwise not be able to express by reconstructing relevant implicit knowledge, or orientations (Bohnsack et al. 2013). Such orientations are identified analyzing *how* informants talk about an issue in several analytic steps. Interview sequences are contrasted with each other to find relevant similarities and differences. Among other aspects, the analyst searches for positive and negative statements marking "horizons" of the orientations (cf. Bohnsack 2014; Nohl 2012).

The elaborate process of data analysis has led to a typology of five different interpretations of the environmental goals. Interpretations describe the orientations within which actors translate the abstract environmental goals into their everyday lives. As an analytical device, the typology exaggerates empirically observable differences. Hardly any informant has interpreted the goal only within one of the social worlds identified or including all aspects of the typology description. Boundaries between types are not persistent in the sense of identifiable groups of specific individuals, even though the following presentation may sometimes suggest that to improve legibility. Orientations are situationally variable. Groups may temporarily agree on orientations, as I clearly observed in one instance.

To prevent biased results, in addition to the procedure to compare passages of different interviews with each other, I discussed parts of my findings, including the typology, with colleagues. I also reflected on in how far the results resonate with the rest of the material of the ethnographic study.

Interpretations of the Environmental Goals

When I asked my informants about the goals to reconcile with creation or to foster the integrity of creation, I was expecting them to talk about problems such as environmental destruction through mining and about organized protest to address such problems. Having read manuals written by the OFM's internal experts, I was also expecting conversations about practices inside the communities such as the separation of waste and the consumption of resources. In the research process, I learned that my expectations had been too narrow. For the informants, the goals were addressed not only through activism, but science and teaching, social work, formation (meaning: internal education), or spirituality, showing potential reasons for disagreement and conflict beyond yes-or-no arguments. Tables 2.1 and 2.2 show the respective typology which I developed by applying the reconstructive approach previously outlined. The first table gives an overview regarding social worlds, horizons, roles, and strategies of the five types. The second table structures them regarding two major lines of difference (the concept of creation and inward or outward action orientation) and the shared basis of all informants. I replenish the results beyond contrasting the types with the discussion of the shared basis, three variants *within* the activism type, and one example of internal tensions.

A major difference between interpretations is the concept of creation implied. Thomas,[4] for example, answers to the question of how his everyday life had been related to reconciliation with creation with activities like "teaching the course . . . ordinary physics." He talks about how the decision of GC 35 influences his work as a scientist at a Jesuit university.

> [We are] trying to set up the environmental science program. When we first started . . . to get SUPPORT from the administration, I would have to bring out in the proposal: Okay, does this fit into the mission . . . particularly with the GC 35? And this is environmental science. So I see: Yeah, it fits in. All the Jesuit superiors would agree.
>
> (Thomas)

Ron is a social worker who works with prisoners and their families. Like Thomas, he is a Philippine Jesuit. Asked how his everyday life relates to reconciliation with creation, he answers, for instance:

> YESTERDAY afternoon we had a formation session FOR the fathers [who are in prison] of the scholars [who are receiving stipends from a program Ron is responsible for, comments JG]. . . . They also have their responsibility to care for their children and care for themselves. So that their children . . . can see that their fathers . . . are doing something about their own (laughing) reconciliation with who they are, what they are. They are families so . . . all creation.
>
> (Ron)

Table 2.1 Types of goal interpretation

Goal interpretation Goal is fulfilled through. . .	*Social world*	*Positive horizons*	*Negative horizons*	*Roles*	*Strategies*
Science and Teaching	Jesuit university	‚Genuine' scientific interest in the common good; scientifically informed approach to solve challenges	Ignorant decisions of political decision-makers; interest in environmental science with only economic interest or/and short-sighted view	Engaged scientist; teacher	Research, teaching, consulting; guidance and promotion of young researchers
Social work	Social organization	Responsibility and care for human beings; inclusion of the excluded; proper hygiene facilities	Egoism and utilitarianism; passivity; bad hygiene facilities	Pastor; supporter	Care for and empowerment of the excluded
Formation	Seminary; internal committee for formation	Awareness about relevance of actions based on (scientific) knowledge; long-term focus; attempt to recreate ecological improvements locally	Apathy and indifference; superficial task fulfillment; short-sighted view; unthinking imitation	Teacher; educator	Integration of JPIC in formation curriculum; best practices in seminaries

(*Continued*)

Table 2.1 (Continued)

Goal interpretation Goal is fulfilled through. . .	*Social world*	*Positive horizons*	*Negative horizons*	*Roles*	*Strategies*
Activism	Local environmental destruction; housing residential community; organizations and committees of the order (for OFM, esp. JPIC groups)	Social movements and civil society; Franciscan (JPIC) spirituality; sustainable lifestyle and (for OFM Philippines) *sapat na*;[3] prophetic claim of religious order	Actors concentrating power and capital, esp. large transnational companies and governments; consumerism; lethargy of other community members; contemplation	Activist; educator; informant; investigator; lobbyist inward and outward; provocateur	Advocacy; civil society engagement; change of lifestyle; lobbying for collective decisions on sustainable procurement; raising awareness; local investigation of irregularities and problems
Spirituality	Community; nature; humanities	Meditation; harmonious interactions; overcoming of boundaries between people e.g., because of religious beliefs; integral way of viewing nature and community; care for creation; beauty of nature; holiness	Rational life directed by others; focus solely on bread-winning tasks; disharmony; pollution through waste and poison; environmentally destructive lifestyle	Community member; mentor for spiritual experiences; theologian	Meditation; meeting in silence; promotion of latitude within the congregation; courses and journeys focusing on creation spirituality; learn from other cultures and religions

Table 2.2 Overview of lines of difference and commonality

Ecological concept of creation				*Human-focused concept of creation*
Action oriented outward	Action oriented inward		Action oriented outward and/ or inward	Action oriented outward or inward
Science and teaching (Thomas, Steve)	*Spirituality* (Leon, Ruben, Bodo)	*Formation* (Cristoforo)	*Activism* (Michael, Martin, Pablo and Camilo, Alon, Bill)	*Social work* (Ron, Karl)
Shared Basis a) Contemporary challenge which *should* influence actions b) Orientation toward "the common good"				

While for the scientist Thomas, the goal is ecological, Ron's interpretation implies that creation is understood with a focus on humans. This he has in common with Karl who fulfills social tasks within the German Jesuit community and with a group of male and female participants of a network meeting in Germany which I observed, who agreed on this orientation for one point on the agenda.[5] Those findings also illustrate that different interpretations cannot solely be explained through geographical context. Main occupations can crucially influence how environmental goals are translated into local everyday practices.

The following contrasts show that there are also crucial differences among those who share an ecological understanding of creation and that those differences do not purely reflect charismatic boundaries of the religious orders. Thomas mostly talks about the "environment" in the narrative passages of the interview and is "teaching a course . . . for example . . . dealing with conservation of energy." The Franciscan Ruben, who had lived in Germany for many years when we met, speaks about "the beautiful nature." During a study tour in India, Ruben asks participants to walk for 15 minutes silently to "just be in touch with the nature. See the beauty." This illustrates differences between the orientation of science and teaching on the one hand and spirituality on the other. What marks a positive horizon in one of the interpretations is seen as negative in the other. For Thomas, scientifically informed approaches are crucial to fulfill the goal to reconcile with creation. For Bodo, "science and technology, rationalism . . . they do not have the final word." Bodo shares with Ruben a strong "spirituality" orientation. He is a Jesuit who had lived in the Philippines for several years when we met.

Next to the concept of creation, an important line of difference between types is if action is oriented inward or outward with respect to the religious community. Thomas, just like Ron, is mostly concerned about fulfilling the goal *outside* of the religious community (with students, researchers, but also companies and politicians he advises). In opposition, Cristoforo, a Franciscan from the Philippines, speaks extensively about internal education, especially the formation program for the young friars. This can be illustrated by his response to the question when he had heard "integrity of creation" for the first time.

> Many years ago. Actually, . . . at formation HERE at [priest seminary], . . . because simply . . . justice, peace and integrity of creation is deeply embedded in our mission . . . as a provincial fraternity. So, when . . . I start[ed] in the formation, many times I heard, I heard about it . . . but my understanding [was] very shallow.
>
> (Cristoforo)

When we met for the interview, Cristoforo was among those responsible for formation. As is true for Thomas and Ron, the interpretation of the goal hence corresponds to his major occupation in his everyday life.

However, this is not necessarily the case. In opposition to Ron, I reconstructed strong signs toward a science orientation for the Jesuit Steve, who was involved in similar social work when we met. Another Jesuit from the Philippines working in the university interprets the goal not as scientist, but as activist, as this small passage from the interview illustrates:

> And if the integrity of the environment is to be challenged by an economic gain, normally the economic gain will prevail. (short pause) For instance, an ONGOING struggle that we have in this university is the struggle against the . . . [COMPANY] mines in [REGION].

The following Table 2.2 structures the types regarding the two major lines of difference and shows the shared basis of all informants.

Shared Basis

The informants shared the view that the goals address a contemporary challenge which *should* influence actions, even though some also shifted responsibilities to experts within the religious orders. This norm cannot be taken for granted for the religious communities globally, as I learned in a focus group discussion with members in formation from different Central and Eastern European provinces.

Further, none of the informants understood the respective goal as invitations for solely individual self-betterment or as a call to conserve nature through shielding it from human doings altogether. As is indicated in many aspects listed in Table 2.2, the orientations point toward some version of "common good" of society.

Activism Type Variants

While neither geography nor charism seem to play a big direct role in determining the orientation for the goal interpretation, the picture becomes more complicated looking into differences within the type of activism. In all variants, the goal is understood politically, as critique of a capitalist societal order and respective lifestyles. The goals to reconcile with or foster the integrity of creation make sense to the informants against the backdrop of a negative status quo. Due to their prophetical assignment, members of religious orders are seen as particularly responsible to engage for changes. Beyond this commonality, there are at least three different "spheres of experiences" which informants who share this orientation refer to, with strong implications for the concrete activities enacted.

The first version is oriented inward. Especially two European informants – one Jesuit and one Franciscan – engage for lifestyle and consumption changes within the local religious communities (e.g., with regard to mobility, waste, or heating). For the Jesuit Michael, changes of practices are predominantly a matter of will. For the Franciscan Martin, it is a complex challenge which he reflects against the backdrop of limited financial and time resources, the latter being dramatically intensified due to the aging of his provincial community. This focus on the local religious community was less present for the activist orientations within the Philippine context.

In the Philippines, for members of both religious orders who express activist orientations, outward is more central than inward orientation. The direct experience of local environmental destruction due to economic activities, especially mining and deforestation, is crucial. Important activities are local and national lobbying and site investigations. However, this regional difference could be either coincidental or relativized through generational differences, as lifestyle choices were a crucial theme in the group discussions of younger Jesuits in formation in the Philippines. Also, in Jesuit universities, ambitious waste management programs exist which are predominantly run by lay university staff and students, but strongly supported by the Jesuit leaders.

Finally, religious communities differ regarding organizational histories and political cultures. Activist Franciscans in both world regions address the collaboration with other religious orders in JPIC groups, especially within the "Franciscan family." Whereas the concrete historical references are different (e.g., martial law and the end of Marcos' dictatorship in the Philippines *versus* NATO Double Track Decision in Europe), two informants describe very similarly the evolution of political themes for those groups, moving from concerns about peace in the 1980s to the awareness of problems of justice and finally the environment. One of the informants also goes further back in his lively narrative than his own time as a community member. Those findings point to organizational knowledge on the history of JPIC engagement on the one hand, and to entanglements of local and global social movement developments on the other. Those orientation frames seem to be more important to Franciscans than to Jesuits.

An Internal Struggle: Bill

Most informants find a way to make sense of the environmental goal of their community and act upon their interpretation in their daily lives. For the Franciscan Bill, this is more difficult. He interprets the goal as one which is to be fulfilled as activist, including protesting with others engaged in JPIC work in the Philippines and worldwide. Throughout the interview, he treats the *trias* JPIC as one goal, often using examples which appear to relate to questions of justice and peace stronger than to those of integrity of creation. However, the province leaders appointed him to guide the postulates during their time in "semi-contemplation," causing an orientation dilemma.

He became a formator unwillingly. While he was on a mission abroad, he volunteered to live in a community in a crisis region in Africa. However, his provincial asked him to return to the Philippines instead to take the new position. When I ask him about this instance, he comments "Yeah. (laughing) Obedience."

His position allows him to sensitize the young friars, as the interpretation "formation" suggests, and to foster changes covered by the goal interpretation "activism" oriented *inward*. In fact, our conversation indicates that he is doing both intensively, for example, by involving the postulates in improvement measures on the outside property and in the clean-up work after storms. However, due to his outward-oriented activist interpretation, his assigned position puts him in an orientation dilemma, meaning that he can hardly act according to his interpretation which would require him to take part in activities such as local protests and demonstrations.

Early in the interview, he talks about this problem and his way to reduce it slightly through online communication:

I: So, if you just think about the last week, in how far has integrity of creation played a role in your life?

Bill: Oh, yeah, yeah. The climate change, yeah. But I, only through facebook. (laughing)

I: Ah, like in which way on facebook?

Bill: No, because I am not really that involved though because I work in the formation that I am taking care of. . . . So I cannot really go out. . . . I can help but help . . . through facebook and other internet, I was able to, (short pause) to update myself and also join.

Bill's orientation dilemma points to possible lines of conflict in the communities with regard to the fulfillment of the goals, but also to a mechanism of local goal translation. On the one hand, "order and obedience" with regard to the major task which the community member should fulfill influences who can act upon which interpretation of the collective environmental goals in which place and with how many resources. On the other hand, facing an orientation dilemma, Bill finds a creative way to still enact his outward activist interpretation on a minimal basis, while also engaging in formation and inward activism, even though these are not his dominant interpretations of the goal.

Final Remarks

Before the discussion, some remarks about the findings are necessary. First, regional differences could be observed which are not reflected in the typology because they did not mount in larger action orientations. Less than one year after the super typhoon Hayan (known as Yolanda in the Philippines) and shortly after several other catastrophes such as landslides, it is not surprising that those events and vulnerability to disasters were much more present in conversations in the Philippines, dramatically turning the goals into pressing issues "quasi-automatically," sometimes without explicit mentioning of this relationship.

Second, due to limitations of the material and my capacities as a single researcher, the five types should not be understood exclusively. Rather, they point to a universe of possibilities just as diverse as the social worlds which the members of the communities are part of. One possible interpretation for which I found some, but not enough, indications to reconstruct a sixth type was to fulfill the goal as a priest animating the Church community to organize local environmental protection initiatives and relief work for victims of environmental disasters.

Discussion

The analysis of the question of how members of the SJ and the OFM interpret their key environmental goals to reconcile with creation or to foster the integrity of creation in the Philippine and German provinces reveals at least five different types of interpretations of the orders' environmental goals. They are addressed through science and teaching, social work, internal education, (political) activism, or spirituality. Those interpretations indicate relevant spheres of experiences for the members of the congregations and they point to possible disagreements regarding the realms of actions related to the goals. Through the reconstructive approach and the documentary method, I found methodological access to different everyday practices as they relate to global religious doctrines.

Except for the interpretation "science and teaching" which was predominantly found among Jesuits, the interpretations are neither a simple outcome of the two different regions nor the membership to one of the orders. For several of the informants, their dominant interpretation of the goal was related to their daily core occupation. For the translation of the goal into local practices in religious orders, the decision about who does what is hence of major importance. In this way, the provincial leadership who takes this decision (mostly in conversation with the person affected), has preponderant influence on the question which types exist and are enacted vitally. As the example of Bill shows, the core task assigned by the province's leadership can be in opposition to the actor's dominant interpretation of the environmental goal. Often, however, the informants integrate the goal into their existing spheres of experience and daily tasks, turning the influence of the leadership into an indirect effect over time. Based on this finding, conflict in the provincial communities is likely to emerge, for example, when orientation and weight given to the goals by consecutive leaders dehisce widely.

The findings improve our knowledge on how the traditions of religious communities influence versions of religious environmentalism of the respective community. Next to political cultures within communities, religious traditions seem to be important for the question how members of religious communities translate global environmental goals into practices *especially if they have an influence on what actors do in their everyday lives* (e.g., religious orders engaged predominantly in schooling *versus* those engaged in poor poverty relief). For religious communities in which leaders do *not* decidc on major occupations of the members – which is the case for most – nonreligious influences can be expected to be even more important for the local translation of global statements.

While this cannot be explicated in depth in this chapter, the results point to several social worlds which reach beyond charisms and regional context, such as that around global JPIC engagement and other social movements. Beyond supplying apparently "neutral" "technical guidance" (Hitzhusen and Tucker 2013, p. 370), professional knowledge (here esp. the natural sciences and social work) importantly influences the goal interpretations, including concrete everyday activities. Theology seems to be part of those interpretations, but mostly on a very broad level, assumingly reflected in the shared basis. This is not irrelevant however, as it marks the common ground which actors might agree on in struggles over interpretations. In opposition to the first hypothesis identified, changes in theological thought, however, do not automatically lead to more ecologically friendly practices locally, as the option to apply a human-focus concept of creation in goal interpretations illustrates.

The findings also question the second hypothesis found in the religion and ecology literature. Only one of the five types includes "classical" social movement activities. However, activism was one of the two types next to that of spirituality which were superimposed on the social world of the major occupation, hence opening up a field of engagement and of struggle beyond or in addition to the core occupations.

Nuanced knowledge on how religious actors interpret environmental goals as it is revealed in this contribution helps to identify potential lines of tensions around the environmental goals in religious communities (such as on inward and outward orientated strategies, on different understandings of creation, or on the leeway which members have to act upon their interpretations within hierarchical organizations). It allows differentiating tensions beyond formal or externally defined boundaries. In my research, I discovered that investigating the orientations of actors in specific situations helps to understand the varying quality of disagreements. Negotiations between actors interpreting the goals within the same orientation often occurred nearly unnoticed, even across formal religious membership boundaries, whereas those involving different orientations were more conflictive, even within religious communities. Especially community members with internal activist orientations reported being sanctioned quite negatively for their engagement sometimes by other community members (Gojowczyk 2020).

Finally, such nuanced knowledge as presented in this contribution enables us to qualify what kind of engagement can be expected from religious communities as a glimmer of hope in times of environmental crises. The activism of the few does not automatically hint toward the potential to mobilize many for the same purpose, contrary to the third hypothesis inherent in large parts of the existing literature as I carved out previously. At the same time, however, the findings reveal no evidence supporting assumptions of inertia. They hint toward dynamism and tensions. Hoping, constructive struggle is the straw to grasp at.

Notes

1 In the social movement literature, religious activists have also gained attention, the environmental movement being one next to others, especially on social justice and the empowerment of marginalized groups. Scholars inquire into the role of religious actors in global, transnational, and local social movements (cf. Boehle 2010; Nepstad 2002; Trigeaud 2012). Mutual exchange between the fields could be much more vital.

2 I elaborate on more results from that study in Gojowczyk (2020). While many informants knew that the Pope was working on a document "on the environment," the data was collected essentially before the encyclical *Laudato Si'* was published.

3 *Sapat na* means "what is enough" or "enough" in Tagalog, one of two official languages in the Philippines. One informant additionally describes this horizon with the following words: "If people would learn to live with what is enough, what is . . . the real need."

4 Names have been changed.

5 It was the *Ökumenisches Umweltforum für OrdenschristInnen*, 2014. They discussed aging convents.

References

Amri, U., 2014. From Theology to a Praxis of "Eco-Jihad". *In:* R.G. Veldman, A. Szasz, and R. Haluza-Delay, eds. *How the World's Religions Are Responding to Climate Change: Social Scientific Investigations*. Abingdon: Routledge, 75–93.

Baugh, A.J., 2019. Explicit and Embedded Environmentalism: Challenging Normativities in the Greening of Religion. *Worldviews: Global Religions, Culture, and Ecology*, 23 (2), 93–112 [online]. Available from: https://brill.com/abstract/journals/wo/aop/article-10.1163-15685357-02301002.xml [Accessed 17 April 2019].

Boehle, J., 2010. The UN System and Religious Actors in the Context of Global Change. *Cross Currents*, 60 (3), 383–401.

Bohnsack, R., 2014. Documentary Method. *In:* U. Flick, ed. *The SAGE Handbook of Qualitative Data Analysis*. London: SAGE, 217–233.

Bohnsack, R., Nentwig-Gesemann, I., and Nohl, A.-M., 2013. *Die dokumentarische Methode und ihre Forschungspraxis: Grundlagen qualitativer Forschung*. Wiesbaden: Springer.

Darlington, S.M., 2012. *The Ordination of a Tree: The Thai Buddhist Environmental Movement*. Albany: SUNY Press.

Dohe, C., 2022. "What Does Religion Have to Do with Nature Conservation?" Investigating the Tensions in an Interreligious Nature Conservation Project in Germany. *In*: J. Koehsen, J. Blanc and F. Huber, eds. *Religious Environmental Activism: Emerging Conflicts and Tensions in Earth Stewardship*. London: Routledge.

Ellingson, S., 2016. *To Care for Creation: The Emergence of the Religious Environmental Movement*. Chicago: University of Chicago Press.

Feldbauer-Durstmüller, B., Sandberger, S., and Neulinger, M., 2012. Sustainability for Centuries – Monastic Governance of Austrian Benedictine Abbeys. *European Journal of Management*, 12 (3), 1–10.

Glaab, K., 2017. A Climate for Justice? Faith-Based Advocacy on Climate Change at the United Nations. *Globalizations*, 14 (7), 1110–1124.

Gojowczyk, J., 2015. Isomorphe Innovation: Umweltmanagement in Evangelischen Landeskirchen in Deutschland. *In:* A. Arnold, M. David, G. Hanke, and M. Sonnberger, eds. *Innovation – Exnovation: Über Prozesse des Abschaffens und Erneuerns in der Nachhaltigkeitstransformation*. Marburg: Metropolis-Verlag, 47–62.

Gojowczyk, J., 2020. *Umweltschutz in katholischen Orden: Interpretieren, Bewerten und Verhandeln als Teilprozesse der Glokalisierung*. Wiesbaden: Springer.

Hall, M., Grim, J., and Tucker, M.E., 2009. Need for religions to promote values of conservation. *Nature*, 462, 720.

Haluza-DeLay, R., 2014. Religion and Climate Change: Varieties in Viewpoints and Practices. *WIRES Climate Change*, 5 (2), 261–279.

Hancock, R., 2018. *Islamic Environmentalism. Activism in the United States and Great Britain*. Abingdon, New York: Routledge.

Herman, L., 2022. Kosher Electricity and Sustainability. Building Block or Stumbling Stone? *In*: J. Koehsen, J. Blanc, and F. Huber, eds. *Religious Environmental Activism: Emerging Conflicts and Tensions in Earth Stewardship*. London: Routledge.

Hitzhusen, G.E., and Tucker, M.E., 2013. The Potential of Religion for Earth Stewardship. *Frontiers in Ecology and the Environment*, 11 (7), 368–376.

Jamil, S., 2022. Halal Wastewater Recycling. Environmental Solution or Religious Complication? *In*: J. Koehsen, J. Blanc, and F. Huber, eds. *Religious Environmental Activism: Emerging Conflicts and Tensions in Earth Stewardship*. London: Routledge.

Jenkins, W., Berry, E., and Kreider, L.B., 2018. Religion and Climate Change. *Annual Review of Environment and Resources*, 43, 85–108.

Jenkins, W., and Chapple, C.K., 2011. Religion and Environment. *Annual Review of Environment and Resources*, 36, 441–463.

Kearns, L., 1996. Saving the Creation: Christian Environmentalism in the United States. *Sociology of Religion*, 57 (1), 55–70.

Moyer, J.M., and Scharper, S.B., 2019. The Fabric of Faith-Based Environmentalism in Canada: Exploring the Strands of Ecological Worldviews. *Worldviews: Global Religions, Culture, and Ecology*, 23 (1), 33–58.

Nepstad, S.E., 2002. Creating Transnational Solidarity: The Use of Narrative in the U.S. – Central America Peace Movement. *In:* J. Smith and H. Johnston, eds. *Globalization and Resistance*. Lanham: Rowman and Littlefield Publishing Group, 133–149.

Nche, G.C., Achunike, H.C., Okoli, A.B., and Diara, B.C., 2017. Challenges of Climate Change and the Culpability of Churches: Towards an Effective Church Climate Change Action in Nigeria. *Missionalia: Southern African Journal of Mission Studies*, 45 (2), 168–187.

Nohl, A.-M., 2012. *Interview und dokumentarische Methode: Anleitungen für die Forschungspraxis*. Wiesbaden: VS Verlag für Sozialwissenschaften.

O'Brien, J., and Palmer, M., 2007. *The Atlas of Religion*. Berkeley, Los Angeles: University of California Press.

Proctor, J.D., and Berry, E., 2005. Social Science on Religion and Nature. *In*: B.R. Taylor, ed. *The Encyclopedia of Religion and Nature*. London, New York: Thoemmes Continuum, 1571–1577.

Ronan, M., 2017. Religion and the Environment: Twenty-First Century American Evangelicalism and the Anthropocene. *Humanities*, 6 (4), 92.

Smith, J.D., 2017. Connecting Global and Local Indonesian Religious Environmental Movements. *Kawistara*, 7 (3), 207–225.

Taylor, B., 2005. Critical Perspectives on "Religions of the World and Ecology". *In:* B.R. Taylor, ed. *The Encyclopedia of Religion and Nature*. London, New York: Thoemmes Continuum, 1375–1376.

Taylor, B., 2010. *Dark Green Religion: Nature Spirituality and the Planetary Future*. Berkeley, Los Angeles, London: University of California Press.

Taylor, B., 2017. Book Review of S. Ellingson, 2016. *To Care for Creation: The Emergence of the Religious Environmental Movement. American Journal of Sociology*, 123 (3), 929–931.

Taylor, B., Van Wieren, G., and Zaleha, B., 2016. The Greening of Religion Hypothesis (Part Two): Assessing the Data from Lynn White, Jr, to Pope Francis". *Journal for the Study of Religion, Nature and Culture*, 10 (3), 306–378.

Trigeaud, S.H., 2012. Christian Prayer for Human Rights and Peace: A Spiritual Or Civic Commitment? *In:* G. Giordan and E. Pace, eds. *Mapping Religion and Spirituality in a Postsecular World*. Leiden: Brill, 99–116.

Tucker, M.E., and Grim, J., 2001. Introduction: The Emerging Alliance of World Religions and Ecology. *Daedalus*, 130 (4), 1–22.

Tucker, M.E., and Grim, J., eds., 1997–2004. *Religions of the World and Ecology Book Series*. Cambridge, MA: Harvard University Press.

Tucker, M.E., and Grim, J., 2017. The Movement of Religion and Ecology: Emerging Field and Dynamic Force. *In:* W. Jenkins, M.E. Tucker, and J. Grim, eds. *Routledge Handbook of Religion and Ecology*. London, New York: Routledge, 3–12.

Veldman, R.G., Szasz, A., and Haluza-Delay, R., 2014. Social Science, Religions and Climate Change. *In:* R.G. Veldman, A. Szasz, and R. Haluza-Delay, eds. *How the World's Religions Are Responding to Climate Change: Social Scientific Investigations*. Abingdon: Routledge, 3–19.

Vincentnathan, L., Vincentnathan, S.G., and Smith, N., 2016. Catholics and Climate Change Skepticism. *Worldviews: Global Religions, Culture, and Ecology*, 20 (2), 125–149.

Witt, J.D., 2016. *Religion and Resistance in Appalachia: Faith and the Fight Against Mountaintop Removal Coal Mining*. Lexington: University Press of Kentucky.

Witt, J.D., and Taylor, B., 2017. Special Issue Introduction: Religion and Eco-Resistance Movements in the 21st Century. *Journal for the Study of Religion, Nature and Culture*, 11 (1), 5–22.

3 Cosmological Tensions

Biodynamic Agriculture's Anthropocentrism and Its Contestation

Stéphanie Majerus

Introduction

In 1924, the anthroposophist Rudolf Steiner gave a series of lectures on agricultural renewals, effectively establishing the field of biodynamic agriculture. Since then, biodynamics has come to oppose the scientization of agriculture and artificial fertilization methods. However, Rudolf Steiner and biodynamic farmers do not call for a return to pre-industrial agriculture. Rather, they propose to work with specific anthroposophical methods, which seek to enhance plant growth and resilience.

The biodynamic milieu is not homogeneous, however; not all farmers in Germany and Switzerland agree with the official anthroposophical premises and Steiner's worldview. In particular, newcomers with no background in anthroposophy have ambivalent opinions about the anthropocentric assumptions underlying their work. Therefore, the question guiding this chapter is: Which possible tensions emerge between the lateral entrants and longtime Steiner followers regarding the anthroposophical anthropocentrism which states that the human builds the core of the cosmos?

In the first section of this chapter, I will outline the historical context in which biodynamic farming emerged and evolved, describe the characteristics and aims of biodynamic farming, and name the most important institutions that shape and support biodynamic agriculture. Through this overview, it will become clear that although a pluralization of the biodynamic milieu is taking place, this process of diversification is limited because stable institutions such as the Demeter label, supermarket chains, anthroposophical schools, anthroposophical banks, and so on support biodynamic agriculture.

I will continue with a summary of the ontological concept of nature in official anthroposophical writings, subsequently comparing the anthroposophical-anthropocentric concept of nature with the ontology proposed by deep ecology since the middle of the 20th century. Deep ecology as well as Dark Green Religious movements are prominent concepts in eco-spiritual milieus and are subtly challenging biodynamics.

DOI: 10.4324/9781003017967-4

The following section will outline the ethnographic methods used to gather data, before an in-depth analysis of current tensions on the basis of an ethnographic fieldwork. The chapter ends with concluding remarks on the tensions between anthropocentric assumptions defended by longtime Steiner followers in biodynamic agriculture, and its contestation by newcomers with no background in anthroposophy. This chapter reveals that ultimately the eco-spiritual milieu is not homogenous, nor is the biodynamic milieu: Cosmological tensions are unfolding in eco-spiritual milieus because the ontology on which they rest is contested, not only from the outside, but also from the inside.

Agricultural Industrialization and Anthroposophical Opponents

The evolution of the field of biodynamics should be understood against a backdrop of broader societal developments related to the agricultural sector. Since the beginning of the 20th century, science, technology, political ideologies, urbanization, agricultural policies, and capital flows have led to fundamental changes in agricultural production. Financial pressure is increasing; political objectives, industrialization, the use of synthetic fertilizers, and the scientization of agriculture mean that new investments are often inevitable. The Common Agricultural Policy (CAP, introduced in 1962 in Europe) brought the guarantee of direct aid, but it also accelerated monocultural agriculture and dairy farming in Europe. In the course of the intensification of agriculture, animals have been increasingly reduced to food-producing creatures. Importantly, the shift from small to large farms undermined everyday contact between farmers and livestock (cf. Mooser 2000, p. 125; Münkel 2009, pp. 61–67; Uekötter 2012, p. 12).

Biodynamic agriculture reacts to these developments. It originated in the 1920s and was mainly influenced by Rudolf Steiner, the founder of anthroposophy. At the time, Steiner turned against artificial fertilizers and the associated scientization of agriculture as well as the decline in seed and food quality after the First World War. He spoke out against soil compaction and salinization, topics which were broadly discussed throughout Germany (cf. Vogt 2000, p. 98). Although Rudolf Steiner thought opposes the scientization of agriculture and promotes more traditional methods, he does not simply call for a return to pre-modern times.[1] Instead, Steiner's biodynamic agriculture offers an alternative ontology and praxis.

Biodynamic agriculture, its monistic ontology and attendant methods, rests on anthroposophical premises. Anthroposophy is a worldview that claims the existence of a spiritual realm beyond the material world; hence, the spiritual and physical realms are entangled, or two sides of the same coin, and this supposed spiritual realm could be perceived by training one's faculty of clairvoyance. In terms of its general practice, anthroposophy suggests that insights gained by such extrasensory means should guide one's daily work (cf. Zander 2019, p. 8).

In June 1924, following a series of lectures by Rudolf Steiner, which are now available as a book under the title *The Agricultural Course* (1999 [1924]), anthroposophical agriculture came into being. At this point, anthroposophy had already established itself in other social fields such as medicine and education. It offered alternatives in these areas, such as the Waldorf schools and Weleda products, which still exist today.

More Than Simply Sowing Seeds: The Characteristics of Biodynamic Farming

In this section the specificities of biodynamic agriculture will be detailed in order to clarify that this form of agriculture builds on specific procedures and a worldview informed by anthroposophy. Likewise, this section will outline premises and practices, which for novices might not be integrated ad hoc and therefore are possible causes for debate and even tension.

Rudolf Steiner gave his lectures at the estate of Count Karl von Keyserlink over the course of a weeklong conference, outlining principles of an anthroposophical agriculture as part of cosmic influences. This contrasted to the materialistic agricultural science focused on biochemical functions. According to Steiner, anthroposophical farmers should optimize the action of etheric and astral forces to counteract the prevailing problems in agriculture. Steiner encouraged farmers to work with *biodynamic preparations*, methods allegedly clairvoyantly elicited by himself (cf. Zander 2007, p. 1587).

Biodynamic Preparations

Biodynamic preparations are synthetizations of mineral, plant, and animal substances intended to stimulate the vitality of the soil and plants. Generally, farmers apply two different processes while working with preparations: spray and compost. Spray preparations optimize the fertility of the soil and plants; compost preparations refine the consistency of the compost and the dung heap and should ultimately revitalize the organic fertilizer produced from them. However, the aim here is not simply to achieve a viable fertilizer – it is also to lead farmers to a deeper insight into nature and put them on a spiritual path, a kind of initiation process (cf. Foyer 2018). For example, when stirring the water during the production of spray preparations, a vortex will manifest itself that should unite the entire power of the cosmos, enabling the farmer to perceive and integrate a bond with all living beings. In such instances, the preparations have an identity-forming effect on biodynamic farmers: they are perceived as "special things"[2] (Taves 2011). Rudolf Steiner describes them as an "extraordinary secret" (Steiner 1999 [1924], p. 3), and this mystery forms the core of anthroposophical agriculture.

According to Steiner, they are a mystery because the effect of the preparations can only be experienced in a "spiritual" realm (Steiner 1999 (1924), p. 3) and cannot be fathomed through conventional natural science.[3]

Human beings are well placed to intervene in the spiritual-physical matrix through the use of preparations because the preparations are a cultural invention of humans to optimize the natural forces. The Anthropos therefore has the ability to transform nature for the better, as humans are incarnated beings with an especially refined consciousness.

Planetary Constellations, Farm-Organism, and Circular Economy

The preparations are not the only key tenet of biodynamics. *The Agricultural Course* also encourages farmers to base their work on planetary constellations. This is because cosmic influences like etheric and astral forces allegedly originate from the planets: If farmers observe planetary constellations, cosmic forces could be transferred to the plants via lime and pebbles (cf. Zander 2007, p. 1586).

The meaning and association of planets and plants is deduced from analogical thinking or correspondence-based thinking. For example, the color red establishes a connection between the planet Mars and the rose: "Then you look at the rose, and in its red color you look at the Martian force" (Steiner 1999 [1924], p. 52). Another example would be the sunflower, whose "yellowness" connects the plant to Jupiter (Steiner 1999 [1924], p. 52).[4]

Another main feature of biodynamic agriculture is the assumption that every farm forms an organism. Biodynamic cultivators see the stable, the forests, the flora and fauna, and the farm community as united: People and animals, technical material, economic reality, and location all form a whole (cf. Besson 2011, p. 51). This holistic conception of the farm is also closely linked to the Demeter ideal of cultivating in circularity: The whole cultivation process should rest on a circular economy, which means, for example, that the number of kept animals is linked to the farm's forage capacity. Thus, the purchase of feed is heavily regulated (cf. Klett 2011, p. 614; Association pour la Biodynamie 2019, p. 20).

Goetheanism and Scientific Studies

As previously mentioned, anthroposophic knowledge is to be generated through suprasensual methods such as the evaluation of the effect of the preparations – a further typical characteristic of biodynamic farming.

One such method spoken of in anthroposophical circles is the "Goethean method."[5] On the AnthroWiki page (anthrowiki.at), an online encyclopedia managed by anthroposophists, this method and the knowledge to be gained through it are described as a holistic method, linked to a phenomenological scientific process. It largely disregards measuring instruments and quantitative evaluations and focuses on qualitative experiences. Although anthroposophists emphasize the phenomenological aspect in their definition on their AnthroWiki page, they assume that this method protects from speculative elements and is therefore objective. Thus, in order to truly recognize a being (such as the constitutive spiritual morphology of a plant), one should apply the "Goethean method." This means that impartiality, openness, and curiosity during sensual observations, coupled with a processual perception of plants, should lead to the recognition of the primordial plant. In Steiner's cosmology, the primordial plant forms the invisible principle that guides growth and life. As such, the primordial plant represents an ideal type that manifests itself in the world in various ways through concrete plants (cf. Choné 2013, p. 21, 2017, pp. 278–279).

In this sense, anthroposophists claim to differ from academic positivism, since they not only register phenomena, but also view them in their ideal-spiritual context. Therefore, true knowledge about the origin of life might be gained by humans in the spiritual world, via anthroposophically trained perception skills.

In the previous description, it becomes clear that anthroposophical knowledge quests differ from academic sciences along their supersensory parameters. Especially in anthroposophy's early days, it claimed that objective knowledge was possible through supersensory approaches. In his last writing phase, however, Steiner partially relativized the claim to objectivity: In particular, he emphasized "empathic considerations," the importance of the imagination, and the metaphorical quality of his language (cf. Clément 2018, pp. cxii–cxiii). Today the notion of objectivity remains contested within anthroposophy. Anna Cecilia Grünn – an agricultural gardener in her early thirties from southern Germany – claims that one's own subjectivity is part of supersensible perceptions. In her book on "extrasensory conversations" with "nature spirits," she writes that all conversations with "nature spirits" are subjectively shaped. These conversations are not a matter of direct, linguistic communication, but of feelings and images that she translates into words (cf. Grünn 2009, p. 11).

Additionally, since the 1970s, research into biodynamic agriculture has been carried out at universities. The research projects are based on system comparison experiments (conventional/organic/biodynamic), and try to measure possible effects of the biodynamic preparations. Methodologically, the studies are conducted in accordance with academic natural science procedures, but in individual cases, they are supplemented by an anthroposophical spiritual science perspective.[6] These studies demonstrate that the biodynamic movement tries to integrate natural sciences but also to exceed them with anthroposophical methods.

In this section I outlined that biodynamic agriculture is derived from anthroposophy, an esoteric-occult worldview, in which nature is conceived as a "spiritual-physical matrix" that can be enhanced by biodynamic preparations (cf. Vogt 2007, p. 19).

In the following section I will deepen the biodynamic understanding of nature and by doing so, once more explain the underlying anthropocentrism of biodynamics. The official anthroposophical ontologie is occasionally contested by novices with no background in anthroposophy[7]; thus, a generation of young lateral entrants that wants to challenge certain features of biodynamics.

Deep Ecology vs. Biodynamics

Steiner perceives nature as alive, as animate and part of a world in which the human as a microcosm reflects the universe, the macrocosm. Anthroposophy assumes that there are four levels of existence with the first level being the physical, to which three further levels are added: the ethereal (a universal force without intentionality but able to animate, especially the growth of plants); the astral (which enables phenomenal consciousness); and the "I" consciousness (a quality of being that enables conscious reflection). The "I" consciousness and ability to understand through thinking is claimed to be an exclusively human capacity (cf. Vogt 2000, pp. 53–54).

In informal and everyday conversations, however, biodynamicists focus on the distinction between the spiritual level and the material level.

In Steiner's monistic conception, the physical and spiritual levels are intertwined. This is evident in *The Agricultural Course*, where Steiner emphasizes that nature forms a whole that is permeated by the forces of the cosmos. Only those who pay attention to this fact could understand nature (cf. Steiner 1999 (1924), p. 175). Like other spiritual worldviews that emerged in the 19th century, Rudolf Steiner speaks in many lectures of a spiritual evolution that aims at betterment; the underlying ideas claim that the humans and the earth go through developmental stages that aspire to continuous perfection (cf. Haller 1995, pp. 95ff). Steiner also combined theories on race with spiritual evolutionism in his earlier work. In this spiritual evolutionism, the Caucasian race is elevated as the most supreme, often through applying a mythological vocabulary. Although the topic remained marginal in his cosmological conception after he left the theosophical society, Steiner employed racist stereotypes to describe non-European cultures until the end of his life (cf. Martins 2012, pp. 133, 142 original quotes in Steiner 1980 (1923)). Moreover, a new understanding of nature developed following Darwin. After Darwin's publications were broadly acknowledged, nature was no longer thought of as unchangeable, but rather as something that could be developed and controlled (cf. Radkau 2005, p. 237). In this sense, it is perhaps not surprising that Steiner's preparations were presented as an extraordinary remedy that helps soils and plants improve continuously, nor that human intervention in nature is a necessary aspect of biodynamics.

Daniel McKanan writes that anthroposophically and biodynamically shaped environmental activism sees itself as part of a cosmic-evolutionary event. As such, biodynamicists rather rarely use terms such as "resilience" or "sustainability," unlike more recent environmental organizations. Anthroposophy also affirms a planetary transmutation or regards it as an inevitable evolutionary reality (cf. McKanan 2017, pp. 238, 241).

In this regard biodynamics differs from other ecological ontologies like, for example, deep ecology, at least if one looks at the statements of its main representatives, Arne Naess (1912–2009) and Joanna Macy (1990). Norwegian philosopher Arne Naess first used the term deep ecology in an essay in 1973. To sum up, deep ecology starts from a biocentric – also called eco-centric – perspective on nature. The emphasis is on the intrinsic value of all life. The human is not thought of as a single, separated entity but as deeply embedded in its surroundings. Therefore the deep ecology milieu links its analysis to a critique of anthropocentrism. In these analyses, Christianity, capitalism, and the dualistic thinking of Western philosophy are regarded as responsible for the decoupling of occidental man from his environment (cf. Taylor 2010, p. 13). Some adherents of Dark Green Religion are influenced by deep ecology. According to Bron Taylor, the movement of Dark Green Religion sees nature as "sacred, imbued with intrinsic value, and worthy of reverent care" (Taylor 2010, p. ix) and mostly rejects anthropocentric worldviews. For Taylor this movement replaces in some way the traditional religions and is distinct from the greening of traditional faiths. For deep ecologists and adherents of Dark Green Religion, nature is a miracle and will never be fully understood by humans, whereas the anthroposophist's claim that objective knowledge is attainable in the spiritual realm.

The German Studies professor Aurelie Choné, for her part, recognizes similarities between deep ecology and biodynamics. She emphasizes that both worldviews focus on cooperation between different organisms (such as bees and plants) and consider the ecological organism of Earth as a whole (cf. Choné 2013, pp. 33f). It is quite legitimate to observe similarities in both milieus, but in this chapter, it has become clear that differences also persist. In biodynamic agriculture, the human, Anthropos, and its unique self-reflexive consciousness play a central role in Steiner's scriptures. This aspect as well as the claim of the existence of "higher realms" as described by Steiner, is largely absent from works of deep ecology and is mostly absent in Dark Green Religious movements.

Some novices in biodynamic farming do however challenge the anthropocentric notion of anthroposophy and biodynamics. Newcomers in the 21st century are influenced by the biocentric tendencies of Dark Green Religion and are challenging the established biodynamic movement. But although these disparities are noticeable, as we will see in the last section of this chapter, anthropocentric-anthroposophical notions are still present among German-speaking biodynamic farmers. Before developing this argument, the next section briefly summarizes how the biodynamic movement already adapted to different societal changes and is shaped by institutional dynamics during the last century.

Developments After Steiner's Death

The characteristics described in the previous sections unfold within a specific historical and cultural context. The biodynamic movement is a response and reaction to other societal changes that have occurred over the last hundred years; thus it is entangled with broader societal dynamics. Furthermore, the movement was able to establish rather stable and far-reaching institutions over the course of the last century, which in some sense prevent a completely heterogeneous pluralization of the movement. In the following section, I will address these aspects.

In 1932 the Demeter trademark and label was patented and biodynamically produced food under the label Demeter was distributed in Germany in *Reformhäuser* (organic food stores), thereby ushering in the life reform movement *(Lebensreformbewegung)* (cf. Hurter 2014, pp. 14f). While non-hegemonic movements were in favor of Demeter agriculture and food, its reception was controversial within conventional farming circles. It was challenging the widely accepted modern narrative of scientific optimism in the interwar period, and the agro-chemistry industry predicted that biodynamic agriculture had no chance to survive (cf. Uekötter 2012, p. 415). During the Nazi regime, an alliance between some biodynamic farmers and the agricultural minister Richard Darré took place since the mid-1930s, which helped biodynamic agriculture spread. However, leading Nazis banned most of the anthroposophical and biodynamic associations in 1941 (cf. Staudenmaier 2014, pp. 135, 143). In the postwar era, organic agriculture remained in the political margins. Not until the mid-1980s, when the environmental damage caused by intensive agriculture became a political issue, did the environmental movement become increasingly interested in understanding alternative agriculture – at least in Germany (cf. Uekötter 2012, pp. 236–237, Zander 2007, p. 1599).

The perception of alternative farming was not the only change in the second half of the 20th century; the social orientation of organic farming and the ecological movements entirely changed as well. They situated themselves more and more on the political left, and environmental movements became increasingly institutionalized and professionalized (cf. Uekötter 2014, p. 7). According to historian Günther Vogt, the publications of biodynamicist Manfred Klett also shifted the focus from the preservation of smallholder structures to environmental issues in the second half of the 20th century (cf. Vogt 2000, p. 184). Furthermore, the 68 generation's criticism of the social structures in the 1970s and 1980s also reinforced the aspiration among young people to move to the countryside and initiate socially inclined projects. In this cultural shift after the Second World War, the desire to regard biodynamic agriculture as a social project was enhanced. Former university students founded new farms, which became an integral part of anthroposophical homes or institutions. In contrast to these developments, conventional agriculture relied progressively on the division of labor.

In the 21st century, organic food and especially biodynamically produced wine are becoming increasingly popular, which causes new paradoxes: Viticulture is a monoculture, which does not implement the ideal of a cycle-oriented agriculture (cf. Hurter 2014, pp. 15f; Von Plato 2003, pp. 49ff; Grandjean 2022). The line between spiritual and economic aspirations might be fuzzy among winegrowers, and it may not be far-fetched to state that sometimes the economic reality takes clear precedence over biodynamic aspects. Nonetheless, some farmers and entrepreneurs in the biodynamic movement seek alternatives to purely capitalist appropriations; for example, a range of farms have implemented CSA models (Community Supported Agriculture). This financing model should ensure that food-producing farms are provided with their revenue before the harvest season. Clients subscribe to a regular purchase in advance, enabling the workers to plan their budgets with greater certainty, to rely less on loans, and to involve customers in production processes (cf. Roquebert 2018, p. 179). Inside the biodynamic movement, the panoply of economic and social orientations has diversified over the last decades, which occasionally causes some internal debates.

Another example of a recent reaction to contemporary topics is the "Farmers for Future"[8] initiative, launched in early 2019. The 33-year-old Demeter farmer Jakob Schererz was inspired to launch this initiative by the "Friday for Future" youth movement. In their statement, Demeter farmers call on politicians to reduce emissions from livestock farming by limiting the livestock farming to feed production capacities of agricultural units. However, this is not the Demeter Association's first appeal to politicians to address climate protection measures: Together with four other associations, the Demeter association published an appeal to act against climate change in 2015.[9]

As previously mentioned, biodynamics is institutionally well rooted. This is not only because the Demeter association manages the distribution of biodynamically produced food, but also because a magazine for members is distributed, conferences are held at the Goetheanum annually, training courses take place all over the year, supermarket chains like Alnatura and DM[10] are managed by anthroposophists and, last but not least, a biodynamic agricultural school is supported by the Demeter association. There are anthroposophically inclined banks like the GLS (Gemeinschaftsbank) founded in 1974, and the Triodos bank that has operated since 1980 in the Netherlands. Since its inception the Triodos bank has spread to France, Belgium, Spain, Germany, and the United Kingdom, and now has more than 700,000 customers. Both banks fund projects in the social and ecological sectors,[11] and many of them – such as Waldorf schools or Demeter farms – have anthroposophical backgrounds. As such, these banks support structures which help diffuse and materialize anthroposophical ideals, though in very subtle ways. As a result of this extensive institutionalization, 5,595 Demeter farms have been established worldwide. There are 1,552 agricultural units in Germany, 100 in Great Britain, and 118 in the United States. Switzerland counts 225 Demeter farms and India 403.[12]

In the highly networked and mobile world of the 21st century, knowledge about biodynamic agriculture circulates nationally and internationally through various channels, which further enhances the presence of biodynamic ideals and the Demeter label. This knowledge influences many different groups: countercultural movements, private companies, environmental NGOs, and other actors of civil society (cf. Choné 2017, p. 274). Even some politicians are in favor of the biodynamic approach or anthroposophy. The Green Ex-Federal Minister Renate Künast, who held the office for Food, Agriculture, and Consumer Protection from 2001 to 2005, was criticized for her proximity to anthroposophy in a widely read German newspaper (cf. Treue 2002, p. 12).

Methodology

The data for this chapter is based on semi-structured interviews, field research, and text analysis.

The fieldwork took place on six different farms in Switzerland and Germany, at three international conferences at the Goetheanum in Dornach (CH), and during workshops. I was involved in many informal talks during coffee breaks at conferences and workshops or during the harvest whilst staying on farms. The overall two months of ethnographic field research from 2017–2020 were documented by means of 50 pages of field notes.

In a first exploratory phase, I conducted ethnographic observations of an array of different events and activities (e.g., harvesting activities; talks during breakfast, lunch, or cooking; work with dairy cows). During the conclusion of this exploratory phase, central categories were identified inductively (e.g., criticism of science and modernity, biodynamic self-understandings, the perspective of newcomers).

During the ethnographic research, I conducted 20 semi-structured interviews. The recorded interviews lasted on average around 60 minutes, whereas informal talks lasted sometimes 3 to 4 hours. The covered questions were around the perception of biodynamics over the course of the farmers' professional life, how they apply preparations, on which level (personal, physical, suprasensual) they situate the effectiveness of the preparations, on anthroposophical assumptions, on their relationality to their ecological surrounding, and so on. My interview technique is mainly informed by Jean-Claude Kaufmann's concept of the "understanding interview" (*entretien compréhensif*) (Kaufmann 2015 [1996]). This means that different interview topics are defined in advance and are adjusted spontaneously depending on the interview partner and the interview situation. The interviews are transcribed and evaluated in accordance with the topics that emerge from the fieldnotes. A successive compression of the material takes place as well, and the data is sorted around certain categories such as "preparations," "anthropocentrism," and "rationality."

Furthermore, I was observing daily interactions on the farm, a process that is usually summarized under the term "ethnomethodology"(cf. Bohnsack 1999, pp. 36, 64, 100f). Its aim is to focus on local and situated practices in an ordinary, and for the group, common context.

The profile of the visited farms was very diverse. Some farms, mainly in Switzerland, are small family farms, cultivating around 15 hectare and owning not more than 30 cows. In Germany, however some farms span over 200 hectares, graze around 50 dairy cows and 50 fattening bulls. These bigger farms are mostly owned by a foundation and organized as a cooperation, and create jobs at different levels of the food-processing scales. The profile of the people working and living on these farms was very diverse as well.

I met women and men who came to biodynamics for very different reasons and who also took different training paths. Some had already had contact with this type of farming prior to their professional career, often because they attended a Waldorf school or because they grew up on a Demeter farm. Others had researched extensively before their training in organic farming and decided that the Demeter Agricultural School had the best teaching offer. Others, again, claimed to have been on a spiritual quest and wanted to reorient themselves after a life crisis. But while some agricultural workers were especially attracted by esoteric elements, for others these were unimportant and completely unknown at the beginning.

Cosmological Tensions

In the first part of this chapter, I stated that biodynamic agriculture is primarily directed against the industrialization and scientization of agriculture. The biodynamic preparations form the primary identity of an agriculture that opposes itself to artificial fertilizers and mass production. For biodynamic farmers, preparations are a cultural product, made by humans and therefore capable to enhance natural processes.

The previous paragraphs reiterate that biodynamic agriculture, which emerges from anthroposophy and was centrally shaped by Rudolf Steiner, is based on an esoteric worldview. Moreover, biodynamic agriculture is not static, but influenced by the historical context in which it unfolds (as has been stated under the last point). To understand its ontological ecological concept and the status of humans in biodynamic agriculture in greater depth, the official anthroposophical ontology was compared to deep ecology. Thereby it has become clear that in the anthroposophical core milieu, an anthropocentric point of view is favored.

This section will explore ongoing dynamics, disparities, and possible tensions within the biodynamic movement, with regard to the notion of anthropocentrism. Anthropocentrism is present in the official writings and brochures of biodynamic agriculture, and surprisingly, some farmers addressed it in interviews. Especially long-term biodynamicists with a background in anthroposophical studies defended the idea of humans being at the center of the universe. However, the biodynamic milieu is not homogenous, and lateral entrants within the biodynamic movement sometimes question anthropocentric notions.

Human intervention is often evaluated in biodynamic writings as an opportunity to steer organic life toward optimization. According to the anthroposophical self-understanding, human mediation, especially through the preparations, can have positive effects. Today this view is not uncommon among long-term farmers and they often adhere to this vision. One elderly woman explained to me:

> So the anthroposophical ideal is that we as human beings start to cultivate, we are mediators . . . [and] inside my mind, I want to cultivate something positive. And I believe nature is waiting for us humans to cultivate it. . . . Nature works for itself, but I think there is a lot waiting for salvation through us humans.
>
> (I-2, 04.05.2017, translated)

This statement was not singular; it was expressed in multiple spontaneous conversations, like on the 8 February 2019 when I was standing in front of the Goetheanum – a building designed by Rudolf Steiner – in Dornach, Switzerland, where an international conference on biodynamic agriculture was taking place. During the lunchbreak, I recognized a farmer I had interviewed a year ago. We started to chat and he reiterated his view that humans are not exploiters of earth; that on the contrary, domestication helps plants and animals to develop. He then provided the concrete example of the apple tree, which through human intervention (grafting) is able to grow tasty apples. Somewhat surprisingly, he asked: "Do you know the story *The Little Prince*?" In this story the fox approaches the little prince and asks the prince to tame him. The farmer reads the relationship between humans and their environment similarly: Through active cultivation and domestication, one can turn plants and animals into more capable beings (fieldnotes 08.02.2019).

A few days later, when I was reading brochures I had picked up during the conference, I discovered a lecture transcribed into English by Jean-Michel Florin[13] that he presented at a conference in 2015. The title of his presentation read, "Please Tame Me," and the introduction revolved around two points of view:

> "Please tame me"; this request from the fox to the Little Prince in the book by Antoine de Saint-Exupéry is puzzling: Why should a wild animal like a fox want to be tamed by a human being? The usual way animal domestication is described is precisely the opposite – human beings took animals against their will and domesticated them for their own use. This understanding has polarised our society. People seeking to protect animals believe that the best thing would be to release them

from human captivity into the wilderness. Others see animals purely as objects to be used for milk or meat production.

(Florin 2015, p. 3)

The lecture concludes with the view that an animal-human relationship beyond exploitation is possible and can release creative forces; through our relationships to animals we create something new and different (cf. Florin 2015, p. 3). In the anthroposophical-anthropocentric tradition, Florin's conclusion underlines the beneficial influence that people can have on their environment.

The annual conferences on biodynamic agriculture, which are held in Dornach at the Goetheanum and attract between 500 and 900 people, are probably some of the most important events for the articulation and negotiation of the self-image of anthroposophic agriculture. These conferences are organized in such a way that, on the one hand, main representatives of biodynamic agriculture (such as Jean-Michel Florin or Ueli Hurter) and experienced farmers present their ideas in official lectures. On the other hand, there is a lively exchange among the farmers in workshops. Due to the formal nature of the official lectures and their subsequent publication on the Internet, the conference may have a trickle-down effect that should not be underestimated, as we can assume from the conversation on *The Little Prince*.

The anthropocentric perspective of biodynamics is presented similarly on the AnthroWiki site. As this page is a central reference for anthroposophists, one can assume that it has a far-reaching effect. According to this online encyclopedia, biodynamics relies on a monistic worldview while simultaneously reiterating the modern nature/culture distinction. According to this encyclopedia the primary activity, which produces this divide, is *agriculture*, an activity, which is guided by the mental ability of humans and which transforms, even elevates, nature.[14]

As in every milieu, however, central assumptions can be contested. Above all, some younger biodynamicists (who were mostly between 25 and 40 years old) did not always agree with the anthropocentric perspective. During the salad harvest in a greenhouse an employee in her mid-thirties told me that she grew up on a small organic winery. Later she lived in San Francisco where she had contact with indigenous peoples; she found their view – that all living beings are animated – more convincing than anthroposophy's subdivisions of astral and ethereal beings. She also believed that nature would be better off without human intervention: Nature could find balance by itself (fieldnotes 20.03.2017). According to Bron Taylor, a kind of Dark Green Religion in which "people feel awe and reverence toward the earth's living systems and even feel themselves as connected and belonging to these systems" will inform the "religious future" (Taylor 2004, p. 1002). It could be that this emerging Dark Green Religion will be even more dominant in the coming years and shape newcomers' ideas about agriculture and create some tension between long-term biodynamicists and apprentices.

In an informal conversation during a greenhouse clean-up, a student completing her training as a biodynamic farmer in Rheinau (CH) made a similar statement. She said that nature conservation was "not an issue" in her training while discussing the specificities of biodynamics. Emphasis would be placed on the idea that humans shape the cultural landscape and "give impulses" to its surroundings. The second generation of anthroposophists (by this she meant people who are about 70 years old) see humans as "saviors" in some respects. She found this mentality to be rather "old-fashioned" (fieldnotes 26.04.2017). Biodynamic practitioners therefore do not share all anthropocentric-anthroposophical assumptions.

Another interviewee, a man in his mid-thirties, who has a master's degree in cultural anthropology took part in the 4-year biodynamic course organized by Demeter in Hessen and co-coordinates today a farm on a CSA-model near Berlin. According to him, there are specific dynamics shaping the relationship between the future farmers,[15] who are often career changers, and the established anthroposophical instructors during the apprenticeship:

> Mostly adults joining the biodynamic school have an alternative and leftist background and have hardly any notion of anthroposophy. During the apprenticeship, they are confronted with a completely different worldview. What they consider problematic then is a perceived dogmatism surrounding the teaching of Steiner, his ideas on races, his references to Christianity, and his ideas on the spiritual heritage and evolution of the earth and finally also the anthropocentrism in anthroposophy. . . . More sympathy is there for the biodynamic ideal of a *wesensgemäss* and not a species-appropriate or animal welfare (*artgerecht*) cultivation. But then again difficult questions came up: why should human beings intervene in the developmental process of non-human beings, be the impulse-giver *(Impulsgeber)*; why should humans dominate other beings? . . . I found these discussions always inspiring. Finally, they gave me the possibility to develop plural perspectives in regard to the question what it means to be a human being.
>
> (I-4, 14.04.2019)

Disparities between lateral entrants, young cultivators, and the more established farmers was a recurrent theme. In March 2017, I was invited to a party for the graduating class of the biodynamic agricultural school in South Germany. I recorded some of the encounters in my field notebook, like one with a woman in her early thirties, who graduated from the biodynamic agricultural school:

> She tells me that she worked as an art teacher and manager in a German city with more than 1 billion inhabitants "until she was tired of it" and

> she realized she wanted to reorient herself and wanted to do "something with nature." Therefore, she registered as an intern at a farm in South Germany. In the beginning, she was not aware that Demeter farms have an anthroposophical background and sometimes she found it somewhat surprising that these spiritual aspirations exist, but at the same time there were not many discussions about them, almost as if there was no language for it. At first, she had her difficulties with some of the ideas. She mentions that for example from an anthroposophical point of view the mouse would be regarded as nervous and the energy of the cow would be considered harmonious. She asked herself whether this is not a purely human ascription. Over time, however, her gaze and attitude would have changed, and she somehow could adhere to the idea that different beings have different qualities. She mentions that, yes, one could say that a green fresh leaf emits a different life energy than a barren, brown one.
>
> (fieldnotes 18.03.2017)

That there were lively debates caused by disparities during the course became clear shortly after this encounter when the students of the second year presented some sketches. Two students disguised as their instructors (who are Demeter farmers) at some point shouted, "XY, how come you are not so into anthroposophy?" The addressed student replied: "Yes, in the beginning I was astonished. Even today, I still am. But I must also say that I learned new perspectives on how to see the world that I no longer want to miss" (fieldnotes 18.03.2017).

Finally, this performance not only shows that sometimes disparities are eased with humor, but also (as in the other examples) that students will selectively integrate the biodynamic worldview, be it the young farmer in Berlin who took the opportunity to reflect on the question as to what kind of responsibilities come with being a farmer, or the young apprentice who started to reflect on her relationship with other beings (such as animals, plants) and what kind of feelings they provoke under observation. Possibly this acceptance was accelerated and channeled through the teaching units at the agricultural school, which include perception exercises, based on an anthroposophical background.

In agreement with the analysis that young biodynamic famers in particular adhere selectively to anthroposophical ideas, a biodynamic farmer in her early thirties (with a master's degree in sociology) said many practitioners of the younger generation would be "cherrypickers"; that is, they first adhere to a practice and not to a worldview. By contrast, she argues that the generation of farmers who are retired today felt more attached to the person, writings, and lectures of Rudolf Steiner, as she wrote in an e-mail (E-mail 09.04.2019).

Conclusion

The first part of this chapter outlined that the biodynamic movement emerged as a counter-reaction to the industrialization of agriculture and especially its usage of synthetic fertilizers and scientific expertise. Even today a tension between conventional agriculture and biodynamic farming is omnipresent. In the second section, this contribution presented the institutionalized worldview of biodynamic agriculture, which emerged from a course given by Rudolf Steiner, the initiator of anthroposophy, in 1924. The interviews and fieldnotes show that the official specificities of biodynamics and the anthroposophical worldview addressed in the first part of this chapter are discussed at conferences organized at the Goetheanum and at schools for biodynamic agriculture. I was also able to show that high esteem for human intervention into nature, as stressed in anthroposophy, is still present in the movement today. Nonetheless, lively debates on anthropocentric assumptions, on dogmatism, Steiner's evolutionistic concept, and race theories take place. This is primarily the case, it seems, when novices or so to say lateral entrants enter the movement. The practitioners' heterogeneous backgrounds lead to multiple, diverse, and sometimes even contradictory contributions to biodynamic agriculture. Furthermore, the Demeter Association cooperates with different groups; it influences and is influenced by other actors in the contexts of NGOs, politics, and environmental movements worldwide. This further enhances internal pluralization to the movement as the biodynamic movement adapts to larger socioeconomic and environmental changes. Nevertheless, specific characteristics can still be discerned in Demeter agriculture in Europe. It relies officially on a monistic worldview in which microcosm and macrocosm reflect each other. At the same time, most biodynamic practitioners affirm the modern distinction between nature and culture and in official contexts, they mostly advocate an anthropocentric environmental perspective. According to anthroposophy, humans are well suited to cultivate land because they have the ability to reflect on the world. But as we have seen, these assumptions do not rest unchallenged, especially as apprentices cultivate a biocentric worldview informed by the emerging "Dark Green Religions" and ontologies such as deep ecology in which people consider themselves belonging to a vast ecosystem. At the same time, the high degree of institutionalization of biodynamics hints at the limits of the impact of the recently emerging "Dark Green Religions." Although the biodynamic-anthroposophical premises are contested among biodynamic newcomers, the Demeter association rests on stable institutions (e.g., official guidelines, a magazine, schools) which prevent completely arbitrary changes in Europe, where the movement is strongly established. Over the years, newcomers will integrate or at least get accustomed to a biodynamic worldview through trainings, the biodynamic schools, working groups, and biodynamic writings. Further research should clarify the precise extent of the diffusion of emerging "Dark Green Religions" and biodynamics on a global scale, and the convergence of biodynamics with other environmental movements.

Notes

1 Rudolf Steiner, for example, claimed during one of his lectures that mankind was trying to fertilize the fields with science, and so will cause vegetal degeneration: "Traditions will disappear. People will fertilize the fields with science. The potatoes, the grain, everything will worsen" (Steiner 1999 [1924], p. 15, own translation).

2 For Ann Taves "special things" and "specialness" are generic attributes for things considered religion-like (cf. Taves 2011, p. 58).

3 During the fieldwork, however, it was sometimes difficult to understand how regularly farmers practiced suprasensual perception, as it is not common to discuss this subject and because it clearly remains an intimate, personal practice for some. Some biodynamic famers and wine producers also even seek to distance themselves altogether from anthroposophy (cf. Foyer 2018).

4 Similar analogies also played a role in deciding the compositions of the preparations. For example, one preparation is composed of nettle, which has a bond with the planet Mars because both entities contain iron.

5 There are some other anthroposophical methods as well, like, for example, the "Bildekräfteforschung": www.bildekraefte.de/

6 The first investigations took place at the University of Giessen under the direction of Professor Boguslawski. A total of 19 dissertations have been published until 2018, as Jürgen Firtz emphasized during his lecture at the Goetheanum during the Agricultural Conference 2018. In addition, the FiBL (Forschungsinstitut für biologischen Landbau) in Switzerland has conducted research since 1978. This institute compares the biodynamic (D), organic-biological (O) and conventional (K) cultivation of arable crops (cf. also www.fibl.org/index.php?id=2018).

7 From all the students joining courses in biodynamic farming, approximately 50% have no knowledge of anthroposophy.

8 See: https://farmers-for-future.de/.

9 See: <www.klimaappell-bauernverbaende.de/>.

10 Alnatura founder Götz Rehn claims to read anthroposophical works every day. The net sales of his company is located around 800 millions in 2018. Götz Werner has established the DM drugstore chain in 1973. Today over 3,500 stores are spread over Europe, counting around 59,400 employees (cf. Zander 2019, pp. 14, 74).

11 See: <www.gls.de/privatkunden/gls-bank/zahlen-fakten/>, <www.triodos.com/about-us>.

12 The figures are from early 2019: <www.demeter.net/statistics>.

13 Jean Michel Florin is coordinator of the biodynamic association in France.

14 "Der Natur steht die Kultur (beginnend mit der *Agrikultur*) gegenüber, als jener Teil der Natur, der durch die von der menschlichen Geistestätigkeit geleitete Arbeit umgeschaffen und durch das menschliche Ich neu geprägt wurde, was im Idealfall nicht zu einer Zerstörung, sondern zu einer Erhöhung und geistigen Vollendung der Natur führt." https://anthrowiki.at/Natur

15 Though many young people joining the Demeter agricultural school often have no background in anthroposophy or biodynamics, half of them do, according to my interviewees. Many either went to a Waldorf School (which are based on anthroposophical principles and do organize internships for pupils at a Demeter farm) or grew up at a Demeter farm.

References

Association pour la Biodynamie, 2019. *Directives pour l'agriculture. Portant sur l'utilisation des marques Demeter et Biodynamique ainsi que des marques dérivées et apparentées* [online]. Available from: https://demeter.ch/wp-content/uploads/2019/01/Anbaurichtlinie_2019_f_definitiv.pdf.

Besson, Y., 2011. *Les fondateurs de l'agriculture biologique. Albert Howard, Rudolf Steiner, Maria & Hans Müller, Hans Peter Rusch, Masanobu Fukuoka*. Paris: Sang de la terre.

Bohnsack, R., 1999. *Rekonstruktive Sozialforschung. Einführung in Methodologie und Praxis qualitativer Forschung*, 3rd revised and expanded edition Opladen: Leske + Budrich.

Choné, A., 2013. Les fondements de l'écologie spirituelle chez Rudolf Steiner. *Politica Hermetica*, 27, 15–35 [online]. Available from: https://unistra.academia.edu/Chon%C3%A9Aur%C3%A9lie [Accessed 14 November 2018].

Choné, A., 2017. Changer le monde par l'agriculture? L'influence des théories et pratiques de l'anthroposophie sur les modèles de pensée écologiques alternatifs. *In:* P. Hamman, ed. *Ruralité, nature et environnement; entre savoirs et imaginaires*. Paris: ERES, 275–304.

Clément, C., 2018. Introduction I to Rudolf Steiner Critical Edition. Writings on Anthropogenesis and Cosmology. *Writings on Anthropogenesis and Cosmology, Rudolf Steiner*. Schriften – Kritische Ausgabe (SKA) [online]. Available from: www.academia.edu/37080193/Introduction_I_to_Rudolf_Steiner_Critical_Edition_vol._8._Writings_on_Anthropogenesis_and_Cosmology_2018_.

Florin, J.M., 2015. Bitte Zähme Mich (Please Tame Me), *Wochenschrift für Anthroposophie*, 12, 7–9 [online]. Available from: www.sektionlandwirtschaft.org/fileadmin/SLW/Literatur/Tagungsdoku/2015/Tiere_Tagungsdokumentation_2015.pdf [Accessed 3 April 2019].

Foyer, J., 2018. Syncrétisme des savoirs dans la viticulture biodynamique: Incorporation dans l'expérience et le sensible et trajectoire initiatique. *Revue d'anthropologie des connaissances*, 12 (2), 289–321.

Grandjean, A., 2022. The Negotiation of Self-Identity in Swiss Biodynamic Wine-Crafting. Facets of a Sentient and Practitioner-based Sustainable Agronomy. *In*: J. Koehrsen, J. Blanc, and F. Huber, eds. *Religious Environmental Activism. Emerging Conflicts and Tensions in Earth Stewardship*. London: Routledge.

Grünn, A-C., 2009. *Ellenlang. Meine Reise mit den Naturgeistern durch Deutschland*. Flensburg: Flensburger Hefte.

Haller, J.S., 1995. *Outcasts from Evolution. Scientific Attitudes of Radical Inferiority 1859–1900*. Carbondale, IL: Southern Illinois University Press.

Hurter, U., 2014. *Agrikultur für die Zukunft. Biodynamische Landwirtschaft heute: 90 Jahre landwirtschaftlicher Kurs Koberwitz*. Dornach Schweiz: Verlag am Goetheanum.

Kaufmann, J.-C., 2015. *Das verstehende Interview. Theorie und Praxis*, 2. Revised ed. Konstanz, München: UVK Verlagsges.

Klett, M., 2011. Landwirtschaft und Anthroposophie. Der biologisch-dynamische Landbau. *In*: R. Uhlenhoff, ed. *Anthroposophie in Geschichte und Gegenwart*. Berlin: Berliner Wissenschafts-Verlag, 613–647.

Macy, J., 1990. The Ecological Self. Postmdern Ground for Right Action. *In:* D.R. Griffin, ed. *Sacred Interconnections. Postmodern Spirituality, Political Economy, and Art*. Albany: State University of New York Press (SUNY series in constructive postmodern thought). 35–48.

Martins, A., 2012. *Rassismus und Geschichtsmetaphysik. Esoterischer Darwinsmus und Freiheitsphilosophie bei Rudolf Steiner*. Frankfurt am Main: info3.

McKanan, D., 2017. *Eco-Alchemy. Anthroposophy and the History and Future of Environmentalism*. Berkeley: University of California Press.

Mooser, J., 2000. Agrargeschichte und Kulturgeschichte im 20. Jahrhundert. *In*: P. Nolte and F.-M. Kuhlemann, eds. *Perspektiven der Gesellschaftsgeschichte*. Münch: C.H. Beck, 124–133.

Münkel, D., 2009. Arbeitsplatz Acker – Umbruch in der Landwirtschaft. *In*: Stiftung Haus der Geschichte der Bundesrepublik Deutschland, ed. *Hauptsache Arbeit. Wandel der Arbeitswelt nach 1945*. Bielefeld: Kerber, 61–67.

Plato, B. Von, 2003. *Anthroposophie im 20. Jahrhundert. Ein Kulturimpuls in biografischen Porträts*. Dornach: Verl. am Goetheanum [online]. Available from: www.gbv.de/dms/faz-rez/FD1200402112205568.pdf.

Radkau, J., 2005. *Max Weber. Die Leidenschaft des Denkens*. München, Wien: Hanser.

Roquebert, C.I., 2018. *La critique écologique radicale dans le contexte de l'entreprise: L'ambivalence des démarches de RSE dans les entreprises en Biodynamie*. Nantes: Audencia Group & Université de Nantes.

Staudenmaier, P., 2014. *Between Occultism and Nazism: Anthroposophy and the Politics of Race in the Fascist Era*. Leiden: Brill.

Steiner, R., 1980. *Vom Leben des Menschen und der Erde. Über das Wesen des Christentums*. GA 349. Dornach Schweiz: Rudolf Steiner Verlag.

Steiner, R., 1999. *Geisteswissenschaftliche Grundlagen zum Gedeihen der Landwirtschaft*. Auflagen der Gesamtsausgabe der Werke Steiners. Dornach Schweiz: Rudolf Steiner Verlag.

Taves, A., 2011. Special Things as Building Blocks of Religions. *In*: R. Orsi, ed. *The Cambridge Companion to Religious Studies*. Cambridge: Cambridge University Press, 58–83.

Taylor, B., 2004. A Green Future for Religion? *Futures*, 36 (9), 991–1008 [online]. DOI: 10.1016/j.futures.2004.02.011.

Taylor, B., 2010. *Dark Green Religion. Nature Spirituality and the Planetary Future*. Berkeley: University of California Press.

Treue, P., 2002. Blut und Bohnen. Der Paradigmenwechsel im Künast-Ministerium ersetzt Wissenschaft durch Okkultismus. *Frankfurter Allgemeine Zeitung*, 13 (3), 2.

Uekötter, F., 2012. *Die Wahrheit ist auf dem Feld. Eine Wissensgeschichte der deutschen Landwirtschaft*. Göttingen: Vandenhoeck & Ruprecht.

Uekötter, F., 2014. *The Greenest Nation? A New History of German Environmentalism*. Cambridge, MA: The MIT Press.

Vogt, G., 2000. *Entstehung und Entwicklung des ökologischen Landbaus im deutschsprachigen Raum*. Bad-Dürkheim: Stiftung Ökologie und Landbau.

Vogt, G., 2007. The Origins of Organic Farming. *In*: W. Lockeretz, ed. *Organic Farming: An International History*. Trowbridge: CABI, 9–29.

Zander, H., 2007. *Anthroposophie in Deutschland. Theosophische Weltanschauung und gesellschaftliche Praxis 1884–1945*. 2nd reviewed ed. Göttingen: Vandenhoeck & Ruprecht.

Zander, H., 2019. *Die Anthroposophie. Rudolf Steiners Ideen zwischen Esoterik Weleda Demeter und Waldorfpädagogik*. Paderborn: Ferdinand Schöningh.

4 The Slow Greening of Established Churches in Switzerland

Tensions Between Local Parishes and Church Head Organizations

Christophe Monnot

Introduction

Switzerland, like many European countries, has a special relationship with one or two churches. In this country, the Reformed and Catholic churches are recognized (Stolz and Monnot 2017). They are the two established churches to be discussed in this chapter. As a result of their special status, these churches are institutionalized with a democratic decision-making system, they are close to state administrations, and above all, they enjoy social recognition. Even in spite of the pressure of advanced secularization in European countries, the established churches still have a voice in the public arena (Willaime 2004). Why, in this fortunate situation, have these two established churches, which still have financial means and political support in Switzerland, not raised their voices for the ecological or climate cause? Why have these established churches, which have long been a very important voice in support of the poor and the needy, not led the way on the ecological or climate issue?

The elites of the established churches, especially at the instigation of the World Council of Churches, were among the first to rethink theology to include environmental concerns. However, if the World Council of Churches (WCC's) process "Justice, Peace and Integrity of Creation" (JPIC), which began in Vancouver in 1983, is remembered as the most iconic ecology program, it is worth noting that in Europe, it was the JPIC Basel Conference (May, 1989) that was significant. This conference, which brought together representatives of 120 European churches in Basel, resulted in a resolution to mobilize for the defense of peace, justice, and the integrity of creation. This important intellectual and theological was almost forgotten until Pope Francis's encyclical *Laudato Si'* (Francis 2015) revived the main features of these eco-theological considerations many years later. Although there has long been in-depth theological work on ecology,[1] the encyclical has raised the concern for the environment among the Catholic Church, but also far beyond (since in Switzerland the Reformed Church also largely participated

DOI: 10.4324/9781003017967-5

in the publicization of the encyclical), placing this topic at the forefront of their broader public agenda.

This chapter considers the case of Switzerland and discusses that established churches in Switzerland, namely, the Reformed and the Catholic churches, are more a follower than an active promoter regarding climate change and environmental concern. Notably, faith-based organizations (FBOs) have initiated actions, for example, several ecumenical (Reformed-Catholic) Lent campaigns (e.g., in 2009, "Curbing global warming to help the poor"[2]), which made significant appeals to the Swiss population; however, these campaigns were short term. Additionally, several synodic decisions have been made, but they were often limited to a particular sector, with the decision to subsidize the installation of solar panels on parish buildings being the most important.

That said, the most visible, concrete initiatives are mainly scattered and local actions. Local parishes have improved the energy efficiency of their buildings in collaboration with Greenpeace, for example, the parish of Biel, and some have created shared gardens for flowers, for example, the parish at Montbillant in Geneva or for vegetables in Saint-Imier or both vegetables and flowers at the parish in Chavannes near Lausanne. Several parishes have organized environmental events or lent their spaces for such events, for example, in Bern for the Green Film Festival, and others have organized fasting for the climate. These examples show that many small local projects have been led by Reformed and Catholic parishes. However, the question remains as to why, at the national level, the head church organizations have not implemented major actions or coordinated environmental actions in joint plans. Indeed, the Swiss Bishops' Conference has no representative for the environment. The Evangelical Reformed Church in Switzerland had only – until 2017 – an ethical advisor whose position included environmental concerns (see Schäfer 2008). There are, on the cantonal level, some actions, but these have been limited such as the following grants by synods (e.g., Aargau or Bern) to help local congregations install solar panels on their buildings. Of these projects dispatched in Swiss territory, none has attracted major public attention; they are neither important nor sufficiently visible to demonstrate that established churches have undertaken pioneering work on environmental concerns. However, on the national level, the Christian church-related FBOs have organized several actions to increase environmental and climate change awareness. However, these campaigns are mainly political advocacy work related to Swiss economic policy and development aid in the countries of the Global South.

The central question of this chapter is therefore: Why are initiatives by local actors and parishes not adopted by head church organizations? This inaction is even more surprising because the established churches, in their attempt to manage the significant decline in their membership, must constantly reinvent themselves to remain stakeholders in society. Instead of exemplifying pioneering models of ecology with their church buildings in the middle of the village, they have missed this opportunity for ecological transformation. On the basis of their recent involvement during the climate strikes in Switzerland, they might be assimilating to the mood of the current time: They have become followers rather than pioneers.

A reason for this can be found in the Reformed and Catholic church bureaucracy. Its multiple levels of decision-making prevent bottom-up innovation processes by local actors (Chaves 1993). In Switzerland, Reformed and Catholic parishes have a certain degree of autonomy. They are, for instance, the institutional level where church taxes are collected, own their buildings, which often benefit from heritage protection. The second level is the cantonal level. The established churches are cantonal churches (Landeskirchen), which have a special status of "recognition." Each canton has a synod or church assembly and an executive that leads to the church in the cantonal territory. On the third level are, of course, national umbrella organizations, but religion in Switzerland is mainly regulated at the cantonal level.

At the national level, the most influential religious organizations are the FBOs. They are linked to the Reformed and Catholic churches and extend their activities to all Swiss parishes, but they are national and mostly independent in their governance. FBOs have a central administration at the Swiss national level, but the established churches have a centralized power at the cantonal level. Regarding the established churches, there is one federation of cantonal churches for Reformed, the Swiss Evangelical Reformed Church,[3] which represents them at the national and international level. For the Catholics, there is the Central Catholic Conference (RKZ), which federates the cantonal churches[4] and, of course, the six dioceses that link the parishes of the Roman Catholic Church hierarchy.[5] At the confederation level, the Swiss central organizations of the established churches are weak, unlike FBOs, which have essentially a central power at the national level.

To understand this multilevel bureaucracy, the chapter draws on the perspective of new institutionalism (DiMaggio and Powell 1983). It has been applied by Chaves in the context of the diffusion of women's ordination in the churches in the United States (1996, 1997a, 1997b). Two empirical cases from Switzerland are based on the different pressures that influence institutions to adopt innovation described by new institutionalism. Both parishes, each part of an established church – a Catholic parish in the German-speaking part of Switzerland and a Reformed parish in Geneva in the French-speaking part of Switzerland – are active in the field of ecology. Using these two cases, I demonstrate the profiles and positions of the innovators and then the institutional barriers that prevent their "green" innovations from spreading within the given churches.

The study assesses the strategies implemented by innovators to circumvent the institutional barriers to promote their actions. These players sustain their innovations over the long term through the efforts of specific FBOs related to the established churches. Historically, these FBOs were founded mainly to sustain missionary activities, but presently, these organizations have large budgets for international development collaborations and humanitarian aid and to provide social assistance in Switzerland. For instance, Caritas, Swiss Catholic Lenten Fund for Catholics, or HEKS/Bread for All for Protestants (Reformed) are among the main Swiss FBOs. These FBOs can be innovators in their fields of activity because they do not depend directly on the established church's authority structure.

This chapter is structured as follows. After giving a historical overview of the important intellectual contributions to establishing an ecological theology especially within the WCC, the chapter will describe the new institutionalist perspective. It will focus on three institutional levels where pressures can slow or implement an innovation in established churches. The following section will present the survey method, then the chapter will emphasize two particularly pioneering and ecologically engaged parishes, one in Thurgau in northern Switzerland and the other in Geneva. These two case studies will show the barriers that appear at the three institutional levels. The discussion underlines the tensions between local parishes and their church head organizations. The parishes are innovative but also bring uncertainty about their future into an institution that has historically obtained its social legitimacy by ensuring peace and social stability to the broad society. Ecological innovation has not become a priority, and this is because the level of uncertainty has increased for established churches with the increase of secularization. To conclude, the chapter will underline two types of tension: one is internal to the established churches and stifles local initiatives, and the other type is external, the church's relationship with society, from which it has obtained legitimacy but faces a continuing loss of church members. The problem of ecology adds to this uncertainty and increases the internal tensions, preventing the diffusion of ecological concern within the churches. On "green" innovation, it is the Faith Based Organizations operating beyond the usual church authority that enable local actors to disseminate their innovations.

The (Bad) Influence of Established Churches' Theology and Values

Lynn White, Jr.'s 1966 lecture (published in 1967) is usually referred to as the starting point for the argument that the ecological crisis is a consequence of Christian theology. Without going into the details of the arguments, White's thesis is much more complex than a simple charge against Christianity (Taylor 2016; Taylor et al. 2016). Few have noted that the renowned historian of science also reflects a sentiment shared by some scholars of the late 1960s. A feeling somewhere between the fundamental critique of the military-technical omnipotence of the United States and the concern about the new spiritualties promoted by the countercultural movements. Lynn

White, Jr. was part of this particular time. He even started his thesis by quoting another contemporary intellectual, Aldous Huxley, who questions the Christian roots of the ecological crisis (White 1967, p. 1203).

The Christianity in question had already initiated the first reconsideration of its theology, at least within the WCC as early as the New Delhi Conference in 1961.[6] The process of reconsidering the creation and human relationship with it then developed from the Upsala Conference in 1968, with the "Message of the West German Synod of the Evangelical Church in Germany (EKD) to the Local Congregations" (1969), and was reinforced by the Justice, Peace, and Integrity of Creation process. This process, although it officially began in Vancouver in 1983, has its roots in the Nairobi Conference in 1975. Notably, although many authors have cited White as the first to stress the link between Christianity and the ecological crisis, I argue that White's lecture was a sign of the context of an awareness of the Anglo-Saxon (and mainly Protestant) elite at that time. This argument also helps to explain why White's thesis had such impact. The question then shifts to why, despite the elites identifying a theological problem at a very early stage, no significant turning point in the established Protestant churches has been noted.

On the Catholic side, there is also an awareness of Christian responsibility toward the environmental crisis. As a highlight of this growing concern, John Paul II designated Saint Francis as the patron of ecologists in 1979. He intervened several times to affirm the importance of protecting the natural environment, highlighting the connection between preserving the environment and a correct understanding of morality (Landron 2008, pp. 340–349; Vaillancourt 1997). Notably, several pastoral letters from bishops, like the one by the Filipino bishops in 1988 ("What is happening to our beautiful land?"), address the ecological question. It is above all the priests and theologians who are committed to the eco-theology in the countries of the Global South that are most widely known to make their voices heard by the Vatican.

Among these pioneers is Sean McDonagh, a missionary priest who witnessed the impact of the destruction of the rainforest on the residents of the island of Mindanao in the Philippines. He has written several books since the 1980s (McDonagh 1986, 1990) and participated in various commissions of reflection with the WCC. His perseverance was fruitful, contributing to the writing of the encyclical "Laudato Si' " of Francis (Roewe 2016; Monnot 2020).

Leonardo Boff is another theologian who influenced the encyclical and was conscious of environmental problems in the Global South, first in Brazil, then globally. In the 1990s, he was already initiating a green shift in liberation theology (Martinez Andrade 2016). Subsequently, in 2015, the work of these theologians from the Global South was acknowledged by the Vatican. Eco-theology has been officially recognized since Pope Francis issued "Laudato Si" (2015), a canonical document on the theology of creation, including implicit references to these theologians. A notable read on this point is the Laudato Si's encyclical, with comments by McDonagh (Francis and McDonagh 2016).

There is a gap between the environmental discourse on the part of the established churches and the actions led by the Reformed and Catholic churches. The well-known success of the Pope's encyclical in the run-up to COP 21 has given new visibility to churches' ecological discourse. However, the release of the encyclical was neither significantly preceded nor followed by wide church climate programs or actions (Becci and Monnot 2016; Grandjean et al. 2018). Both on the Catholic and Protestant (Reformed) sides, the elite theologians were critical early on of the inaction on environmental causes of the churches and their members. However, although these voices represent the highest levels of the current theology, the translation of these theological reflections into programs of action has not fulfilled expectations. This gap between discourse and action is the starting point for the institutional analysis that I present.

This paralysis of the established churches is partly because of their emphasis on social programs. On the agenda of the JPIC process, it is especially the first two (justice and peace) that have occupied the churches. These priorities were observed by Koehrsen (2015) in a field survey of the small German town of Emden. The churches were far more engaged in social justice activities than in environmental preservation. The leaders of the three local churches, Lutheran, Reformed, and Catholic, testified that they also identified with the "integrity of creation" and with sustainability and energy transition. However, Koehrsen observed competition between the church priorities of social justice and ecology. In Emden, the churches did not influence the other nonreligious actors involved in the energy transition in the town.

Another quantitative study focused on student environmental engagement in Chile according to religious affiliation. In this quantitative study, Parker noted that again,

> It can be argued that the religious factor provides a generally favorable, but weak, orientation to the energy transition. The religious factor, as a performative social representation, does have an impact, although it is neither decisive nor unequivocal.
>
> (Parker 2015, p. 357)

This study shows again that the process of greening Christian theologies does not affect the actions of churches and their members.

This difference between pro-environmental discourse and church inaction can be illuminated by an analysis based on quantitative data from the 30-year-long longitudinal survey: the Swiss House Panel (Becci et al. 2021). Conservative members of the churches would neutralize the progressive engagement of others. Indeed, the study shows that individuals in Switzerland who define themselves as spiritual, meditative, politically left-wing, and progressives are significantly more committed to environmental protection than the rest of the population. This analysis confirms what Sherkat and Ellison (2007) observed in the United States. For them, religion

> influences political orientations that often inform environmental beliefs and actions. . . . Political conservatism influences environmental orientations mostly through calling into question the seriousness of environmental problems. Political conservatism also dampens environmental political activities.
>
> (Sherkat and Ellison 2007, p. 82)

Thus, conservative members are informed by beliefs about the environment that negate those of progressive members within established churches, which in Europe are not defined by politics but by an affiliation inherited from parents. Looking at the values on the environment by denominational affiliations, the pro-environmental stance of progressive members is counterbalanced by the commitment of conservative members.

In the ecclesial domain, it would therefore seem that there is a pro-environmental theology defended by progressive elites, but that their ideas are then often hindered by conservative actors. As such, churches experience internal tensions between pro-environmental and anti-environmental actors and groups. These tensions are all the more notable because this topic has received little attention in the literature, since it is almost invisible from a social perspective. This tension would even partly explain why studies show contradictory results on the relationship between religion and environmental attitudes. On the one hand, studies have demonstrated that the non-affiliated are more pro-environmental than the Christians (Clements, McCright et al. 2014; Clements, Xiao et al. 2014). On the other hand, studies have suggested a slightly positive influence of religion on environmental attitudes (Tucker 2003; McDuff 2010; Wilkinson 2012; Johnston 2013; Stoll 2015).

State of the Art: Progressive Actors vs. Institutional Pathways

From the perspective of new institutionalism (see Powell and DiMaggio 1991), the disjunction between a normative discourse, such as the theology of creation supported by the established churches, and practice, such as a practical commitment to the environment that is rather small, stems from the pressure for change that is mainly from outside institutions. As Scott and Meyer (1994, p. 2) stated: "The dependence of organizations on the patterning built up in wider environments – rather than on a purely internal technical and functional logic – produces organizational forms that are often rather loosely integrated (or decoupled) structures." That practice only vaguely reflects the discourse of an institution is an important clue to understanding this discourse more as a response to the pressures of the societal environment than as an internal regulation to govern the organization.

In this context, Francis' encyclical letter can be understood primarily as a response to society rather than an exhortation to the Catholic hierarchy to start all it can to protect the integrity of creation. Moreover, Francis makes no secret that he addresses his letter to the whole society, believers or not. Thus, this loose coupling between an external discourse very favorable to the environment and a negligent practice of the churches suggests that the spread of an eco-theology or theology encouraging the preservation of the environment should be studied from a new institutionalism perspective.

This concept is what Chaves suggested in his US study of the diffusion of women's ministry in the US churches (Chaves 1996, 1997a, 1997b). According to him, three levels of institutional pressures would explain the variations in the diffusion of women's ordination in the United States. The first level is the parish level, with its internal pressure (the pressure of the members or the staff). The second level is the institutional pressure outside the denominational field, such as the pressure from society on church head organizations. The third level is the pressure from the internal organizational structure (synods, church parliament of the church head organizations) and the pressure from the FBOs that are part of the established churches milieu but autonomous in their decision and action. These three levels are described while applying them to the diffusion of the care of creation in churches in Switzerland.

Regarding ecology, its diffusion among the established churches recently started in Switzerland. The first environmental interdenominational FBO, Oeku "Church and Environment," was founded in 1986 in Bern by Lukas Vischer, then a WCC staff member, and other pioneering theologians such as Christoph Stückelberger. This work is a direct result in Switzerland of the JPIC process of the WCC. Oeku[7] is especially known in the various parishes for its liturgical books,[8] which it produces annually in September, a liturgical month for the creation for Christians at the initiative of the European Christian Environmental Network since the 1990s.[9] Its action is developed on three levels. First, it informs and trains parishes in ecological commitment (e.g., building insulation, energy saving, worship services on creation).[10] Second, it sets up ecological programs and certifications in line with federal directives on energy (e.g., the certification of the "Green Rooster," which includes numerous federal norms on energy saving and building standards).[11] The third level is political and includes the leitmotif "Churches for a climate of change." This document is about lobbying for ecology and adding the voice of the established churches into the various political debates on climate and the environment.[12] After 30 years of Oeku work, increasing theological awareness, lobbying, and training, concrete programs are being conducted, and the parishes of the established churches are, slowly but surely, improving their carbon footprint. The first synod fund (to help parishes improve the energy efficiency of their buildings) was established in 2004 (Catholic cantonal Church of Aargau). The first parish to obtain its public "Green Rooster" certification was in a small town on the Swiss-German border, on 8 November 2015.[13] This parish is a focus of this chapter. Currently, 22 parishes have been certified (out of ca. 2,800 Catholic and Reformed parishes in Switzerland; Monnot and Stolz 2020, p. 140). Moreover, Oeku estimates that a little less than 10% of the parishes are active in environmental conservation in Switzerland.

The institutions are far from their innovation process moving from discourses to action, especially in the case of the established churches, from eco-theology to churches' active programs for ecology. Drawing on new institutionalism, Chaves (1996, 1997a, 1997b) outlined three levels of pressures that promote innovation. For the first level, this scholar noted – on women's ordination – two types of internal pressure: claims from the members and the clergy shortage. This second pressure is ambivalent because

> Neoinstitutionalist logic, . . . while not denying the market forces that lead congregations faced with a shortage of male clergy to draft women to do the work, emphasizes the likelihood that such market forces will not directly translate into formal rules. In this case, [new] institutionalism counterintuitively predicts that a clergy shortage will not raise the likelihood that a denomination will begin to officially permit women's ordination.
>
> (Chaves 1996, p. 847)

For the members, the new institutionalism perspective shows that individual preferences are guided by the institutional context in which they are embedded. In this regard, internal pressures may have an influence locally but are weak regarding having a rule adopted by the whole church. Consideration of the ecological dimension can be treated in the same terms as the context of the clergy shortage, as long as the context affects concrete, practical local dimensions such as the cost of maintaining a building, its heating, or its energy efficiency. This level is critical, however, because it allows local innovators to start pressing the institution for change. This phenomenon is what Geels in the multilevel perspective called "the niche level, [where] actors in precarious networks work on radical innovations" (Geels 2002, p. 1262). These innovators have tension with other institutional levels.

The second level described by Chaves is institutional pressures outside the denominational field. For ecology, this pressure has been expressed recently by, for example, climate strikes. The churches of the main cities of German-speaking Switzerland reacted positively by ringing the bells for the strike rally or some bell towers stopped at 11:55 to recall that it is "minus five for the climate." Another axis that Chaves raised is the diffusion by interdenominational networks. Diffusion networks have been studied, such as "contagion" (Galaskiewicz and Burt 1991), and show that organizations influence each other. The closer organizations are to each other, the more likely they are to take over the innovations of others. Applied to churches by Chaves, he suggested: "There is good reason to expect that denominations will similarly be influenced more by the denominations with which they are most closely tied" (1996, p. 850). This level is also crucial because it is the level where different "linkages" can be made. This level is where "radical innovations may also gradually stabilize into dominant design," as Geels suggested (2002, p. 1262).

Regarding the environment, the major Protestant and Orthodox denominations in the context of the WCC have influenced each other. This environmentally friendly position has been adopted by the Catholic Church. In his encyclical Laudato Si', the pope mentions Patriarch Bartholomew as a role model in the first article of the letter. This transfer occurs especially if organizations think they have similarities with each other.

> We begin with the observation that linkages may be cultural as well as relational. That is, the cultural understanding that social entities belong to a common social category constructs a tie between them. Such ties, while easily represented in graph theoretic terms, invoke a different substantive imagery from that of direct relations like friendship and exchange. We argue that where actors are seen as falling into the same category, diffusion should be rapid.
>
> (Strang and Meyer 1994, p. 102)

The observation is that flows are increased where the actors involved are perceived as similar (by themselves, and others, and within social institutions more generally). Most obviously, perceptions of similarity provide a rationale for diffusion. Transposed to the denominational field, an innovation taken by the Greek Orthodox may be of interest to Catholics. Bartholomew, Patriarch of Constantinople, is one of the five great traditional patriarchs of the primitive church of which the Bishop of Rome (the Catholic pope) is a part. There are therefore far fewer symbolic barriers to be crossed to justify a theological innovation (with the requirement for churches such as Catholicism and Orthodoxy to inscribe it in a long line of tradition).

The third level is the internal organizational structure. The first aspect is to assess the structure of the organization. As suggested by Brown Zikmund (1986, p. 34), decentralized denominations allow more freedom to congregations to ordain women, and more centralized denominations have more rules and policies for their local bodies. In Switzerland, the parishes mainly own their buildings. They can renovate them and install heat pumps and solar panels. Regarding this point, three cantonal head church organizations, the two in Aargau and the Reformed Church of Bern-Jura-Solothurn (the largest Protestant church in Switzerland), created (small) ecofunds to help parishes install solar panels. The decision is made by the parishes that take care of their buildings.

The second aspect of this level is the FBOs. They work parallel to established churches, such as missions, social welfare organizations, and development nongovernmental organizations (NGOs). In Switzerland, the aforementioned ecumenical organization Oeku "Church and Environment" is campaigning for eco-theology to move from theory to the routine practice of churches. Since 1986, this organization has built links with both secular and religious actors and organizations in the Swiss landscape. It is from Oeku that the most demanding and recognized certification "Green Rooster" can be obtained for churches. Such organizations have political and social influence. Inevitably, this influence puts pressure on churches to practice what they preach.

Methods

The data was collected through an ongoing ethnographical study: "Towards a spiritualization of ecology?"[14] The first step was a field observation of the ecological scene in Switzerland. Previously, the same research team was involved in research to assess if spirituality and religion are energy transition enablers in Switzerland.[15] Cultural and political events, ecological events, and festivals enabled the research team to identify a network of spiritual or religious actors. This was all the more dense because our survey began in September 2015, in the middle of the preparations for COP 21 in Paris, a particularly active period for individuals involved in environmental protection and climate concerns. For churches, it soon became clear that the core actors were structured around two Christian and ecumenical FBOs – the Laboratory of Inner Transition (supported by Protestant "Bread for All" and "Swiss Catholic Lenten Fund") and Oeku.

We also identified local actors who promoted shared gardens on the grounds surrounding the church building in the center of the cities, parishes seeking green labels, and groups organizing religious reflections and events around ecology. A search for church documents, church groups, or publications of important actors made it possible to trace the genealogy of a network of church actors.

The network of these local initiatives was then consolidated through participation in two eco-spirituality weekends organized by the Laboratory of Inner Transition. These weekends brought together each time approximately 100 actors involved in the parishes and in the networks of new spiritualties. In addition to field observations, the research team undertook semi-direct interviews with key actors in the ecological and spiritual milieu. Eighteen were conducted with respondents from religious organizations, that is, their involvement was linked to a church or church-related (faith-based) organization. The respondents were involved – beyond the discourse and good spiritual intentions – in concrete, observable, and often pioneering actions for the institutionalized churches or interdenominational organizations.

Two of them were especially relevant for this paper because they represented two of the pioneering actions that involve a whole parish.

In addition, each of them wants to provide their innovations to the Cantonal Church. Each canton has unique laws regulating churches and religious communities. These differences are why the churches in Switzerland have historically been organized into cantonal entities that are almost independent of each other. Each canton has one Catholic and one Protestant organizational structure with its own administration. The recognized churches receive money from the taxpayers of the canton.

The first is the Protestant parish of a wealthy suburb in Geneva, and the second is the Catholic parish in a town on the German border in Canton Thurgau. These two cases are therefore representative of a marginal but growing movement. They are also special because they represent pioneering cases in Switzerland. I present their initiatives as closely as possible to the point of view of the actors, to highlight their reaction to the immobility they feel from their churches. Though based on interviews with the actors, for whom I attempted to communicate the logic of their actions, my analysis considered the institutional context of my ethnographic observations and the interviews I conducted with individuals active at other institutional levels.

Results: Two Environmentally Engaged Parishes Managing Institutional Barriers

For the parish in Geneva, the pastor took advantage of the impressive resources available to him by the members of his parish. In this important, committed parish were many former high-level WCC employees, former university professors, PhDs, engineers in environmental sciences, musicians, philosophers, and ethicists. He set up a working group that worked on a climate justice charter that was approved by the Parish Council and published in French and English (Piguet et al. 2016). Moreover, the parish has remarkable vitality. In a country where participation in Protestant worship is approximately 70 individuals per congregation (Monnot 2012), this parish gathers more than 150 individuals of all ages every Sunday. Its strong vitality and financial health are supported by the parish's ownership of several buildings, providing to the congregation some independence from its cantonal head church organization. The cantonal head church organization, the Protestant Church in Geneva, for the past 20 years, has responded to its financial crises with layoffs of pastors and the sale of many buildings. Although the city is one of the cradles of Protestantism, the church is witnessing massive disaffiliation of members and the loss of many donors, such as local bankers (Stolz and Ballif 2010).

Since 2015, the parish in Geneva has been organizing congregation festivals around the theme of ecology. The event I attended began with the screening of the film *Tomorrow* (2015), which was a big hit in the French-speaking world. The projection was followed by a debate and exchange over a drink. The next day, a service on the theme of ecology took place with an ordinary liturgy while articulating moments of biblical questioning and a more relaxed time for the younger individuals. Then, a convivial afternoon with more than 120 participants continued until 4:00 p.m. If the success of the weekend for the climate appears to be total, the charter nevertheless came up against a wall on the side of the cantonal church.

One of the contributors to the Climate Justice Charter reported his frustration during the interview that such a document could not be disseminated within his head church organization. The leaders of the parish first attempted to introduce their charter to the organizational level above the parish, the council of the pastoral unit (bringing together the four – wealthy – parishes of this area of Geneva). However, they were never able to put it on the agenda of a pastoral unit meeting. The president of the unit pointed out that the council was constitutionally only concerned with the management of the region's employees. My interlocutor wondered why then must this council meet monthly if it is to manage the staff, which has remained almost unchanged during the years of the drafting of the Climate Justice Charter. A hope suddenly arose with a new church's consultation on ecology that was first initiated by the cantonal church newspaper *La Vie protestante* in the autumn of 2015 and then by the continuation of the consultation within the church synod. The leaders presented their charter to the synodal council, who found it "much too complicated and demanding," according to the respondent.

Finally, the head church organization did not take the green turn hoped for by the bearers of the Climate Justice Charter. They therefore resolved to pursue two other strategies. The first strategy was to disseminate the charter through the Globe Ethics platform founded by the former theology professor in Basel (and also co-founder of Oeku), Christoph Stückelberger, where the respondent works as a collaborator. The second was to ally the neighboring Catholic parish around the climate charter and pressure the municipality to adopt several ecological resolutions, particularly on the management of sorting and waste bins. Five years later, these strategies have not led to success. Currently, the parish is considering entering the process of "Green Rooster" certification provided by Oeku. This strategy leads us directly to the second case presented in this chapter.

The Catholic parish in Thurgau is among the first parishes in Switzerland to have been certified "Green Rooster." This certification owes nothing to chance in Switzerland because it is the fruit of a theologian and pioneer of the Christian ecological cause: Ursula Aebi.[16] In Switzerland, for many years, in the absence of a priest, female theologians have been able to officiate in a Catholic parish. Ursula has been officiating in this parish for approximately

10 years, "But if a priest were to arrive, it is clear that I would lose my position," she confessed during the interview. She has always relayed the ecumenical Lenten campaigns,[17] and she has been aware of ecological concerns for more than 30 years. Since 2010, she attempted to initiate a practical action for the environment supported by the church.

At a "Kirchentag"[18] in Germany in 2010, namely, at a workshop, she first heard of the church certification of "Green Rooster."[19] The "Green Rooster" label is a certification compatible with other secular building certifications established in Germany that certifies parishes that conform to environmental management with a 10-step protocol that allows parishes to improve their ecological efficiency. Its added value is that the process requires a strong ecological commitment of the parish with an active local group, which allows the renewal of the certification every 3 years. Before the workshop at the "Kirchentag," Ursula Aebi had been involved in several events and used the Lenten campaigns of the Swiss Catholic Lenten Fund and other campaigns of FBOs to remind individuals of their Christian responsibility for the environment. For 30 years, she had also been involved in a group of theologians who discussed liberation theology. Every year, this group held retreats lasting several days with theologians from South America. At the parish level, a group specifically dedicated to ecology had been formed since 2000. However, here, it was different – the certification was methodical and concretely concerned all the areas of the parish.

For Ursula Aebi, the challenge remained considerable. First, parishes in Germany and Switzerland have very different financial means to bring their buildings up to energy efficiency standards; second, building laws and standards differ between countries; third, a Swiss certification body had to be established. It was by finding an ally, Georg Baumann,[20] a former pastor who made a professional conversion to environmental and sustainability consulting for Church Institutions. It was at this "Kirchentag" workshop that she met him.

She then met with Oeku's leaders to convince them to take the necessary steps to become the Swiss certification body for the "Green Rooster" label. However, Oeku leaders faced a great dilemma: The office had only three staff members, all part-time. With very limited financial resources, they did not know how to get involved in the certification of the "Green Rooster" label. Nevertheless, Ursula Aebi continued to pressure them because she believed that this certification was a unique means for churches to move from words to deeds in a serious manner. For her part, she began the process of adopting all the certification standards to Switzerland with a working group she set up with G. Baumann. At first, she did not think it would be such an important task.

In her parish, she brought the group dedicated to ecology and the building commission to work together to draw up a building renovation project. Together they were able to lead the process involving renovations and the installation of heavy infrastructure such as heat pumps and solar panels.

A few years of effort later, on 8 November 2015, the parish in Thurgau received the "Green Rooster" certification by Oeku during a worship service. The pugnacity of the pioneer enabled her congregation to have parish buildings with the Swiss "Minergie-A" standard (energy independence). Additionally, during this service, several other neighboring Catholic parishes were also certified. Ursula Aebi had succeeded in getting four other parishes in her region to go through the certification process. In 2017, two more parishes in the region and the administration offices of the Cantonal Catholic Church (Thurgau) were certified.

However, although Ursula Aebi's outcome is positive for her region, she regrets that her diocese neither certified its buildings nor encouraged all parishes to become certified. She then requested a meeting with her bishop. When she was finally granted a one-hour meeting, she went with Georg Baumann to present the "Green Rooster" certification to the bishop. The reaction she received a few days later in a letter was that it was not yet the time for that.

To circumvent the bishop's negative reaction, Ursula Aebi's strategy was to have all the parishes in her canton certified. The idea was to have a certified cantonal church and thus be able to set an example for other cantonal churches and perhaps make the bishop bend. However, in this case, too, the strength of the theologian was somewhat limited because if the six certified parishes and the cantonal administration were included, there remained more than 35 Catholic parishes to be certified in this canton.

Another barrier for Ursula Aebi on the local level is the indifference her project encounters with the Reformed parish of the town and other Reformed parishes in the region. Thus, as I observed in the parish in Geneva, the concrete actions of the parishes for the environment are very local. They come up against a glass ceiling (Bryant 1984) or "stained glasses ceiling" (Purvis 1995; Adams 2007), which prevents their actions from spreading further. The theological discourse of these pioneers takes up the arguments advanced and made known by the elites of the WCC and the Pope. This discourse of the theological elites is obviously relayed by the pioneers, but not just that. The established churches in Switzerland generally accept that God's mandate described in the Bible book of Genesis is no longer "to exploit nature" but "to be careful stewards of creation." This point is critical to consider because despite this common knowledge, promoters of concrete ecological action are confronted with a crucial obstacle. Therefore, it is useful to assess institutional functioning and blockages to understand how a shared discourse has not borne the expected institutional fruits.

Discussion

Drawing on the new institutional perspective, the three levels of pressure outlined by Chaves are discussed. In terms of internal pressures, both cases present local spiritual leaders having attracted skilled members around them to conduct their local project in a distinct bottom-up process. Members have thus entered into a process of commitment to the environmental cause through parish practice.

At this level, one point that Chaves identified as ambivalent for innovation in his study field was the rational or contextual needs. In the case of environmental standards, an important issue is the upgrading and renovation of the church's building. Heating is a costly matter for a parish, although renovation in accordance with the Swiss "Minergie-A" environmental standards seems more expensive. The Swiss government also offers subsidies that do not cover the costs. In the case of the parish in Thurgau, the need to renovate parish buildings was a trigger for entering the "Green Rooster" certification. However, the theologian was a long-time member of a very progressive circle of Catholic theologians, with groups of members active in the field of ecology locally. Notably, this set of internal pressures allowed local initiatives to be implemented but not diffused beyond the local group.

The second level is the pressure outside the denominational field. There are, of course, environmental organizations and the Green Party. In both cases presented in this chapter, the ecological commitment is presented as a concern for the local parish. This internal pressure is coupled with an external pressure to take advantage of them, for example, in the case of the Catholic parish in Thurgau, to renovate buildings according to the high criteria of the Swiss government. For the Protestant parish in Geneva, it directs its actions on the basis of a charter and was able to convince its Catholic sister; however, in terms of dissemination, it did not even succeed in influencing the first institutional level above the parish in the church: the pastoral unit, gathering of a couple of parishes around a single team of ministers.

For the parish in Thurgau, dissemination to nearby parishes has had some effect but has not generated a wave of enthusiasm from all the surrounding parishes to convince other cantonal churches and the bishop. Moreover, and intriguingly, the neighboring Reformed parish in the same town has never shown an interest in the process initiated by the Catholic sister parish. Notably, there has been some resistance at this second level. However, the large popular movements to protest for ecology have just begun. This led to a wave of green deputies for the last renewal of the Swiss Federal Parliament. Notably, these movements have not influenced national policies on the topic. Their impact has been insufficient to stabilize a new regime of environmentally friendly institutional innovation (Geels 2002, p. 1262).

The third level highlights two institutional levers – that of the cantonal church and those of the church-related Christian FBOs – which are financially supported by the churches.

The strategy of Ursula Aebi seems pragmatic. Faced with the bishop's refusal, she attempted to win over the cantonal church to then convince other churches and the diocese. For the parish in Geneva, all the institutional doors within the church head organization were closed. Every level of the institution has resisted innovation. The level above the parish, the pastoral unit, refused to deliberate on the proposals of the Climate Justice Charter, and the synod did not consider a discussion of the charter a necessity. This also shows a certain autonomy of the parish, which has many resources. It has sufficient internal resources to produce a Climate Justice Charter of remarkable quality without outside help. However, this characteristic is also its weakness. Its decentralized power prevents it from reaching the centralized government of the cantonal church. However, its members with social capital have been able to publish the charter in French and in English and disseminate a PDF[21] version on a global platform of an ethic and theological think tank.

For the parish in Thurgau, Ursula Aebi did not have all the necessary resources in the congregation. She had to ask another pastor known for his consulting company and Oeku. It is through these organizations that certification can take place. The pressure can be structured. Instead of maintaining the pressure from the parish in Thurgau, Ursula Aebi, by asking Oeku for help, also shifted and structured the pressure from an organization that is not church-based but transversal in the denominational field. Through Oeku, it exerts some "internal-external" pressure.

Despite the small number of certified parishes in Switzerland (one third of which are the result of Ursula Aebi's dissemination work), it is an observable long-term success because it provides a recognized certification. This result can structure the pressure on the churches through Oeku. Moreover, because of the failure of the diffusion of its charter inside the church, it is also through the "Green Rooster" certification of Oeku that the parish in Geneva considers continuing on the path of ecology.

The environmental pioneers of the parishes are in a tense relationship with a church bureaucracy that prevents them from disseminating their innovations. The cantonal church legitimacy earned through several political struggles provides them a special status. Although this status is not frontally questioned, the bureaucracy of the churches have interest in conforming to the desires of political and secular powers because these powers provide churches with legitimacy in society. The political context in Switzerland reinforces a conservative attitude toward the cause of the environment. Except for a few cities, the public authorities have not prioritized ecology. The established churches therefore have no social advantage in choosing this path.

This tension with the bureaucracy is continually rekindled by the aforementioned pro-environmental theology of the WCC and Pope Francis, which has widely spread. Moreover, the contexts of increasing climate protests and social pressures are increasing regarding this issue, and the churches in cities are sensitive to and respond positively to them. Several churches have expressed sympathy for the movement. However, again, Oeku plays a key coordinating role in major public actions. Two emblematic actions for the Swiss population are, for instance, a large inter-religious worship service which was held in a spacious church building just before the national climate strike[22] in Bern on 28 September 2019, and the stopping of the clocks at "minus five for the climate." At the local level, various parishes have provided church buildings or meeting rooms for free to climate strike meetings, and several parishes have also hosted the fasting for the climate actions of their areas. For the public, one of the most spectacular actions of the churches in support of youth climate strike initiatives is the ringing of bells at the start of the Friday demonstration, especially in the cities of Basel, Bern, and Zurich. In addition to this pressure from the youth and the particular circumstances of the climate strikes, I observed that globally, the cantonal churches are afraid of a tense relationship with the public authorities if they become too green. The churches have responded favorably to societal pressure from the youth on the climate strike. However, simultaneously, because the political sphere is not conquered by ecological demands, the churches are confused: They want to be favorable to the environmentally friendly youth movement without engaging in political decisions or actions that would cause them to lose part of their social legitimacy among decision makers.

Another interesting tension concerns the clashes between these local pioneers and the church hierarchy's routine. The topic of ecology makes it possible to observe that (1) churches are not homogeneous and (2) parishes have a certain freedom of action. However, individuals who innovate or develop alternative perspectives are ignored by the main decision bodies of the established churches. Throughout history, the church bureaucracy has learned how to control internal tensions by establishing all types of administrative and "democratic" bodies to prevent internal tensions from diffusing. Thus, the bottom-up movements, such as the recognition of women's ordination and taking care of creation, cannot diffuse in the church without a complex set of internal and external pressures. In this case, FBOs such as Oeku, which are external to the churches but supported by them, can become levers structuring the diffusion of ideas and maintaining the transformative pressure in the very long term.

Conclusion

In the debate on whether religion can be a vector for ecological transition, I focused on the barriers that prevent established churches (Reformed and Catholic churches in Switzerland) from acting effectively through different internal institutional levels. Other research has shown how the cultural context may block rising environmental concern in the church. As such, Bertina (2016) has demonstrated how the cultural context prevented collective appropriation of the environmental cause by churches. The cultural context constitutes, as observed in this chapter, one important element of external pressure that influences the decisions of the churches at the macro level. For the local level of congregations in the United States,[23] Djupe and Hunt (2009) noted that members and clergy held pro-environmental views but that "religious beliefs have little to no effect once social communication is controlled." The ethnographic survey presented brings nuance to this observation. In the particular case of clergy and members who organize themselves around a pro-environmental project, they manage to put internal pressure that facilitates pro-environmental decisions at the congregational level. Beyond this level, however, the effects are limited. I demonstrated that the local level was insufficient to illuminate the limited commitment of established churches in the environment. As Koehrsen (2018) pointed out, it is necessary to consider the different institutional levels to understand the role of religion in the energy transition. For him, "The 'greening' implies that religious organizations experience internal sustainability transition processes" (Koehrsen 2018, p. 8). I observed in Switzerland that these processes are implemented in several local niches. Their implementation is difficult because of the different institutional levels. Notably, the pressure maintained by national FBOs provides an opportunity for change in established churches (see Johnston 2006; Hawken 2007).

In the case of Switzerland, I demonstrated how a parish was able to disseminate, albeit in a limited manner, its "green" innovation because of a certification from a church-related FBO that operates beyond the usual church authority structures. In the other case, despite the resources and social capital of the members, the innovation remained stuck at the local level. Here, I observed that internal pressure from the members and elite is insufficient to change the action of the church, remaining mostly limited to single actions in time such as open support for climate strikes. If their action cannot change the church, it is at least protected from disqualification by the theology of the elites. Moreover, the history of the church relates a few individuals who have taken care of creation, such as Saint Francis of Assisi, priests, or botanical monks, offering a reference and role models for current environmental innovators.

This very slow progress of the ecological cause has also allowed me to show that the tension within the established churches is not so much between pro- or anti-ecology but between the bureaucracy that attempts to maintain its legitimacy with the public authorities and the progressives who could decrease it. The churches are losing credibility and need this acquired legitimacy now more than ever. The pioneers with their actions create uncertainties that the church's bureaucracy is denying because it can.

In history, established churches have managed different pressure groups and theological currents. In Switzerland, Reformed and Catholic churches have been driven to put democratic bodies in place to temper and negotiate these tensions. These bodies then become barriers that prevent innovation from reaching the decisive power in the church head organizations. This process has been sharply observable in the case of the parish in Geneva, which has never been able to cross the institutional gateway above the parish level for administrative reasons.

The topic of ecology allows pinpointing two types of tension: one type is contained within the established churches and stifles local initiatives, and the other type is external, that is, the church's relationship with society, from which it has obtained a privileged status that assures it of financial support. The tension is all the more diffuse as the members disperse and because the societal demands for ecological concerns are not unanimous.

Notes

1 For instance, the importance of the theologian Jürgen Moltman in the Protestant theological faculties. See: Moltmann, J., 1985. *God in Creation: A New Theology of Creation and the Spirit of God*. San Francisco: Harper & Row.
2 In collaboration with Oeku-Church and environment, see later.
3 See: www.evref.ch [Accessed on 11 May 2020]
4 See: www.rkz.ch/wer-wir-sind/mitglieder [Accessed 10 May 2020].
5 See: www.eveques.ch/nous/suisse/dioceses [Accessed 10 May 2020].
6 Especially by the Lutheran key speaker Joseph Sittler, see later.
7 See: www.oeku.ch [Accessed 30 June 2020].
8 https://oeku.ch/de/schoepfungszeit.php [Accessed 30 June 2020].
9 On 1 September 1989, the Greek Orthodox metropolis Demitrios I declared the first of September as the liturgical day for the protection of the environment. At the instigation of Lukas Vischer, the European Christian Environmental Network (ECEN) adopted on 31 October 1999 that September would be a "Season for Creation," see: ecen.org [Accessed 30 June 2020].
10 https://oeku.ch/de/prakt_schoepfungsb.php [Accessed 30 June 2020].
11 https://oeku.ch/de/gruener_gueggel.php [Accessed 30 June 2020].
12 https://oeku.ch/de/klimakrise.php [Accessed 30 June 2020].
13 On 6 December, a second parish in the Zurich area publicly celebrated its certification obtained in September of the same year.
14 Irene Becci from Universty of Lausanne is the PI. The research is supported by the Swiss National Science Foundation (2017–2020). Alexandre Grandjean is PhD candidate and the author of this chapter is Senior researcher. Later Salome Okoekpen joined the research team as Student researcher.
15 Irene Becci and Christophe Monnot were the PIs of the research and belong to the program "Volteface" at the University of Lausanne. See: Niwa, N., and Frund, B., 2018. *Volteface. La transition énergétique: un projet de société*. Lausanne; Paris: Editions d'en bas; Editions Charles-Léopold Mayer.
16 Fictitious name.
17 The Lenten campaigns are organized by the Reformed and the Catholic churches' agencies (Bred for all and Swiss Catholic Lenten Fund), identified as major drivers of ecological consciousness in the established churches and Swiss population.
18 The Kirchentag is a lay movement founded in 1949 in Germany. It is a five-day festival gathering ca. 100,000 German Protestants with a faith-based forum for, for example, democracy, human rights, and ecumenism.
19 851 parishes are certified in Germany, see: www.kirum.org/tl_files/kirum-files/content-pics/Einrichtungen%20und%20Gemeinden/Zertifizierte%20Einrichtungen%20und%20Gemeinden%20Stand%2011.06.19.pdf [Accessed 05 May 2020].
20 Fictitious name.
21 http://hdl.handle.net/20.500.12424/238448 [Accessed 10 May 2020].

22 The strike gathered approximately 100,000 individuals.
23 Also see: Cousineau, M., 1997. Communautes de base, terre et environnement en Amazonie. *Social Compass*, 44 (3): 429–441; Gottlieb, R., 2006. *The Oxford Handbook of Religion and Ecology*. Oxford: Oxford University Press, 467–509; Koehrsen, J., Huber, F., Becci, I. et al. 2019. How Is Religion Involved in Transformations Towards More Sustainable Societies? A Systematization. *Historia Religionum*, 12 (1), 99–116.

References

Adams, J., 2007. Stained Glass Makes the Ceiling Visible: Organizational Opposition to Women in Congregational Leadership. *Gender Society*, 21 (1), 80–105.

Becci, I., and Monnot, C., 2016. Spiritualité et religion : nouveaux carburants vers la transition énergétique. *Histoire, monde et cultures religieuses (HMC)*, 2016/4 (40), 93–109.

Becci, I., Monnot, C., and Wernli, B., 2021. Sensing 'Subtle Spirituality' Among Environmentalists: A Swiss Study. *Journal for the Study of Religion, Nature and Culture*, 15 (3), 344–367.

Bertina, L., 2016. Le catholicisme et la question environnementale en France: les raisons culturelles d'un retard. *Revue Française d'Histoire des Idées Politiques*, 44 (2), 127–155.

Brown Zikmund, B., 1986. Winning Ordination for Women in Mainstream Protestant Churches. *In:* R. Radford Ruether and R. Skinner Keller, eds. *Women and Religion in America. Vol. 3: 1900–1968*. San Francisco: Harper & Row, 339–350.

Bryant, G., 1984. *The Working Woman Report: Succeeding in Business in the 80s*. New York: Simon & Schuster.

Chaves, M., 1993. Denominations as Dual Structures: An Organizational Analysis. *Sociology of Religion*, 54 (2), 147–169.

Chaves, M., 1996. Ordaining Women: The Diffusion of an Organizational Innovation. *The American Journal of Sociology*, 101 (4), 840–873.

Chaves, M., 1997a. *Ordaining Women: Culture and Conflict in Religious Organizations*. Cambridge: Harvard University Press.

Chaves, M., 1997b. The Symbolic Significance of Women's Ordination. *The Journal of Religion*, 77 (1), 87–114.

Clements, J.M., McCright, A.M., and Xiao, C., 2014. Green Christians? An Empirical Examination of Environmental Concern Within the U.S. General Public. *Organization & Environment*, 27 (1), 85–102.

Clements, J.M., Xiao, C., and McCright, A.M., 2014. An Examination of the "Greening of Christianity" Thesis Among Americans, 1993–2010. *Journal for the Scientific Study of Religion*, 53 (2), 373–391.

DiMaggio, P.J., and Powell, W.W., 1983. The Iron Cage Revisited: Institutional Isomorphism and Collective Rationality in Organizational Fields. *American Sociological Review*, 48 (2), 147–160.

Djupe, P.A., and Hunt, P.K., 2009. Beyond the Lynn White Thesis: Congregational Effects on Environmental Concern. *Journal for the Scientific Study of Religion*, 48 (4), 670–686.

Evangelische Kirche in Deutscland (EKD), 1969. Christian Love Means Forestalling Disaster. *The Ecumenical Review*, 21 (1), 71–73.

Galaskiewicz, J., and Burt, R.S., 1991. Interorganization Contagion in Corporate Philanthropy. *Administrative Science Quarterly*, 36 (1), 88–105.

Geels, F.W., 2002. Technological Transitions as Evolutionary Reconfiguration Processes: A Multi-Level Perspective and a Case-Study. *Research Policy*, 31 (8), 1257–1274.

Grandjean, A., Monnot, C., and Becci, I., 2018. Spiritualités et religions: des "facilitateurs" pour la transition énergétique? *In:* N. Niwa and B. Frund, eds. *Volteface. La transition énergétique: un projet de société.* Lausanne, Paris: Editions d'en bas; Editions Charles-Léopold Mayer, 157–175.

Hawken, P., 2007. *Blessed Unrest: How the Largest Movement in the World Came into Being, and Why No One Saw it Coming.* New York: Viking.

Johnston, L., 2006. The "Nature" of Buddhism: A Survey of Relevant Literature and Themes. *Worldviews: Global Religions, Culture, and Ecology*, 10 (1), 69–99.

Johnston, L., 2013. *Religion and Sustainability: Social Movements and the Politics of the Environment.* Sheffield: Equinox.

Koehrsen, J., 2015. Does Religion Promote Environmental Sustainability? – Exploring the Role of Religion in Local Energy Transitions. *Social Compass*, 62 (3), 296–310.

Koehrsen, J., 2018. Religious Agency in Sustainability Transitions: Between Experimentation, Upscaling, and Regime Support. *Environmental Innovation and Societal Transitions*, 27, 4–15.

Landron, O., 2008. *Le Catholicisme vert. Histoire des relations entre l'Eglise et la nature au XXe siècle.* Paris: Le Cerf.

Laurent, M., and Dion, C., dir., 2015. *Tomorrow. Take Concrete Steps to a Sustainable Future.* France: Under the Milky Way.

Martinez Andrade, L., 2016. *Ecologie et libération. Critique de la modernité dans la théologie de la libération.* Paris: Van Dieren.

McDonagh, S., 1986. *To Care for the Earth: A Call to a New Theology.* London: G. Chapman.

McDonagh, S., 1990. *The Greening of the Church.* London: G. Chapman.

McDuff, M.D., 2010. *Natural Saints: How People of Faith Are Working to Save God's Earth.* Oxford, New York: Oxford University Press.

Moltmann, J., 1985. *God in Creation: A New Theology of Creation and the Spirit of God.* San Francisco: Harper & Row.

Monnot, C., 2012. Mesurer la pratique religieuse. Différentes mesures, différents taux? Analyse comparative à partir de la Suisse. *Archives des sciences sociales des religions*, 158 (1), 137–156.

Monnot, C., and Stolz, J., 2020. The Diversity of Religious Diversity. Using Census and NCS Methodology in Order to Map and Assess the Religious Diversity of a Whole Country. *In:* A. Körs, W. Weisse, and J.-P. Willaime, eds. *Religious Diversity and Interreligious Dialogue.* Cham: Springer, 133–151.

Niwa, N., and Frund, B., eds., 2018. *Volteface. La transition énergétique: un projet de société.* Lausanne, Paris: Editions d'en bas & Editions Charles-Léopold Mayer.

Parker, G.C., 2015. Religion et transition énergétique: une étude en Amérique du Sud. *Social Compass*, 62 (3), 344–361.

Piguet, F.-P. et. al., 2016. *Climate Justice Charter: Statement and Commitments on Climate Justice and Respect for the Biosphere.* Geneva: Arve et Lac Publications.

Pope Francis, 2015. *Laudato Si'. Encyclical on Climate Change & Inequality: On Care for Our Common Home.* Brooklyn: Melville House.

Pope Francis, and McDonagh, S., 2016. *On Care for Our Common Home: The Encyclical of Pope Francis on the Environment, Laudato Si'.* Maryknoll, New York: Orbis Books.

Powell, W.W., and DiMaggio, P.J., eds., 1991. *The New Institutionalism in Organizational Analysis*. Chicago: University of Chicago Press.

Purvis, S.B., 1995. *The Stained Glasses Ceiling: Churches and their Women Pastors*. Westminster: John Knox Press.

Roewe, B., 2016. Eco-Theologian Fr. Sean McDonagh: Don't Let This "Laudato Si" Moment Pass. *NCR Online* (March 9, 2016).

Schäfer, O., 2008. *Ethique de l'Energie. Vers une nouvelle ère énergétique. Perspectives durables pour l'après-pétrole*. Berne: FEPS.

Scott, W.R., and Meyer, J.W., 1994. *Institutional Environments and Organizations: Structural Complexity and Individualism*. London: Sage Publications.

Sherkat, D.E., and Ellison, C.G., 2007. Structuring the Religion-Environment Connection: Identifying Religious Influences on Environmental Concern and Activism. *Journal for the Scientific Study of Religion*, 46 (1), 71–85.

Stoll, M., 2015. *Inherit the Holy Mountain: Religion and the Rise of American Environmentalism*. Oxford, New York: Oxford University Press.

Stolz, J., and Ballif, E., 2010. *Die Zukunft der Reformierten: Megatrends – Wirkungen – Reaktionen*. Zürich: TVZ.

Stolz, J., and Monnot, C., 2017. The Established and the Newcomers. A Weberian-Bourdieusian View of Congregations in the Swiss Religious Field. *Revue d'histoire et de philosophie religieuses*, 97 (1), 69–90.

Strang, D., and Meyer, J.W., 1994. Institutional Conditions for Diffusion. *In:* R.W. Scott and J.W. Meyer, eds. *Institutional Environments and Organizations: Structural Complexity and Individualism*. London: Sage, 100–112.

Taylor, B., 2016. The Greening of Religion Hypothesis (Part One): From Lynn White, Jr and Claims That Religions Can Promote Environmentally Destructive Attitudes and Behaviors to Assertions They Are Becoming Environmentally Friendly. *Journal for the Study of Religion, Nature and Culture*, 10 (3), 268–305.

Taylor, B., Van Wieren, G., and Zaleha, B. (2016). The Greening of Religion Hypothesis (Part Two): Assessing the Data from Lynn White, Jr, to Pope Francis. *Journal for the Study of Religion, Nature and Culture*, 10 (3), 306–378.

Tucker, M.E., 2003. *Wordly Wonder: Religions Enter Their Ecological Phase*. Chicago: Open Court.

Vaillancourt, J.-G., 1997. Les prises de position du Vatican sur les questions d'environnement. *Social Compass*, 44 (3), 321–331.

White, L., 1967. The Historical Roots of our Ecologic Crisis. *Science*, 155 (3767), 1203–1207.

Wilkinson, K.K., 2012. *Between God and Green: How Evangelicals Are Cultivating a Middle Ground on Climate Change*. Oxford: Oxford University Press.

Willaime, J.-P., 2004. *Europe et religions: les enjeux du XXIe siècle*. Paris: Fayard.

Part II

Interdenominational Tensions

5 Halal Wastewater Recycling

Environmental Solution or Religious Complication?

Sofiah Jamil

Introduction

The existing Islam and ecology literature – in line with broader eco-theology scholarship – generally argues that Islamic scholars and clerics can play a significant role in interpreting religious norms on the environment and influencing their congregations to adopt environmental practices. The validity of these assertions has been strengthened by the emergence of faith-based environmental activities, notably between the 1990s and 2000s. Many of these activities were under the auspices of international organizations such as the World Bank, IUCN, and the Alliance for Religions and Conservation, and sought to engage local Islamic leaders as partners to implement conservation or environmental projects (Schwencke 2012; Dinata et al. 2013). Indonesia – home to the world's biggest Muslim population – has been involved in several of these Islamic environmental initiatives, both at the national and local level.

Existing empirical studies on these Islamic environmental initiatives, however, have not comprehensively evaluated the effectiveness of engaging Islamic clerics in addressing environmental challenges. In particular, these studies have ignored the tensions that exist within Islamic societies, which in turn potentially jeopardize the effectiveness of such activities. For instance, while there are cases of Islamic clerics and organizations being more involved in environmental activities (Mangunjaya 2011; Mangunjaya and McKay 2012; Amri 2014), there is also evidence where Islamic organizations or the pursuit of Islamic objectives have inadvertently contributed to environmental degradation in Indonesia (Lahiri-Dutt 2004; Chozin 2008; Resosudarmo et al. 2009; Blacksmith Institute and Yayasan Tambuhak Sinta 2013; Varkkey 2015). Such conflicting views thus validate arguments that religious environmentalism is "an *interpretation of tradition* rather than a *traditional interpretation*" (Tomalin 2002, p. 15), and that the relationship between religion and the environment is not static, but rather is dependent on specific contexts (Kent 2010).

One specific context in which the relationship between Islam and the environment becomes subjective is when the issue of *Halal*-ness is concerned.

DOI: 10.4324/9781003017967-7

Defined in Islamic jurisprudence as being "permissible," the term is generally understood in layman terms as permissibility in relation to food consumption and based on technical issues relating to cleanliness and being free of impurities. Questions over what is considered halal have thus, at times, revealed tensions among some Islamic schools of thought and communities, such as the opposition to driving biofuel powered vehicles, as ethanol contains alcohol (Al Arabiya 2009), and debates over halal products that are not organically produced. In this regard, while Islamic clerics and organizations unanimously agree on the importance of environmental protection in Islam, they differ on specific approaches on how this should be done.

Given these contrasting depictions, this chapter asks the following question: What impact do ideological and historical differences have on Islamic organizations' ability to address water challenges in Indonesia? To be specific, this chapter examines how Indonesia's three main Islamic organizations interpret and respond to the state-proposed solution of wastewater recycling. As will be demonstrated, while their differences reveal intra-religious tensions that hamper collective consensus on specific water-related initiatives, and jeopardize institutional legitimacy, these organizations respond to the water issues in ways that are suitable and relevant to their respective members.

This chapter is divided into the following sections. First, an explanation of the methodology adopted in this chapter. Second, a brief introduction to Indonesia's three main Islamic organizations. Third, these organizations' deliberations over the permissibility of recycled wastewater. In particular, it highlights the tensions that rose in the formulation of the MUI's fatwa on recycle wastewater, and organizational reactions after the fatwa was issued. This chapter then concludes with an evaluation of these Islamic organizations' significance in water conservation initiatives, and prospects for future research.

Methodology

The organizations examined in this research study are the Indonesian Council of Islamic Scholars (*Majlis Ulama Indonesia* or MUI for short), *Nahdahtul Ulama* (NU), and *Muhammadiyah*. NU and Muhammadiyah are the two biggest Islamic organizations that were established during the Dutch colonial period and are commonly described in the existing literature as the traditionalist and modernist camps of Indonesian Islam respectively. MUI is a state-supported organization that was established during the Suharto era as a means of officially representing all major Islamic organizations in Indonesia.

This study is part of a broader PhD study on Islamic environmentalism in Indonesia and was conducted through a combination of semi-structured interviews, participatory observation, and desk research. While the broader research includes more than 40 interviews (conducted between 2013 and 2014), 12 interviews are specifically relevant to the issue of wastewater recycling, and how it sits in the broader context of Islamic environmental initiatives in Indonesia. Among these 12 interviewees are members of MUI's fatwa committee (who are also members of either NU or Muhammadiyah), staff of MUI, representatives of secular environmental non-governmental organizations (ENGOs), and officials from the environment ministry. Sourcing these interviewees was undertaken through a mix of pre-existing contacts (namely, the NGO and governmental contacts) whom I had known professionally. Contacts to the Islamic organizations were primarily through mutual contacts, through networking at Islamic events, or cold calling directly to specific individuals.

The duration of each interview was approximately 2 hours, with follow up interviews with particular representatives from the three Islamic organizations. As a way of building trust and familiarity with my interviewees – particularly those from Islamic organizations – I conducted the interviews in Indonesian and shared my personal family background, that is, having roots from two Indonesian provinces (East Java and West Sumatra) and a grandfather who is an Islamic cleric. Additionally, I also "gave back" to the community by sharing my expertise and experiences with students and members from the Islamic schools and organizations.

The questions posed to the Islamic organizations were framed in a way that would allow interviewees to share their broader ideological perspectives and range of organizational activities, rather than a narrow focus on their environmental activities. This has allowed for a more comprehensive understanding of the different – and at times – competing organizational objectives and interests. To prod further in terms of the challenges and tensions these organizations have faced, I framed my questions in a way that would allow them to further "let their guard down" – in particular I related my own real experiences and challenges in engaging Singaporean Islamic organizations and communities in being more active in environmental activities. Participatory observation during some of these organizations' environmental activities allowed me to ask more informal questions and gain insight from their members and local communities that they served.

To avoid a conflict of interest, interviewees from MUI made clear when their opinions reflected their position as MUI officials, or as members of Islamic organizations. An Islamic scholar from Indonesia's outer islands was also interviewed as a means of triangulating information and perspectives on MUI from within Java. These interviews have also been triangulated with Islamic organizations' official documents relating to water challenges and wastewater recycling (e.g., Fatwa, advisories, and meeting notes), as well as governmental reports on public infrastructure, and news reports on the development of water infrastructure in Indonesia.

Indonesia's Islamic Organizations and the Environment

As mentioned in the introduction, Indonesia has received substantial attention with regard to Islamic environmental initiatives. According to Fazlun Khalid – founder of the Islamic Foundation for Ecology and Environmental Sciences, and regarded as one of the pioneers of the global Islamic environmental movement (Johnston 2012) – his observations of Indonesia's pre-existing Islamic environmental norms and initiatives in some local communities and *pesantren* (a traditional Islamic boarding school), and participation by leaders of major Islamic organizations in discussing environmental challenges, were testimony of a thriving Islamic environmental culture in the world's largest Muslim country. Speaking in an interview on the sidelines of the International Conference on Muslim Action on Climate Change in Indonesia, he noted:

> Indonesia has something special to offer. . . I seem to come here more often than I go to any other country. . . we find that here there is an interest and motivation . . . the Ulama (Islamic scholars or clerics) here in Indonesia are far more far-sighted than any that I have been to anywhere else. They are developing special teachings on Islamic environmentalism, and they can give the world these teachings.
>
> (Fazlun Khalid, Interview on Metro TV, 15 April 2010)

Regardless of their organizational backgrounds, Indonesia's Islamic organizations are generally focused on three areas – maintaining or promoting religiosity (Latief 2012; Bush and Fealy 2014; Saat 2016), socio-economic development (Fuad 2002; Fernandez 2009; Candra and Ab Rahman 2010; Al Banna Choiruzzad and Nugroho 2013), and education (Zuhdi 2006; Marongo 2012; Bryner 2013). Environmental activities have only become more prominent in recent years, in tandem with national and global environmental developments.

These developments broadly fall into two categories. The first includes the efforts of Indonesian environmentalists that sought to engage Islamic organizations in sustainable development activities. One key personality is Fachruddin Mangunjaya, a conservation biologist who has facilitated much of the externally funded Islamic environmental activities, such as those organized by Khalid through IFEES and the ARC (Muhammad et al. 2006; Mangunjaya et al. 2010, 2015; Dinata et al. 2013). The second development is the Indonesian government's environmental agenda, specifically during President Susilo Bambang Yudhyono's term (2004–2014), which placed emphasis on demonstrating Indonesia's efforts in responding to climate change. In particular, the introduction of a new carbon trading mechanism, Reduction of Emissions from Deforestation and Forest Degradation in developing countries (REDD) in 2005 was a golden opportunity for Indonesia to utilize its forests as a carbon sink and gain access to international funding. A year later in 2009, Indonesia made a voluntary pledge to reduce its emissions by 26% by the year 2020 with domestic resources (Murdiyarso et al. 2011).

In both these developments, Islamic leaders and their respective organizations have been engaged to facilitate the activities – notably MUI, NU, and Muhammadiyah. Not only are these Indonesia's three biggest and most established Islamic organizations, but their differentiated backgrounds and capabilities allow them to engage varying segments of Indonesia's Muslim population. NU and Muhammadiyah, for instance, are engaged to help promote environmental initiatives at the local level, by mobilizing their respective networks to carry out activities such as tree planting or recycling campaigns, while making reference to Islamic principles pertaining to environmental stewardship (Amri 2011, 2014. See also Table 5.1 for a sample of Quranic references and Prophetic sayings on the environment, which are often referred to by Islamic organizations and environmentalists). MUI, as a national-level council of Islamic scholars, has occasionally been engaged to issue environment-related fatwa (religious pronouncements). Additionally, in 2010, MUI established its Institute for Awareness on Environment and Natural Resources (*Lembaga Pemuliaan Lingkungan Hidup dan Sumber Daya Alam*, LPLHSDA) to facilitate the promotion of Islamic perspectives on the environment, and awareness of MUI's environment related fatwa.

It is important to note, however, that the existing Islamic environmentalism discourse is arguably apologetic in nature (Foltz 2003; Johnston 2012; Bagir and Martiam 2017; Afzaal 2012). In fact, much of the work on Islamic environmentalism in Indonesia has not fully taken into consideration

Table 5.1 Examples from the Quran and Hadith on environmental responsibility

Resource conservation
"but waste not by excess: for Allah loves not the wasters" (Qur'an 6:141, and Qur'an 7:31)
Prophet Muhammad's (PBUH) noted limiting the use of water during ablution (*wudhu*) and ritual cleansing (*ghusl*) to 1 *madd* (amount of water in 2 cupped hands) of water and 1 *sa'a* (equivalent to 4 madd) respectively. (Related by al-Bukhari and Muslim; Ahmad, al-Bazaar and at-Tabarani in al-Kabeer)
Prophet Muhammad was also noted for saying that water should be conserved even when taking ablution at a flowing river. (Related by Ahmad and Ibn Majah)
Charity
"There is none amongst the Muslims who plants a tree or sows seeds, and then a bird, or a person or an animal eats from it, but is regarded as a charitable gift from him." (Al-Bukhari Vol. 3: Hadith 513; Vol. 8 Hadith 41)
Abu-Hurairah reports a long hadith from the Prophet in which he said, "Removing harmful things from the road is a charity."
Cleanliness
Abû Mâlik Al-Hârith bin Âsim Al-Ash'ari reports that the Prophet said "Cleanliness is half of faith." (Related by Muslim)

existing literature on the politics within Indonesia's Islamic organizations (Saat 2016; Bush 2009), strong patron-client relations in Indonesian politics and society, and the prevalence of short-term projects as a tool for distribution of economic resources (Hicks 2012; Aspinall 2013; Mietzner 2013). Additionally, my conversations with other Indonesian environmental practitioners (Interview with two secular ENGO representatives, Jakarta, 11 December 2013; Interview with Government official, Jogjakarta, 9 November 2013) have revealed the lack of confidence these individuals had in the presumed significance of Islamic clerics – a contrast to the opinions of proponents of Islamic environmentalism. Conversely, as will be shown in the following paragraphs, some sections of Islamic organizations also express their lack of confidence in the state, and in each other, to effectively meet the needs of local communities. The next section will introduce the three Islamic organizations in question.

Nahdlatul Ulama (NU)

NU represents a federation of kiai and pesantren with a strong rural base, particularly in Java, and has approximately 40 million members nationwide, commonly referred to as *Nahdliyin*. Although officially established as an organization in 1926, NU has its roots in the traditional form of Islam that has been practiced since the coming of religion in the archipelago. In line with Sufi traditions, the Islamic boarding school (*pesantren*) is the node of religious learning from the Islamic leader or *Kiai*. The kiai is understood to be well-versed in the study of the *Kitab Kuning*, which refers to a series of classical Islamic books that encapsulate the historical development of Islamic teachings and practices (Isbah 2012). These topics include Islamic history and literature, Sufism, and the importance of moral behavior, as well as the science behind Quranic interpretations, Prophetic traditions, and Islamic jurisprudence (Geertz 1960; Bruinessen 2009).

In terms of the values espoused in a pesantren, their formation is influenced by Islamic jurisprudence, while their implementation is through Sufism. The students in the pesantren (*santri*) follow a tutelage system, with the kiai teaching and leading by example (Faisal 2011). This traditional mentorship relation between the santri and kiai is reflective of the Sufi practice of gaining knowledge from a revered leader. Some empirical research has examined the significance of this santri-kiai relationship in promoting sustainable development at the local level (Manshur 2009; Mangunjaya et al. 2010; Gade 2012).

Muhammadiyah

As for Muhammadiyah, it is the second largest organization after Nahdlatul Ulama with approximately 29 million members. Muhammadiyah was founded in 1912 by Ahmad Dahlan, a court official in the royal palace in Yogyakarta, as a reformist and progressive platform that combined religious and secular education. This was done as a means of equipping Indonesian Muslims to overcome contemporary challenges – namely Dutch colonization – while maintaining their Islamic identity. Unlike the traditionalists who rejected anything that signified Dutch colonial culture, Muhammadiyah promoted an education system that incorporated the secular Dutch curriculum alongside religious education. This is represented by their modern madrasah, as opposed to the traditional pesantren. To date, the organization is known for its vast network of education and tertiary institutions, as well as its healthcare facilities, as crucial resources for mobilizing collective action.

Muhammadiyah, however shares a love-hate relationship with Nahdlatul Ulama, which is well-documented in existing literature on traditionalist and modernist interpretations of Islam in Indonesia (Howell 2010; Burhani 2013; Aljunied 2016). NU was officially established as a response to the growth of Islamic modernism (represented by Muhammadiyah), which they felt threatened the traditional kiai and pesantren order. This was because Dahlan was greatly influenced by the ideals of Egyptian reformist Muhammad Abduh and rejected the syncretic forms of Islamic culture practiced by the traditionalists, which are perceived to be tainting the purity of Islamic beliefs and practices. In this regard, Muhammadiyah refers strictly to the Quran and Sunnah as sources of Islamic knowledge. The organization also utilizes *ijtihad* (independent reasoning), as a secondary source of Islamic knowledge – particularly in terms of ways of incorporating technological advancements in daily life. This is in contrast to the traditionalists, who tended to defer to the traditional interpretations by Islamic clerics and scholars – in particular Sufism in accordance with the teachings of Al-Ghazali. This NU-Muhammadiyah rivalry is well documented in the existing literature, and continues until this day, through different approaches in determining critical dates in the Islamic calendar and differences in opinion over wastewater recycling (which will be discussed later).

Majlis Ulama Indonesia (MUI)

MUI was established in 1975 by the state as an umbrella organization of Islamic clerics and scholars from various Islamic organizations in Indonesia. As such, MUI serves to unify or standardize legal opinions that may differ among Islamic organizations, in consultation with state agencies and wider society (Hosen 2004). In addition to NU and Muhammadiyah, other smaller Islamic organizations are also represented in MUI. The national level MUI in Jakarta is the central node of leadership and is assisted by religious councils located in the major cities of provinces and districts (known as MUI Daerah).

Although MUI comprises 12 commissions and 4 institutes, its Fatwa Commission is the most active with 42 members while the other commissions have approximately 8 members each. Indeed, much of the emphasis of MUI's work has been to clarify and provide legal opinions on Islam as experts in Islamic law (*Mufti*) in the form of fatwa. As a rule of thumb, fatwa are formulated by coming to a consensus based on a combination of various considerations – theological grounds, scientific factors, and societal needs. From 1975 to 2011, MUI issued 137 fatwa – 14 related to piety and religious denominations, 37 on religious rituals, 51 on social and cultural issues, and 35 related to food, drugs, and science and technology (Majelis Ulama Indonesia 2011). These fatwa are issued based on public or state queries on religious opinions regarding a specific matter.

The formulation of some of these fatwa is in fact where part of the tensions lie. In addition to the differences in opinion between the traditionalists and modernists, the MUI, particularly the central body in Jakarta, has often been criticized as being a religious rubber stamp to support state policies. This will be discussed in greater detail in the following sections.

In terms of MUI's environmental activities, the MUI representatives and Islamic environmentalists that I have spoken to noted that three environment-related fatwa had been issued at the national level since 2008 – sustainable mining, the use of formalin in the fisheries industry, and on wildlife protection. Interestingly, however, these interviewees did not mention the fatwa on wastewater recycling as an environment-related fatwa. This is likely because, as far as MUI is concerned, the fatwa was issued based on whether recycled wastewater is *halal* (permissible) or pure enough for consumption and religious ablution. Hence, while a factor warranting the adoption of wastewater recycling on the part of environment authorities is water conservation, Islamic clerics are more concerned that the sanctity of wider Islamic values not be impinged on. The following section will discuss this in further detail.

Water Conservation Through Wastewater Recycling

Indonesia's water crisis is a major challenge. Despite possessing nearly 6% of the world's freshwater supply, the country faces water-stress issues, which have been exacerbated by high rates of deforestation and groundwater overextraction. Approximately 27 million Indonesians, representing 10% of the population, lack access to clean, safe water and close to 20% (51 million) lack access to improved sanitation. The situation is particularly acute in urban areas such as Jakarta, where groundwater extraction has resulted in the city "sinking" and increasing the city's vulnerability to rising sea levels. Groundwater extraction coupled with rising sea levels has resulted in Jakarta sinking at an average rate of 7 cm per year (CNN Indonesia 2018), thus increasing the likelihood of flooding.

Jakarta's urban poor inevitably bear the brunt of these urban vulnerabilities, as their access to water is either of poor quality or unaffordable and they are often living in areas most susceptible to flooding with little insurance or social safety nets. Moreover, the poor maintenance of water infrastructure and sewages exacerbates the spread of water-borne diseases (APIP et al. 2015). In 2004, for instance, *E. coli* was detected in all 13 rivers in Jakarta monitored by the Jakarta Environmental Management Agency due to the excessive amounts of sewage produced in the city.

Adopting wastewater recycling technology has thus been seen as a possible way of easing the water crisis in the city. Although advancements in technology are able to ensure a high degree of water purity, the government is conscious that ensuring public acceptance of using recycled wastewater would be challenging. In this regard, the Ministry of Environment requested that the MUI issue a fatwa to determine the permissibility (*halal*) of recycled wastewater. Deliberating over the formulation of the fatwa, however, required just as much acceptance from Islamic clerics of varying ideologies.

Tensions in Fatwa Deliberations

A workshop was organized in March 2009 for MUI's Fatwa committee to understand the issue holistically. The case for deliberating the permissibility of wastewater recycling was made on three main grounds. First, that advancements in technology make it possible to remove impurities from water. Second, that the use of recycled water in the community is increasing along with the rapid increase in water demand and a decrease in the quality of water sources due to the increase in population, the pace of urbanization, and the development of industry. Third, that there is currently no standard for halal use of recycled water so that questions arise regarding the law of benefits. As such, it would be necessary to stipulate a fatwa on the use of recycled water to be used as a guideline.

In addition to referring to Islamic scholarship and jurisprudence on what determines water purity from a religious perspective, members of MUI's fatwa committee also heard from experts on other social and scientific perspectives surrounding the issue. These experts included academics from the Institute for Agriculture in Bogor who spoke about the use of recycled wastewater in other countries, the State Health Department on standards for safe drinking water, and members from the Ministry of Environment who spoke about the governmental efforts in managing water resources and recycled wastewater.

It was also noted that while recycled wastewater has been used in other Muslim countries to cope with water shortages, their use has been limited to agricultural production, and not for direct consumption (Abderrahman 2001; Al Khateeb 2001). This was the case despite a special fatwa that was issued in 1978 by the Council of Leading Islamic Scholars in Saudi Arabia which noted that wastewater could "theoretically be used for ablution and drinking, provided that it presents no health risk." Even so, Saudi Arabia has limited the use of wastewater for agriculture. Singapore, however, stands out as an exception where their Islamic Council issued a fatwa stating the permissibility of recycled wastewater produced in the city-state (known as NEWater) for consumption or domestic use (MUIS 2001).

Despite these comprehensive arguments for wastewater recycling, two types of tensions arose during these deliberations. First, a tension in terms of differing religious interpretations on what sort of wastewater recycling mechanisms are permissible. This was particularly seen in the rift between the traditionalist and modernist camps. Although the modernist sections of the MUI were amenable to mechanized wastewater technology, several older NU members in the committee rejected it in favor of a more traditional notion of wastewater recycling, that is, natural water cycle processes (Interview with Environment Ministry official, 1 July 2013). The latter group was also concerned that the existing poor track record of proper maintenance of public infrastructure in Indonesia could jeopardize the quality of recycled wastewater regardless of the technological advancements in producing high-quality recycled water (Interview with NU representative, 13 July 2013). This is particularly critical given the fact that one of the conditions for the permissibility of wastewater recycling is that the process (and appliances used in it) are free from impurities. There was also the perception among some members of MUI that Indonesia has sufficient water resources and, unlike Singapore, is not in a dire state of emergency and need not resort to issuing a fatwa on the permissibility of wastewater recycling (Interview with NU representative, 13 July 2013).

The concern about proper maintenance of public infrastructure also revealed a second tension about MUI's institutional legitimacy. Should there be contamination in the wastewater recycling process, MUI as an organization would inevitably suffer a backlash from society by being blamed for condoning the permissibility of wastewater recycling, and thus tarnish its credibility (Interview with NU representative, 13 July 2013). Being a state-sponsored organization, MUI was by design meant to issue fatwa that are generally supportive of government policies; this was particularly so since its inception under the Suharto government, leading many to view the MUI as a government mouthpiece (Ichwan 2013). These include other controversial fatwa such as allowing vaccines which contain derivatives of pork (Tehusijarana 2018).

The same could be said about the wastewater recycling fatwa. The MUI was obliged to issue the fatwa, as it was primarily in the state's interest to roll out its water sustainability policies. Deliberations spanned a period of 10 months, which by normal MUI fatwa committee standards was much longer than usual. The fatwa was finally issued in January 2010 and in a bid to protect the MUI's interests, was carefully worded in two ways. First, it highlighted the conditions that allow for the wastewater's permissibility (see Table 5.2). Like other fatwa, it begins with references to relevant Quranic verses and hadith – in this case verses on the purity of water. It also included four Hadith relating Prophetic traditions of removing impurities in water. It also noted that by convention, the default position of the Islamic law

Table 5.2 Conditions for determining *halal* recycled wastewater

Recycled wastewater is suitable for purification in view of the following criteria:

1. Removing the impurities in the water, to the point that the remaining body of water has not changed in its physical properties – namely, smell, taste, and color.
2. A body of water that has a quantity of less than two *qullah* (270 litres) becomes impure by contact with impurities, but as soon as the amount is increased to the minimum of 2 qullah, the water becomes suitable for purification (such as ablution or total ablution).
3. Changing water that has come into contact with impurities or has changes in its physical properties by using tools that will return the original water properties by meeting two conditions: first that the volume of water is more than 270 litres, and the tools and appliances used are free from impurities.

in matters (excluding worship) is one of permissibility, and that a state of emergency must always be avoided or eliminated (Majelis Ulama Indonesia 2011, p. 756). By meeting one of these criteria, the fatwa notes that recycled wastewater would be suitable for ablution, bathing, cleaning impurities, as permissible for consumption, so long as there is no harm to health – a stance that is similar to the 1978 Saudi fatwa as mentioned previously.

Second, the fatwa concluded with two recommendations in ensuring the cooperation of other critical stakeholders managing wastewater recycling infrastructure. The first is a request to the government to include halal standards for water as part of the requirements for clean water and drinking water, in addition to their health standards. Second, a request was made to government authorities and stakeholders involved in managing wastewater recycling to consistently improve and maintain the quality of the wastewater facilities and tools, thus ensuring the validity of the fatwa as a guideline (Majelis Ulama Indonesia 2011, p. 762). These recommendations were thus a caveat to ensure that MUI's credibility is not jeopardized in the event of a fault in the wastewater recycling system. The recommendations calling for cooperation from other stakeholders are also important to note, as while MUI may be seen as the highest level of authority in providing religious guidance to Indonesian Muslims, its role is merely advisory, and it does not necessarily have the capacity to monitor or control the implementation of policies or initiatives related to the fatwa that they have issued.

Resolving Tensions With Diverging Reactions

Following the issuing of the MUI's fatwa on wastewater recycling, the varying stakeholders responded in accordance with their own interests. With the MUI fatwa, the government agencies proceeded to install wastewater recycling systems in specific areas. One such area was Istiqlal Mosque in Central Jakarta, whose management was supportive of the proposed installation

(Detik 2012). The mosque, which is the largest in Southeast Asia, has a capacity of 200,000 people, and spends IDR 2 billion annually for clean water (approximately USD 137,500) for the ablution purposes of its visitors (The Jakarta Post 2018). Installing a wastewater recycling system would therefore ensure more efficient use of mosque funds and reduce the stress on the city's limited water supply.

For Muhammadiyah, the fatwa was a win as it not only reflected their modernist stance on societal issues, but also allowed them to integrate technological advancements as part of their broader sustainability measures. In particular, the organization set out to include wastewater recycling systems as part of retrofitting their existing mosques into Eco-Mosques (Taylor 2017). Further steps were taken to create societal awareness through their networks, as well in the mass media. In particular, the issue has been raised several times between 2012 to 2017 in *Republika* – a newspaper closely aligned with Muhammadiyah – explaining the permissibility of recycled wastewater (Ruslan 2012, 2013; Republika 2015, 2016; Wulandari 2015a, 2015b; Damhuri 2017).

As with other fatwa that are deemed controversial, however, MUI has often had to bite the bullet in the face of public criticism. Given its reputation as a state puppet, some have noted that MUI potentially devalues the importance of religious clerics (as indicated in the previous quote), so much so that Islamic organizations will choose to follow their own organizations' fatwa rather than that of the MUI. As one Islamic cleric in West Sumatra relayed to me:

> The job [of Ulama] is to give advice and direction, not to be given advice and directions. These days it is the latter . . . In some regions, MUI is considered lower than NU, Muhammadiyah. Muhammadiyah members trust the outcomes of their Majlis Tarjih (Council) more than that of MUI's fatwa. Same too for NU members who would trust the outcomes of their NU. By right, this should not be the case . . . politics! Why has this happened? Allah knows, but in my opinion, it's about interests.
>
> (Interview with Islamic cleric, 16 December 2013)

Tensions remain within and around MUI due to the institution's limited resources to effectively socialize the fatwa at the local level, and thereby increase their legitimacy – a point agreed by Islamic scholars irrespective of their organizational background (Interview with MUI and Muhammadiyah representative, 10 September 2013; Interview with Islamic cleric, 16 December 2013). Rather, their ability to socialize fatwas is largely dependent on the availability of external funding. Even MUI's environment institute, which is meant to be promoting environment-related fatwa, is dependent on donor funding (Interview with MUI representative, 3 May 2013). In fact, since its inception, it was the head of the institute himself who paid for administrative expenses out of his own pocket (Interview with MUI staff,

17 June 2013). Other MUI Fatwa committee members noted that they disseminate Islamic environmental perspectives in their own capacity, such as through lectures (Interview with MUI representative, 14 September 2013). In this regard, it is the Islamic welfare organizations themselves, such as NU and Muhammadiyah, that have greater agency to operationalize or disseminate religious pronouncements.

As for NU, varying opinions among NU members have revealed tensions between older and younger generations (Interview with MUI Fatwa committee representative, 18 May 2013). Similar to the stance of NU representatives in MUI's fatwa committee, some leaders of local-level mosques, particularly in rural areas, did not agree with the permissibility of wastewater recycling technology (Republika 2015). Moreover, given NU's strong rural support base, wastewater recycling technology is not as relevant and much more costly than existing traditional practices of accessing water. It is also possible that the latter point on costs has reduced willingness to implement such technology in smaller rural-based mosques. In this regard, the mosques equipped with wastewater recycling technology are more often than not those with significant access to financial resources, such as the centrally located Istiqlal Mosque in Jakarta.

For mosques and communities that lack access to such financial and technical resources, alternative traditional water saving/harvesting practices are thus more attractive. Examples of these practices can be traced as far back as the 1970s. Pesantren An-Nuqayah in Madura, is one such example, whose tree planting activities in arid land areas have successfully raised groundwater levels (Ghazali 2001). More interesting, however, are the motivations for doing so. According to Prof. Emil Salim, former Indonesian Minister for the Environment:

> I asked Kiyayi Haji Basith [of Pesantren An-Nuqayah], how is it possible to have a forest with tall, matured trees in the midst of a bare and salty [land area like in] Madura? . . . He said, "We are obliged to pray 5 times a day, and when we pray, we have to take ablution (wudhu) to be clean, so that we will face God in a clean state. Wudhu requires water. When there is no water, God gives us the means of how to have water. And that is by creating forests . . . planting trees." So Kiyayi Haji Basith was planting trees for 5–6 years continuously with his santri. Planting the trees to better absorb the rainfall resulted in the creation of a creek amidst the trees, which after 10 years became a solid river with clean water. So, the motivation for planting trees is for the perfection of prayer (kesempurnaan solat). That was the driving force [back in the 1970s], and all the santri up until now, are dedicated to do this.
>
> (Interview with Emil Salim, 6 December 2013)

There are two points in Kiai Basith's responses that are worth noting. First, the solution to water scarcity/stress was tree planting, an act which is also documented as a Prophetic tradition in the *Hadith*. This is reflective of broader conventional thinking within NU, where solutions to contemporary problems are available in existing Islamic scholarship – whether it be the Quran, Hadith, or *Kitab Kuning* (Interview with NU representative, 13 July 2013). Second, the motivation to harvest water was primarily linked to fulfilling religious obligations, rather than environmental protection *per se* – similar to observations made in other chapters in this volume (see Herman 2022).

Given these traditional practices which continue to exist in NU heartlands, there is little need or relevance in adopting expensive water technology, especially when issues such as drought are more pertinent. Raising water tables through tree planting is therefore preferred. Moreover, there is a perception that state agencies tend to consult religious knowledge and organizations only when it furthers their agendas and provide only limited support to religious institutions to effectively implement environmental programs at the local level (Interview with NU representative, 13 July 2013).

Yet, with the increasing severity of water challenges in Indonesia, some arguably younger members of NU have supported the idea of wastewater recycling and disseminated the information on NU's official media platforms (Kholil 2017; NU Online 2019). These intra-religious tensions that have surfaced over time are therefore not simply based on religious jurisprudence, but also on socio-political affiliations, geographical needs, and generational differences.

Conclusion

This chapter has highlighted how ideological and historical differences not only impede collective decision making among a country's religious elite, but also potentially tarnishes their institutional legitimacy. Contrary to the optimistic nature of existing Islamic environmentalist literature (Mangunjaya 2011; Mangunjaya and McKay 2012; Amri 2014), Islamic clerics' opinions may not always be in line with ideal environmental actions. Moreover, rather than simply pursuing environmental sustainability, considerations relating to religious credibility, and socio-political transparency and equity are significant factors that determine how a religious organization responds to a given environmental challenge.

Intra-religious tensions are exacerbated by the degree to which members within Islamic organizations accommodate state interests. As far as the state is concerned, Islamic organizations are valued primarily for their utilitarian purposes. Although environmental practitioners may engage Islamic organizations in environmental initiatives, there is often limited interest in considering the non-environmental concerns expressed by the Islamic actors. For the Ministry of Environment, the modernists' religious perspectives were more in line with their policies than the traditionalists', and were thus keen to see a fatwa being amenable to wastewater recycling. The traditionalists not only questioned the religious permissibility of wastewater recycling, but also the capability of government agencies in securing the proper maintenance of water and sewage facilities.

Despite the varying perspectives, Islamic organizations' responses after the fatwa was issued reflect efforts to maintain institutional legitimacy. MUI sought to mitigate the existing tensions that it faces within and with wider sections of the Islamic community by including recommendations for further action by government agencies to ensure the efficient and credible use of wastewater recycling systems. Muhammadiyah provided a boost to the state's promotion of wastewater recycling by incorporating it as part of their wider environmental sustainability strategies. As for NU, the internally disparate views on the feasibility of wastewater recycling may cause some degree of confusion for its supporters. Nevertheless, the organization continues to focus on enhancing its local sustainable development projects within its realm of traditional Islam.

It is interesting to note that the findings from this research study parallel developments in other religious communities such as those mentioned in other chapters in this book, namely Herman (2022) on Kosher electricity. Future research could explore Islamic (as well as other religious) organizations' role in relation to other environmental challenges, such as large-scale mining and palm oil plantations. Based on some preliminary studies (Chozin 2008; Mukaddar 2013; Varkkey 2015), similar tensions are likely v isible – that is, MUI fatwa on sustainable mining vis-à-vis small-scale mining activities by local religious leaders.

References

Abderrahman, W.A., 2001. Water Demand Management in Saudi Arabia. *In:* N.I. Faruqui, A.K. Biswas, and M.J. Bino, eds. *Water Management in Islam*. Tokyo: United Nations University, 68–78.

Afzaal, A., 2012. Disenchantment and the Environmental Crisis: Lynn White Jr., Max Weber, and Muhammad Iqbal. *Worldviews: Global Religions, Culture, and Ecology*, 16 (3), 239–262.

Al Arabiya, 2009. *Saudi Scholar Warns Alcohol in Bio Fuel Is a Sin* [online]. Available from: www.alarabiya.net/articles/2009/02/19/66803.html [Accessed 4 April 2022].

Al Banna Choiruzzad, S., and Nugroho, B.E., 2013. Indonesia's Islamic Economy Project and the Islamic Scholars. *Procedia Environmental Sciences*, 17, 957–966.

Al Khateeb, N., 2001. Sociocultural Acceptability of Wastewater Reuse in Palestine. *In*: N.I. Faruqui, A.K. Biswas, and M.J. Bino, eds. *Water Management in Islam*. Tokyo: United Nations University, 79–84.

Aljunied, K., 2016. Reorienting Sufism: Hamka and Islamic Mysticism in the Malay World. *Indonesia* (101), 67–84.

Amri, U., 2011. Divine Power: Faith-Based Mass Groups Take on Climate Change. *Strategic Review*, 1 (1), 95–108.

Amri, U., 2014. From Theology to a Praxis of "Eco-Jihad": The Role of Religious Civil Society Organizations in Combating Climate Change in Indonesia. *In:* R.G. Veldman, A. Szasz, and R. Haluza-DeLay, eds. *How the World's Religions Are*

Responding to Climate Change: Social Scientific Investigations. 1st ed. London: Routledge, 75–93.

APIP, Sagala, S.A., and Pingping, L., 2015. Overview of Jakarta Water- Related Environmental Challenges. *Water and Urban Initiative Working Paper Series*, 4, 1–5.

Aspinall, E., 2013. A Nation in Fragments: Patronage and Neoliberalism in Contemporary Indonesia. *Critical Asian Studies*, 45 (1), 27–54.

Bagir, Z.A., and Martiam, N., 2017. Islam: Norms and Practices. *In*: W.J. Jenkins, M.E. Tucker, and J. Grim, eds. *Routledge Handbook of Religion and Ecology*. New York: Routledge, 79–87.

Blacksmith Institute and Yayasan Tambuhak Sinta, 2013. *Mitigating Mercury Emissions from Artisanal and Small Scale Gold Mining in Indonesia* [online]. Available from: www.unep.org/chemicalsandwaste/Portals/9/Mercury/Final Report Narrative_June2014.pdf [Accessed 4 April 2022].

Bryner, K., 2013. *Piety Projects: Islamic Schools for Indonesia's Urban Middle Class*. Columbia: Columbia University.

Burhani, A.N., 2013. Defining Indonesian Islam: An Examination of the Construction of the National Islamic Identity of Traditionalist and Modernist Muslims. *In*: J. Burhanudin and K. Van Dijk, eds. *Islam in Indonesia: Contrasting Images and Interpretations*. Amsterdam: Amsterdam University Press, 25–47.

Bush, R., 2009. *Nahdlatul Ulama and the Struggle for Power Within Islam and Politics in Indonesia*. Singapore: ISEAS.

Bush, R., and Fealy, G., 2014. The Political Decline of Traditional Ulama in Indonesia. *Asian Journal of Social Science*, 42 (5), 536–560.

Candra, H., and Ab Rahman, A., 2010. Waqf Investment: A Case Study of Dompet Dhuafa Republika, Indonesia. *Jurnal Syariah*, 18 (1), 163–190.

Chozin, M., 2008. *Illegal but Common: Life of Blast Fishermen in the Spermonde Archipelago, South Sulawesi, Indonesia*. Ohio: Ohio University.

CNN Indonesia, 2018. *Anies Ingin Masjid di Jakarta Daur Ulang Limbah Air Wudu* [online] Available from: www.cnnindonesia.com/nasional/20180525 141744-20-301255/anies-ingin-masjid-di-jakarta-daur-ulang-limbah-air-wudu [Accessed 4 April 2022].

Damhuri, E., 2017. Hukum Mengonsumsi Air Daur Ulang, *Republika* [online]. Available from: www.republika.co.id/berita/dunia-islam/fatwa/17/10/13/oxqr5o440-hukum-mengonsumsi-air-daur-ulang [Accessed 4 April 2022].

Detik, 2012. *Pengelola Masjid Istiqlal Tak Persoalkan Air Wudhu Bekas Daur Ulang* [online]. Available from: https://news.detik.com/berita/2108391/pengelola-masjid-istiqlal-tak-persoalkan-air-wudhu-bekas-daur-ulang [Accessed 4 April 2022].

Dinata, Y. et al., 2013. *Integrating Religion Within Conservation: Islamic Beliefs and Sumatran Forest Management*. Canterbury: DICE, University of Kent.

Faisal, B, 2011. *Local Knowledge and Education: Developing Ecopesantren, Green Islam in Indonesia* [online]. Available from: https//vimeo.com/45589145 [Accessed 21 March 2022].

Fernandez, J.M., 2009. From Charity to Social Investments and Social Justice: A Study of Philanthropic Institutions in Indonesia. *In*: *Asian Transformations in Action: The Work of the 2006/2007 API Fellows*. Bangkok: The Nippon Foundation, 26–35.

Foltz, R.C., 2003. Islamic Environmentalism: A Matter of Interpretation. *In*: R.C. Foltz, F.M. Denny, and A. Baharuddin, eds. *Islam and Ecology*. Cambridge: Harvard University Press.

Fuad, M., 2002. Civil Society in Indonesia : The Potential and Limits of Muhammadiyah. *Sojourn : Journal of Social Issues in Southeast Asia*, 17 (2), 133–163.

Gade, A., 2012. *Indonesia's Islamic Environmental Boarding Schools, Inside Islam: Dialogues and Debates* [online]. Available from: http://insideislam.wisc.edu/2012/02/indonesias-islamic-environmental-boarding-schools/ [Accessed 4 April 2022].

Geertz, C., 1960. *The Religion of Java. Glencoe.* Glencoe, IL:The Free Press.

Ghazali, M.B., 2001. *Pendidikan Pesantren Berwawasan Lingkungan: Kasus Pondok Pesantren An-Nuqayah Guluk-Guluk, Sumenep, Madura.* Jakarta: Pedoman Ilmu Jaya.

Herman, L., 2022. Kosher Electricity and Sustainability: Building Block or Stumbling Stone?" *In*: J. Koehrsen, J. Blanc, and F. Huber, eds. *Religious Environmental Activism: Emerging Conflicts and Tensions in Earth Stewardship.* London: Routledge.

Hicks, J., 2012. The Missing Link: Explaining the Political Mobilisation of Islam in Indonesia. *Journal of Contemporary Asia*, 42 (1), 39–66.

Hosen, N., 2004. Behind the Scenes: Fatwas of Majelis Ulama Indonesia (1975–1998). *Journal of Islamic Studies*, 15 (2), 147–179.

Howell, J.D., 2010. Indonesia's Salafist Sufis. *Modern Asian Studies*, 44 (5), 1029–1051.

Ichwan, M.N., 2013. Towards a Puritanical Moderate Islam: The Majelis Ulama Indonesia and the Politics of Religious Orthodoxy. *In*: M. Van Bruinessen, ed. *Contemporary Developments in Indonesian Islam: Explaining the 'Conservative Turn'*. Singapore: ISEAS, 60–104.

Isbah, M.F., 2012. Religiously Committed and Prosperously Developed : The Survival of Pesantren Salaf in Modern. *Review of Indonesian and Malaysian Affairs*, 46 (1), 83–104.

The Jakarta Post, 2018. *Istiqlal Mosque to Have Wastewater Plant* [online]. Available from: www.thejakartapost.com/news/2018/11/25/istiqlal-mosque-to-have-wastewater-plant.html [Accessed 4 April 2022].

Johnston, D.L., 2012. Intra-Muslim Debates on Ecology: Is Shari'a Still Relevant? *Worldviews: Global Religions, Culture, and Ecology*, 16 (3), 218–238.

Kent, E.F., 2010. A Road Runs Through It: Changing Meanings in a Sacred Grove in Tiruvannamalai, Tamil Nadu. *Journal for the Study of Religion, Nature and Culture*, 4 (2), 213–231.

Kholil, L., 2017. Daur Ulang Air Mutanajjis, *Syiar NU* [online]. Available from: http://nahdlatululama.id/blog/2017/10/19/daur-ulang-air-mutanajjis/ [Accessed 4 April 2022].

Lahiri-Dutt, K., 2004. Informality in Mineral Resources Management in Asia: Raising Questions Relating to Community Economies and Sustainable Development. *Natural Resource Forum* (28), 123–132.

Latief, H., 2012. *Islamic Charities and Social Activism: Welfare, Dakwah and Politics in Indonesia.* Utrecht: Utrecht University [online]. Available from: http://dspace.library.uu.nl/handle/1874/253590 [Accessed 21 March 2022].

Majelis Ulama Indonesia, 2011. *Himpunan Fatwa MUI Sejak 1975.* Jakarta: Erlangga.

Mangunjaya, F.M., 2011. Developing Environmental Awareness and Conservation Through Islamic Teaching. *Journal of Islamic Studies*, 22 (1), 36–49.

Mangunjaya, F.M. et al., 2010. Muslim Projects to Halt Climate Change in Indonesia. *The Journal of Islamic Perspective*, 3, 116–130.

Mangunjaya, F.M. and McKay, J.E., 2012. Reviving an Islamic Approach for Environmental Conservation in Indonesia. *Worldviews*, 16 (3), 286–305.

Mangunjaya, F.M. et al., 2015. Faiths from the Archipelago. *Worldviews: Global Religions, Culture, and Ecology*, 19 (2), 103–122.

Manshur, F., 2009. *Entrepeneur Organik: Rahasia Sukses K.H. Fuad Affandi Bersama Pesantren dan Tarekat 'Sayuriah'-nya*. Bandung: Nuansa.

Marongo, U., 2012. Islamic Education in Indonesia and Malaysia. *At-Ta'dib*, 7 (2), 363–377.

Metro TV Indonesia, 2010. *Islam dan Perubahan Iklim (Islam and Climate Change)', Green Lifestyle* [online]. Available from: www.youtube.com/watch?v=ExXdWqdH8wY [Accessed 4 April 2022].

Mietzner, M., 2013. Fighting the Hellhounds: Pro-democracy Activists and Party Politics in Post-Suharto Indonesia. *Journal of Contemporary Asia*, 43 (1), 28–50.

Muhammad, A.S. et al., 2006. Fiqh Lingkungan (Fiqh al-bi'ah). *In*: A.S. Muhammad, ed. *Pertemuan Menggagas Fikih Lingkungan (Fiqh al-Bi'ah) oleh Ulama Pesantren di Lido, Sukabumi, 9–12 Mei 2004*. Jakarta: Indonesia Forest and Media Campaign (INFORM).

MUIS, 2001. *Jawatankuasa Fatwa Majlis Ugama Islam Singapura* [online]. Available from: https://www.muis.gov.sg/officeofthemufti/Fatwa [Accessed 16 August 2022].

Mukaddar, M., 2013. *Eko-Tarbiyah: Aplikasi Pendidikan Islam Berwawasan Ekologi di SMK Kehutanan Wali Songo Tuban*. Jakarta: PKBM Ngudi Ilmu.

Murdiyarso, D. et al., 2011. *Indonesia's Forest Moratorium: A Stepping Stone to Better Forest Governance? Working Paper 76*. Bogor: CIFOR.

NU Online, 2019. *Hukum Bersuci dengan Air Daur Ulang Limbah* [online]. Available from: www.nu.or.id/post/read/104885/hukum-bersuci-dengan-air-daur-ulang-limbah [Accessed 4 April 2022].

Republika, 2015. *Pemerintah Didesak Batalkan Kerja Sama Penguasaan Air* [online]. Available from: www.republika.co.id/berita/nasional/umum/15/02/23/nk864p-pemerintah-didesak-batalkan-kerja-sama-penguasaan-air [Accessed 4 April 2022].

Republika, 2016. *Air Daur Ulang, Halalkah?* [online]. Available from: www.republika.co.id/berita/dunia-islam/fatwa/16/09/06/od2lax313-air-daur-ulang-halalkah [Accessed 4 April 2022].

Resosudarmo, B.P. et al., 2009. Socioeconomic Conflicts in Indonesia's Mining Industry. *In*: R. Cronin and A. Pandya, eds. *Exploiting Natural Resources: Growth, Instability, and Conflict in the Middle East and Asia*. Washington, DC: The Henry Stimson Center, 33–46.

Ruslan, H., 2012. Air Daur Ulang, Bagaimana Hukumnya? *Republika* [online]. Available from: https://republika.co.id/berita/dunia-islam/fatwa/12/02/26/lzymaq-air-daur-ulang-bagaimana-hukumnya [Accessed 4 April 2022].

Ruslan, H., 2013. Benarkah Air Daur Ulang Halal? *Republika* [online]. Available from: https://khazanah.republika.co.id/berita/dunia-islam/khazanah/mpccie/benarkah-air-daur-ulang-halal [Accessed 4 April 2022].

Saat, N., 2016. Theologians "Moralising" Indonesia? *Asian Journal of Social Science*, 44 (4–5), 546–570.

Schwencke, A.M., 2012. *Globalized Eco-Islam: A Survey of Global Islamic Environmentalism*. Leiden: Leiden University.

Taylor, M., 2017. Indonesia Unveils Plan to Roll Out 1000 Eco-Mosques by 2020. *Reuters* [online]. Available from: www.reuters.com/article/us-indonesia-climatechange-religion/indonesia-unveils-plan-to-roll-out-1000-eco-mosques-by-2020-idUSKBN1DG1J8 [Accessed 4 April 2022].

Tehusijarana, K.M., 2018. MR Vaccine Contains Haram Substance, but Muslims May Use It: MUI. *The Jakarta Post* [online]. Available from: www.thejakartapost.com/news/2018/08/21/mr-vaccine-contains-haram-substance-but-muslims-may-use-it-mui.html [Accessed 4 April 2022].

Tomalin, E., 2002. The Limitations of Religious Environmentalism for India. *Worldviews*, 6, 12–30.

van Bruinessen, M., 2009. *Kitab Kuning Pesantren dan Tarekat*. Mizan: Jakarta.

Varkkey, H., 2015. *The Haze Problem in Southeast Asia: Palm Oil and Patronage*. New York: Routledge.

Wulandari, I., 2015a. Daur Ulang Air Bekas Wudhu Masjid Az-Zikra Belum Terlaksana. *Republika* [online]. Available from: www.republika.co.id/berita/dunia-islam/islam-nusantara/15/05/05/nnuuv9-daur-ulang-air-bekas-wudhu-masjid-azzikra-belum-terlaksana [Accessed 4 April 2022].

Wulandari, I., 2015b. Umat Islam Perlu Memikirkan Daur Ulang Air Wudhu. *Republika* [online]. Available from: www.republika.co.id/berita/dunia-islam/islam-nusantara/15/04/23/nn99nv-umat-islam-perlu-memikirkan-daur-ulang-air-wudhu [Accessed 4 April 2022].

Zuhdi, M., 2006. Modernization of Indonesian Islamic Schools' Curricula, 1945–2003. *International Journal of Inclusive Education*, 10 (4–5), 415–427.

6 From "Why Should?" to "Why Do?" Tensions in the Christian Context While Acting for the Environment[1]

Julia Blanc

Introduction: The Divergence Between Requirement and Reality – An Unbridgeable Gap?

"Drinking wine while preaching water" is only one of many reproaches most religious communities face regularly. Probably all people who belong to a religious community know situations in which reality is fundamentally opposed to the noble aims and particular guidelines of their religious community. This text focuses on the tensions that arise between ambition and reality in the field of ecology and environmental protection.

Theological contributions to the environmental debate focus primarily on a normative approach to the question of why one should protect creation (Deane-Drummond and Artinian Kaiser 2018; Vogt 2019). Often, the answer is quite theoretical, supported by some quotations from various books of the Bible and ecclesial/community traditions. When drawing upon these theoretical perspectives, most religious actors seem open to an "environmental turn." My contribution goes beyond the theoretical perspective and explores the empirical undertaking of this "environmental turn." Based on qualitative interviews, I will investigate the question of what reasons representatives of Christian communities give for their environmental (non-) engagement.

The article addresses (a) the status quo of the enviro-ethical debate in the Christian milieu as well as (b) its real-life application in Christian churches focusing on the congregational level. This reveals multiple tensions in how the topic is addressed. The dissimilarities between theory and practice seem typical for all engagement that goes along with the hierarchical structure of most Christian denominations: Different levels treat different issues differently.

This chapter begins with an overview of the development of Christian environmental ethics, as represented in official papers and communications with a focus on the German-speaking European context from which the case studies are taken. The subsequent section introduces the research methods that were undertaken to gather and analyze the empirical data. The following empirical part describes reasons that Christian actors give for

DOI: 10.4324/9781003017967-8

their environmental (non-)engagement and links them back to these teachings. Moreover, this section also provides insights into the actions and their rationale: It explores the place of theological argumentation in Christian communications and shows that Christian communities often prioritize other fields of activity.

Finally, the conclusion relates the findings to the divergence between requirement and reality highlighted in the opening of this article: The Catholic and the Protestant/Lutheran/Reformed churches in particular (the "established" ones, cf. Huber 2022, can rely on a longer tradition in the environmental ethical context at a higher level, such as the umbrella organization or the diocese). These reflections, however, are not simply passed down to the congregation and applied. Contrary to assumption, ecological reflections at a higher level rarely lead to significantly more sustainable behavior on the ground.

The Development of Christian Environmental Ethics

In the aftermath of World War II, many actors in politics, society, and church(es) realized that the time of "doing as we have always done" was over. This led to fundamental upheavals in Christian communities, such as the admission of women as pastors in various Protestant and Reformed communities, but also big shifts in the Catholic context: The Second Vatican Council (1962–1965) reformed the liturgy and led to numerous other changes. The "aggiornamento," initiated by the election of John XXIII and Paul VI as popes, led inter alia to the integration of human rights discourse into the ecclesiastical domain. In this context, Lynn White formulated in 1967 his fundamental critique of Christianity, especially the Roman Catholic Church: For him, it was due to medieval monks that the idea of the submission of the earth was so present and influential (White 1967). This text, however, was primarily influential in the US-American context. Its widely precepted German equivalent was written and published by the German environmental activist Carl Amery under the title "Das Ende der Vorsehung. Die gnadenlosen Folgen des Christentums" ("The End of Providence. The Merciless Consequences of Christianity") (Amery 1972).

During the same time, the Club of Rome published its position paper on the fact that the world in which we live is not without limits (Meadows 1972). Unlike when similar positions had been espoused in the 1930s or earlier, humanity – as well as religious communities, which was new – seemed to have reached a point where ecological concerns were becoming more widely shared, and a serious discussion of environmental matters started even within churches on a visible level.

The more recent publication of the encyclical Laudato Si' (Pope Francis 2015) can be described as the culmination of Christian (or even religious) environmental commitment to date; however, it is certainly not the only evidence of the emergence of a Christian interest in ecological (and wider) concerns (Blanc 2017; Peppard 2015). John Paul II had adopted his own approach based on the inter- and transdisciplinary study of "human ecology," which concentrates on the relationship between humans and their natural, social, and built environments (Blanc 2021; Vogt 2013). This philosophy-based approach has a diverse history of advancements in various fields of the social sciences and humanities, but remains strictly within the limits of anthropocentrism. John Paul II's predecessors Pius XII, John XXIII, and Paul VI had also shown an interest in ecological issues, although in ways that were shaped by the conventions of their time. The latter in particular demonstrated a very progress-friendly position in his encyclical on development, Populorum Progressio (Paul VI 1967), which was written in light of the belief that every challenge facing humanity could be resolved within the upcoming years. Even if this position now sounds outdated, it was very modern and open-minded when published; and it was the first time that environmental issues were approached, albeit obliquely, in an encyclical letter (Vogt 2013). Clearer references were made in official statements and speeches by the different popes, for example, by Paul VI during various meetings in the early 1970s (Santos 2019). Bartholomew, the patriarch of Constantinople and therefore a point of reference in the Orthodox world, also has to be named in the list of Christian figures central to the environmental cause (Koehrsen et al. 2019). He is one of only a few living figures on which Pope Francis relies in Laudato Si' (older examples that underline his importance and application concerning ecological questions are given in McGreevy and Hamlin 2006; McCright et al. 2014; Konisky 2018).

These reflections demonstrate that faith can help to address environmental sustainability (Dilmaghani 2018; Grim 2011; Palmer 2013): Even though Christian membership numbers are declining in Western Europe, churches can still motivate people to commit themselves to environmental protection (Gardner 2002).

Environmental Ethical Reflections in History

> And God blessed them, and God said unto them, "Be fruitful, and multiply, and replenish the earth, and subdue it: and have dominion over the fish of the sea, and over the fowl of the air, and over every living thing that moveth upon the earth."
>
> (Gen 1:28)

This biblical passage is the starting point for most "traditional" criticism of environmental degradation caused in the name of Christianity – or at least Christians. Since the translation and interpretive problems of this quotation are not within the scope of this article (on the first theological responses, see Münk 1987; Rappel 1996), I will instead focus on its possible influence, as demonstrated in Lynn White's 1967 piece, "The Historical Roots of Our Ecological Crisis." Though perhaps unintentionally, this article formulates the basic tenets of modern Christian environmental commitment, starting with medieval monks' enthusiasm and affinity for technology. Motivated by the Biblical passage just quoted, these monks set out to subjugate the world, and according to White's interpretation, they were successful (Taylor et al. 2016). The fact that today theological research has found out that these interpretive difficulties are mostly due to tendentious or even wrong translations does – as indicated previously – not change much about the influence this passage has had so far.

The Christian treatment of ecological questions might always be seen in the interaction of immanence and transcendence: The connection between religion and environment is always shaped through the point of view of the individual who connects both fields. His or her interpretation of God and the world has an enormous impact on the position they choose to adopt. Generalizing this approach, one could probably even say that organizing life in tandem with the (God-given) environment is the core task of religion. These same aspirations of living in harmony with the earth are nowadays quite often supposed to be more present in indigenous religions than they are in the allegedly modern "book religions" (i.e., Judaism, Islam, and Christianity). Yet even if the latter seem more distant from the environment (and therefore often less caring toward it), they also reveal a coming to terms with an environment that reaches beyond human explanation.

During the first centuries of Christianity, ecological issues remained topical, but primarily because of people's living conditions: The assumption was that society was rural, and therefore it had to have faith. Dealing with environmental challenges like droughts, storms, and food scarcity was easier within a framework that made reference to faith. The theological guidelines gave meaning to people's suffering and hope for life after death. In this context, of course, the criticism formulated by White must be acknowledged.

Nowadays, many tend to refer to Saint Francis of Assisi (very probably also because of his elevation to the position of the patron saint of ecology by John Paul II in 1979) as one of the first Christian actors who cared for creation beyond humans (Carroll 2001; Nothwehr 2019). The uncritical reception and (re)interpretation of his deeds, however, ignores the historical context in which they happened and prevent a full understanding of the Christian relationship to the environment. These interpretations emphasize Francis's care for the environment, but omit the fact that in the 12th century his care was a form of worship of God, to whom he referred as the creator. St. Hildegard of Bingen, a German Benedictine abbess, can be read in contradistinction to this, as she accumulated enormous knowledge in relation to plants and their active substances. For her, the blessings of the animal and vegetal prove the value that humans have endowed upon their environment without taking too much out of it (Roth 2000; Philipp 2009; Dadosky 2018).

In the following, I will focus on the modern development of Christian ecological arguments, as the actual debate does not – at least not explicitly – go further back in history. It might be seen as a parallel to the philosophical development, where issues going beyond humans arrived only in the mainstream during the last decades. As this article is based on case studies in the German-speaking European context, which was (and still is) dominated by the Roman Catholic and the mainly Lutheran, Protestant Church in Germany (EKD) and the Reformed Church in Switzerland, my focus is on theologians who were influential in this field. There were undoubtedly other fore-runners in other places (Nasr in the eco-Islam; Walter Rauschenbusch as a representative in the Baptist tradition), but their influence remains scarce.

The French Jesuit and paleontologist Pierre Teilhard de Chardin, in contrary, is a central figure in the Christian turn to the non-human environment in the European context. The majority of his eco-theological writings were made available to the general public only after his death in 1955 and led to a real trend in the following years. Especially "Le phènomène humaine," published in 1957, has to be named as a central work (Teilhard de Chardin 1958). The Catholic priest was followed in the Germanophone field by various theologians from both traditions: namely, from the Protestant/Lutheran/Reformed field Günter Altner (1975), Jürgen Moltmann (1985) and Wolfhart Pannenberg (2000), and on the Catholic side Alexandre Ganoczy (1983) (Ostheimer/Blanc 2021).

Environmental Ethics in WCC, the Conciliar Process and the Catholic Hierarchy

As already indicated, many things changed in both religious and secular settings from the late 1960s. These (fundamental) shifts often begin on a grassroots level and have to overcome various challenges – especially tradition and hierarchical structures – on their way to creating broader-scale change, their influence only becomes apparent after a certain time. What is more, religion does not think in short time spans. For instance, Catholic ecclesial life today is governed by the "new" CIC (codex iuris canonice), considered the "fundamental body of ecclesiastical laws for the Latin Church" (John Paul II 1983). It was, however, promulgated in 1983, nearly 40 years ago. The same applies to changes in liturgy starting in the aftermath of the Second Vatican Council (1962–1965). These examples might illustrate the velocity with which shifts can expect to find purchase in this long-term religious context. The non-Catholic but mainline Christian churches also underwent fundamental changes during the last century. Even before World War II, early attempts were made to connect both on inter-confessional as well as international levels. This is how the World Council of Churches (WCC) came to see the light of day in 1948 in Amsterdam, where 147 member churches signed its founding papers. It was built upon the mainly orthodox desire to create a "League of Churches" in parallel to the League of Nations. At the end of 2013 (the most recent general meeting at the time of writing, as the 2020 meeting was delayed due to Covid-19), the membership stood at 345 churches that practice baptism and eucharistic fellowship. The WCC has also established a joint direction on environmental ethics. Although the Catholic Church is not a member of the WCC, it maintains close contact through observation. Thus, progress made at the WCC influences the Catholic Church and vice versa. Even if decisions are not made in unison, they often respond to each other.

As Vogt (2013, pp. 180–183) highlights, the WCC's concept of a Sustainable Society makes it a pioneer in the religious as well as the applied ecological context. In 1974, at the WCC conference in Bucharest, sustainability had already become a keyword and a guiding concept toward a future society, which was expressed in a communiqué under the title "Sustainable and Just Society." The authors were committed to the idea that ecological limits had to be respected in order to create a society that was also economically sustainable over the long term. In 1975, during the Nairobi conference, this paper was adopted as an important complement to the ecumenical and social thinking of the WCC. Vogt underlines, however, that the follow-up study "Just, Participatory and Sustainable Society" opposes economic and theological concepts in a too fundamental way. In interpreting "sustainability" in an exclusively God-focused way (sustainability is used as a synonym for God's care for the world), economic terms get lost in favor of an exclusively religious vocabulary, which is also illustrated by the later renaming of the field as the "integrity of creation." This much more theological labeling might be seen as a loss of inner-world orientation, as Vogt mentions. On the other hand, it re-inscribes a Christian approach to the theme and leaves potentially more space for religious reflections and justifications.

The reimagination of environmental concerns is nevertheless significantly encouraged during the "Conciliar Process," which strongly shaped the environmental ethics of Christian institutions in Western Europe. This process officially started in 1983 at the 6th World Assembly of the World Council of Churches (WCC) in Vancouver, Canada. It is known as the conciliar process for Justice, Peace and the Integrity of Creation (JPIC) (Blanc 2017; Blanc and Ostheimer 2019; Philipp 2009; Rosenberger 2001). This shift in nomenclature – from "sustainable society" to "integrity of creation" – also underlines the shift in the concept, which becomes even more theological and less political. The first publication mentions, inter alia, the importance of ideas like (over)population and birth control, which were abandoned at the urging of various members of the WCC in favor of a religiously "more acceptable" approach to societal demands (Best 1990; Cobb 1992; Rosenberger 2001; Vogt 2013). In this context, it is important to remark that the ecological crisis had become a central focus for the Northern Hemisphere. Many Southern churches, however, perceived this as a distraction from their "real" problems: for example, the treatment of poverty as a core issue of (Christian) justice in the new era of post-colonialism. Today this line of argumentation remains prevalent among some fundamental Christian Evangelical actors who have labeled the religio-environmental movement "the green dragon" (Taylor 2016).[2]

Discussions concerning the "right" positioning continued in the WCC in the years following the Vancouver meeting, for example, at the 1990 assembly in Seoul. As a result of the WCC's internal quarrels concerning the organization's ongoing positioning in relation to sustainability, it did not participate in the 1992 United Nations meeting on the environment in Rio de Janeiro, Brazil. One might say that a number of the initiators of the sustainability discourse yielded to other, more political and NGO-affiliated actors (Kopiec 2016). Churches were, however, present at the United Nations Conference on Environment and Development (UNCED), also known as the Rio de Janeiro Earth Summit. Unfortunately, research concerning their influence is still scarce. David Hallmann underlines, however, the importance of the ecumenical exchanges that took place during this conference, especially in regard to how these signaled an awareness of "the inter-relatedness of environment and development" (Hallmann 2002). Sustainability as a central term became unfortunately less important during the following years on the WCC level. A Christian delegation organized through the WCC, but including some Catholic actors, tried to participate in the World Summit on Sustainable Development (WSSD) in Johannesburg in 2002, but was only admitted to background discussions and therefore had only a very limited impact (Vogt 2013).

The WCC grew very fast and different questions seemed more or less important to different world regions. Therefore, religious actors from Europe considered it useful to pool Christian confessions in the Council of European Churches (CEC), which was founded in 1959 and today acts as a regional sub-section of the WCC. As for the worldwide version, its members come from Orthodox, Protestant/Lutheran/Reformed, Anglican, and Old Catholic Churches, all ecclesial communities that reject the Roman Catholic idea of primacy. This opposition between the two largest established Christian blocks, however, was (and might still be) overcome in the European Ecumenical Assembly and its focus on (environmental) sustainability. The first meeting was held in 1989 in Basel, Switzerland. Here, sustainability-related matters forced their way into the ecumenical discourse, which now included Roman Catholic actors as well, thereby placing sustainability and ecological considerations at the center of theological debates in the European context for the first time. With this first step, a new field was not only conquered but also brought its opportunities to the ecumenical discourse.

Eight years after Basel, a second meeting of the European Ecumenical Assembly was organized in 1997 in Graz, Austria. It focused on the theme of "Reconciliation – a gift of God and a source of new life." The meeting's concluding document formulated four central recommendations focusing on the field of sustainability, which are much more inner-world oriented than one might imagine from the overall theme. These are: (1) the recognition of sustainability as part of ecclesial life; (2) a plea for a sustainable lifestyle and economy; (3) a position on Agenda 21 as a political framing for the potential role of the Conciliar Process; and (4) the generation of ecclesial networks. All these suggestions were supported by practical and concrete information, which led to significant changes in the daily life of the churches concerned: Among other things, the celebration of the Day of Creation, which originally used to be an exclusively Orthodox celebration (initiated in 1989 at the beginning of the Orthodox liturgical year[3]), now became an event solemnized in all Christian denominations on the first Sunday in September. Following this prelude, the initiators proposed that Creation be celebrated for a season that lasted from that day until 4 October. An even more tangible result was the establishment of the European Christian Environmental Network (ECEN), which brought together representatives in charge of environmental issues from the Anglican, Orthodox, Protestant, and – which is new in this ecumenical context – Catholic churches, as well as other Christian environmental networks (Deutsche Kommission Justitia et Pax 1997).

The third meeting of the European Ecumenical Assembly took place in 2007 in Sibiu, Romania, under the motto "The light of Christ shines upon all." This meeting took the form of an ecological pilgrimage, taking in various stations all over Europe, with the purpose of positioning the care of creation at the core of different denominations in different places. In addition to this, the meeting set standards to be followed throughout its organization, which followed the world's strictest demands for environmental management, the EMAS (Eco-Management and Audit Scheme) regulations.

The Conciliar Process advanced in parallel to the secular movement that followed the publication of the Brundtland report, "Our Common Future" (World Commission on Environment and Development 1987), and the World Summit in Rio de Janeiro 1992. The parallels between the sustainability triangle – social welfare as aim, economy as means, and ecology as natural constraints – and the Conciliar Process – peace, justice, and the integrity of creation – are obvious: Both stress the relationality of life on Earth. Instead of separating different fields or zones ("the ecology," "the economy," . . .), the connection is stressed, one might even say put into the center of all attention. Both have in common to see life in its various dimensions. Social welfare as aim might be seen as a vector for peace, economy of means must be organized in a just way, whereas the parallels between ecology as natural constraints and integrity of creation are quite obvious (Sachs 2017).

Various actors from the secular as well as the religious field therefore undertook the shaping of the "modern ecological approach," as well as the concept of sustainability that is prominent today. This broad participation from different areas, however, risks being quickly forgotten: The early impact of Christian actors on environmental questions on a societal and political level is still an open question and needs to be investigated.

The Conciliar Process encouraged activities within the individual Christian churches that participated in the process. An example of this might be the Roman Catholic Church. A Europe-wide survey on the ecological engagement of Catholic bishops' conferences (and therefore dioceses) was undertaken during the first years of the new millennium. Published in 2007 under the title "Responsibilità per il Creato in Europa. L'impegno delle Conference Episcopali" (Mascia et al. 2007), it gives a complete overview of already realized or planned actions, resources, and limits. Beneath the enumeration and presentation of environmental bureaus and staff in each bishops' conference, as well as their financial support and the titles of all official writings referring to environmental topics, it names, for example, reforestation projects and awareness seminars, youth challenges and liturgical propositions. The publication is doubly impressive in that it assigns an enormous importance to the ecological question within the Catholic Church and collects an abundance of answers, figures, and numbers. Thus, the "environmental question" became inextricably linked to the field of pastoral work and environmental commitment; likewise, lobbying for ecology-related questions became recognized as an immanent responsibility of the Church. If one considers that this survey took place during the time of Pope Benedict XVI, under whose leadership different sides emphasized the ecological oblivion of the Catholic Church (Vogt 2009), these self-assurances are even more surprising.

In 2012, various Christian actors participated in the Rio+20-Conference and thus contributed to the development of the SDGs (Sustainable Development Goals) that were "born" there. These 17 aims associate 169 targets and 232 indicators which cover a wide range of social, ecological, and ecumenical challenges that must be faced and addressed in common until 2030 to permit an equilibrate progress of the world community (Ostheimer and Blanc 2021).

Since then, many things have changed: Jorge Mario Bergoglio was elected Pope of the Roman Catholic Church in 2013 and became the first Pope in history to choose Francis as his papal name, foreshadowing his later concern for ecological themes. This presentiment was confirmed when he published his encyclical Laudato Si' in 2015, thereby bringing environment, ecology, and sustainable behavior into the focus of ecclesiastical reflection and action. Significantly, the sustainability encyclical Laudato Si' was formulated in the same year the SDGs were published. Together with the Paris Climate Conference at the end of the very same year, it must be seen as a central turning point in the shift from development to post-development (Sachs 2017).

Methods and Data

This chapter draws on the analysis of documents and interviews undertaken in the context of the SNSF-funded project "Urban Green Religions" at the University of Basel. The empirical part of the research focuses on four different ecological benchmark cities of less than 100,000 inhabitants in Germany and Switzerland. We have conducted 67 interviews. My focus here is on the responses of the interviewees from the established churches (Huber 2022): the Roman Catholic and the Protestant/Lutheran/Reformed Churches. These represent 21 interviews: 7 representatives of the Catholic Church, 12 of Protestant/Lutheran/Reformed Churches, and 1 of an intrareligious Christian environmental umbrella organization. Of the interviewees, 15 were located in Germany and 6 in Switzerland.

Semi-structured interviews were used to collect the information. The use of an interview guide ensures a joint approach to the selected topics while leaving time and space to ask additional questions and delve further into topics, and – importantly – to create a relationship with the interview partner. Thus, interviewees seem to feel more comfortable, and therefore more willing to share their knowledge when they are addressed as individuals, speaking about their own experiences and background, than if they were asked pre-formulated questions in an anonymous setting.

The questions asked during the approximately 60-minute interviews focused on the position of the interviewees, their experience, and their environmental collaborations with other actors. Many of them used the opportunity to inform themselves about the research project either while making the appointment or after having spoken with us. All interviews were transcribed and coded using MAXQDA.

When interviewees gave information material to us or pointed explicitly to homepages, we considered them as background information. Especially a printed brochure promoting environmental behavior within the Muslim community as well as the homepages of different Christian environmental actors have to be enumerated in this context. They helped us to understand in a broader context what "greening" meant to the communities and congregations. For this understanding, it was also necessary to know the "official" positionings as formulated in policy papers, pastoral letters, and so on, which we studied carefully already in the preparation of the interviews.

The Application of Theological Findings on the Ground: A Field of Tensions

The development of ecological interest, eco-sustainability, or care for creation as part of the churches' body of activities, as shown in the first part of this chapter, could lead one to think that this field has found a place at the core of parish activities. This, however, turns out to be wrong. Even if there are congregations that do show quite a high level of engagement (in most cases distinguished with the green rooster,[4] a label created for and from ecologically engaged congregations, cf. Monnot 2022), this is not the mainstream position. Putting this together with research that has shown that ecological engagement on the part of religious actors can be summarized via the three potential functions of "public campaigning," "materialization," and "value dissemination" (Koehrsen 2015, 2018) makes various insights possible. The first function, "public campaigning," concerns the fact that religions play an important role in shaping public opinion and can therefore be impactful. The success of this, however, depends on the interest of the community as well as the amount of information on the subject that is accessible to (and accessed by) the people being addressed from other sources. The second item, "materialization," highlights traditional measurable output, such as installing solar panels on a church roof or applying EMAS. On the one hand, this function is the easiest to measure. On the other hand, while it can potentially show people what they can and should be doing, it does not convey any theological content underlying the fact that Christians do care about the environment. The third aspect, "value dissemination," deals with the implementation of ecological concerns in religious discourse and an ethical approval of sustainability claims. Examples of this function do effectively apply to the calls of the various WCC publications as well as environment-encyclical Laudato Si'. It seems, however, that this aspect is "less attractive" for religious actors from mainline Christian congregations to be active in. This could be due to the "outdatedness" of theological arguments, as society presumes them to be. It could also be due to the still widespread under-representation of environmental issues in the training of clergy and religious ministers.

Environmental Engagement and Theological Reasoning

What reasons do religious actors give for their environmental engagement? Is it an inherent need to do something for creation or is it a societal demand that has to be satisfied? Are there personal experiences from their youth and lived reality that are evoked, or is the mention of influence and pressure from their surroundings dominant?

Many of the statements that qualify for this subsection, especially regarding intrinsic spiritual motivation, are typical representations of public theology (Bedford-Strohm 2008; Blanc 2017; Kim and Day 2017). For instance, a church official responsible for environmental questions states:

> On the basis of this common creation of all being, we have this responsibility, which has also been entrusted to us in this stewardship, to stand up for God's creation.
>
> (Interview 23, 20.02.2019)

He is a key participant in preparing theological communiqués that are intended to be read in churches as well as behind the scenes. Furthermore, he gives advice to parishes that want to become "greener" and composes "green" liturgical texts. This biblical, creation-focused approach is also employed by a pastor who is known in his diocese for his ecological engagement:

> From my understanding of the church, we have a mandate to be active in the world, for the good of humanity, well, and the whole of creation, but this is also founded on my biblical background.
>
> (Interview 7, 20.07.2018)

Most of the statements that I summarize under this point refer to the more obvious biblical reasons for environmental protection. There are, however, also positions that refer to the specific field of social ethics, as well as moral teaching. This is the case for a Catholic pastoral worker who quoted the European Ecumenical Assembly meeting in Basel (see earlier in this text) as a source of motivation:

> In the ecumenical world we have a social doctrine. This debate interweaves political and social aspects, as the assembly of Basel has done, which focused on these three deeply linked words: peace, justice, and the integrity of creation. And I think that the commitment to environmental issues is actually based on THAT.
>
> (Interview 10, 23.07.2018)

All actors who give theological reasons for their environmental engagement refer to their long-standing connection to environmental values, which can be traced back to childhood or is described as just "having always been there" (Interview 7, 20.07.2018). Even if the question was explicitly asked, nobody referred to a turning point that converted him or her to Christian environmentalism.

As mainstream Christian churches consider themselves as part of the society they live in, it is evident that "worldly" issues have to be addressed. For this reason, extrinsic motivations should not be perceived as less valuable. Many pastoral issues were not in the "inner circle" of pastoral work in the beginning but went on to become quite important fields of engagement, as for example, modern development aid did. One can state, however, that in general those actors who give more personal reasons for their environmental engagement and who link it back to their religious convictions try more actively to include this field of pastoral work in their congregations.

The Difficulty of Theological Reasoning in an Increasingly Secular Society

Other actors who are environmentally engaged, especially in materialization activities, do not refer directly to religious or spiritual motivations. The existence of these statements is particularly surprising as the present research relies exclusively on interviews with religious actors. It is, however, the dominant way that those active in this field speak about environmental commitments. This dominance might also be due to the fact that mainline Christian churches in Germany and Switzerland have to respect ecological policies of public administrations as well as religious umbrella organizations. As most of the quoted examples also come from contexts where building renovations were taking place, it is self-evident that a focus is placed on the field of materialization. Nevertheless, many interviewees give, at the end of the interview, a more intimate description of their religious motivations.

These interviewees were very careful not to use religious speech when speaking about their engagement in the first place. It is beyond the scope of this chapter to consider whether their avoidance of religious speech is a product of a "distance to theological reasoning," their conception of their professional role, or their consideration of non-religious arguments as more convincing.

One example is an environmental church official who campaigns for sustainable building refurbishments and the application of up-to-date knowledge in the art of construction. Examples of this refurbishment include replacing the insulation in buildings that are centuries old with ecological alternatives, as well as the planned installation of solar panels (Interview 10, 23.07.2018). The same parish also recently introduced palm-oil-free cleaning agents. A colleague of the interviewee referred to previously stated:

> We just started a project: palm-oil-free cleaning agents. I wasn't aware of the problem at all but, like in food, there are also a lot of detergents that incorporate palm oil. So, when a company approached us about replacing palm oil with local oils, we were happy to test them and now we use them. . . . We think that we as a church should be a role model and we must live up to this responsibility. Since we have funds available, we really should invest them into our long-term environment and not only see the short-term financial aspects.
>
> (Interview 17, 11.10.2018)

In an interview that lasted over an hour, this is the only reference that goes beyond exclusively technical questions to talk about the impact of the church on civil society.

What is surprising is that these non-religious arguments are put forward not only by non-theological employees of the parish but also by staff trained for liturgical roles. One of these interviewees worked together with the aforementioned officer but forwent nearly all theological speech when explaining his engagement. Instead, he pointed to the societal impact of his faith, in a tradition linked to the Böckenförde Dilemma: This means that the state builds on preconditions that it cannot create itself (Böckenförde 1976). Without explicit reference to this quotation, he underlines the importance of the role his denomination plays in civil society, but avoided the use of more explicitly theological speech.

The same problem appears in many theological position papers from bishops' conferences and so on. The German bishops' conference as well as the EKD (Evangelische Kirche in Deutschland, which is the federation of 20 Lutheran, Reformed, and United Protestant Regional Churches in Germany) are leading players in the field of theological publications on ecological questions. This tradition started in the 1980s (EKD: Ratserklärung der EKD zur Energiediskussion (Council declaration of the EKD on the energy discussion), 1977; DBK: Zukunft der Schöpfung, Zukunft der Menschheit (Future of creation, future of humanity), 1980; EKD und DBK (joint publication); Verantwortung wahrnehmen für die Schöpfung (Take responsibility for the creation), 1985) and continued ever since (Vogt and Numico 2007; Grevel 2015). These German pastorals are recognized for their strong scholarly foundation, which is due to strong cooperation between bishops and university researchers, and are therefore often translated into other languages. It appears, however, that in most cases, exclusively theological content only forms part of the pastorals' framework at the beginning and the end of the publications. The central part focuses quite often on scientific solutions (e.g., fracking, use of water, animal care) (Blanc 2017, 2021). It lacks the more theological argumentation that terms like "integrity of creation" (used by the WCC since the late 1970s; for a more critical ethical approach, see Rasmussen 1995) or "sister earth" (Pope Francis 2015) might lead one to expect.

The Prioritization of Engagement in Other Pastoral Fields

Quite often, interviewees excuse or explain their lack of commitment to the care of creation by indicating that, until now, it was not an issue that was of concern to their members. It appears that other issues were more pressing for some congregations and their members, such as the field of justice. One theologically trained parish pastoral worker manages, however, to link focused services from the ecclesial "one-world" working group to the field of ecology:

> Recently [we] had a service . . . about "Dios o el Oro," "God or gold." One cannot serve both God or Mammon, that is already a biblical instruction, which of course only then makes room for ecological action. If one is not primarily Mammon-oriented.
>
> (Interview 19, 23.10.2018)

Others see environmental engagement as a task that goes along with other church activities. It has to be done, but is not the core issue of the interviewed person. The pastor of a Reformed parish underlines the connection between environmental engagement and the societal importance of environmental issues:

> Whereas, you know, when I look at my audience, I don't have to "preach" environmental issues there anymore. They all actually have a sensitivity to sustainability issues.
>
> (Interview 12, 27.09.2018)

He cares about the implementation of these issues but does not seem to be as personally involved as others.

The interwoven- and interconnectedness of the different issues that are part of the triangle of sustainability – ecology, economy, and social – make it difficult to regard only one aspect. Especially the first example shows that justice appears in different forms and contexts. Ecological justice is just one aspect and yet its boundaries are still today not clearly defined. In a more general approach one can say that the Catholic and the Protestant/Lutheran/Reformed Churches referred to in this chapter consider themselves (still) as "Volkskirchen." "Volkskirchen" might be translated as popular churches, but means primarily that these congregations see themselves as pillars of society and actors (also) in the political field. This self-image is based, among other things, on European history but also on the financial and legal framework conditions traditionally resident congregations (still) face today. For them, it is "normal" to position themselves on various topics of daily life. As any other actor in civil society, they have to make choices concerning this prioritization – and as any other actor in civil society, these choices are also influenced by personal preferences and interests.

Conclusion: Tensions Between Christian Ethics and Environmental Action

The "ecological cause" is treated differently in different Christian denominations. These differences, however, do not only manifest between Christian churches. Strong – indeed probably even stronger – differences concerning the reasons underpinning environmental commitment, the application of this commitment, and the forms that environmental engagement takes, emerge today within individual Christian denominations on an international, national, regional, or local level. In other words: The frontiers between ecological and "not-so-ecological" reasoning and behavior are rarely the same as the borders between different denominations. Especially the last decades have led to a departure from strict discipline in regard to this issue. This finding contrasts with the overview Kearns proposed more than 25 years ago (Kearns 1996). Today the actual situation has become potentially even less clear. Additionally, research and ethical reflections are often regionally limited (by language but also by teaching resources, for example) (Blanc 2022).

The great admiration and esteem with which Protestant/Lutheran/Reformed interviewees speak about Laudato Si' is particularly testimony to the abovementioned blurry boundaries. This vanishing of frontiers can also be seen in the large number of ecumenical communications in regard to ecological questions as published by ÖKU in Switzerland and DBK and EKD in Germany.

This commonly approachable basis, however, is quite often forgotten on the way to the application of environmental commitments in particular parishes. As was shown previously, there is environmental action taking place, especially in the field of materialization, but it is not clearly linked back to a theological basis, and therefore might be perceived as an action that is not intrinsically motivated but rather stems from societal expectations. The enviro-ethical debate that was discussed at the beginning of this article seems to have little effect on environmental engagement.

Nevertheless, established Christian actors produce a variety of different approaches, explanations, and actions. Despite their joint ethical basis, there is not one single way of engaging religiously for the environment, but many. Environmental commitment appears as an individual issue shaped by personal experience, rather than collective theological reasoning.

The interviews, as well as the existence of a common ethical basis (formulated in the various publications), show that numerous ideas in the field of ecology can be shared and applied in new contexts. One example might be the Austrian "Autofasten"[5] campaign that encourages people to reduce their car use during Lent as a way of reducing pollution and raising awareness of creation. It has existed since 2007.

The importance of non-theological argumentation, especially with regard to communities that are engaged in the environmental field, emerges in this context as potentially one of the most fundamental questions. Is ecology really perceived as a part of the theological core task, or does it merely play a role in regard to churches' societal responsibility? This search for the appropriate localization of environmental engagement is decisive concerning the greening-of-religion hypothesis (Grim 2011; Dilmaghani 2018): Either ecology is part of the theological teaching and therefore it must occupy a much more important place in the formation of religious actors or it is "just" part of the societal engagement of churches. In that case, ecological arguments could be left behind in the formations of multiplicators. This self-restriction, however, would extremely limit the Christian potential. Finally, it is a, perhaps even the, unique selling point of theological discourse that ultimate reasons are offered.

Notes

1 I want to thank Jens Koehrsen and Fabian Huber for the great cooperation we had in the SNF-funded project Urban Green Religions.

2 www.resistingthegreendragon.com [Accessed 6 January 2020]. For an analysis of this group from a liberal political perspective, see People for the American Way, "The 'Green Dragon' Slayers: How the Religious Right and the Corporate

Right Are Joining Forces to Fight Environmental Protection": www.pfaw.org/press-releases/the-green-dragon-slayers-how-the-religious-right-and-the-corporate-right-are-joining-forces-to-fight-environmental-protection/ [Accessed 24 August 2019].

3 For more information, see Season of Creation [online] https://seasonofcreation.org/about/ [Accessed 7 February 2022].

4 Further information see www.kirchliches-umweltmanagement.de/ [Accessed 22 June 2020].

5 Autofasten: Heilsam in Bewegung kommen [online]. www.autofasten.at/ [Accessed 01 June 2020].

References

Altner, G., 1975. *Zwischen Natur und Menschengeschichte. Perspektiven für Eine Neue Schöpfungstheologie*. München: Kaiser.

Amery, C., 1972. *Das Ende der Vorsehung. Die gnadenlosen Folgen des Christentums*. Berlin: Rowohlt.

Bedford-Strohm, H., 2008. Öffentliche Theologie in der Zivilgesellschaft. *In*: Gabriel, I., ed. *Politik und Theologie in Europa. Perspektiven ökumenischer Sozialethik*. Ostfieldern: Gruenewaldverlag, 340–357.

Best, T., ed., 1990. *Vancouver to Canberra, 1983–1990: Report of the Central Committee of the World Council of Churches to the Seventh Assembly*. Geneva: WCC Publications.

Blanc, J., 2017. *Ökokatholizismus. Sozialethische Analysen zu ausgewählten Ländern und Institutionen in Europa*. Marburg: Metropolis.

Blanc, J., 2021. Arme Tiere: die Option für die Armen als möglicher Anschlusspunkt einer christlichen Tierethik. *In*: Martin Lintner, ed. *Mensch – Tier – Gott: Interdisziplinäre Annäherung an eine christliche Tierethik*. Baden-Baden: Nomos, 219–240.

Blanc, J., 2022. Believe in the Future: Wie sich religiöse und nicht-religiöse Akteure glaubensmotiviert für eine nachhaltige Entwicklung einsetzen. *In*: *epd Dokumentation* (in print).

Blanc, J., and Ostheimer, J., 2019. Religion as Transmission Belt for Promoting the Sustainable Development Goals. *In*: Leal W. Filho, A. Azul, L. Brandli, P. Özuyar, and T. Wall, eds. *Partnerships for the Goals. Encyclopedia of the UN Sustainable Development Goals*. Cham: Springer.

Böckenförde, E.-W., 1976. *Staat, Gesellschaft, Freiheit. Studien zur Staatstheorie und zum Verfassungsrecht*. Berlin: Suhrkamp.

Carroll, J.E., 2001. Catholicism and Deep Ecology. *In:* David L. Barnhill and Roger Gottlieb, eds. *Deep Ecology and World Religions: New Essays on Sacred Ground*. New York: State University of New York Press, 169–192.

Cobb, J., 1992. *Sustainability: Economics, Ecology, and Justice*. Eugene: Wipf and Stock Publishers.

Dadosky, J., 2018. The Original Green Campaign: Dr. Hildegard of Bingen's Viriditas as Complement to Laudato Si. *Toronto Journal of Theology*, 34 (1), 79–95.

DBK, 1980. *Zukunft der Schöpfung, Zukunft der Menschheit* [Future of creation, future of humanity] [online]. Available from: www.dbk-shop.de/media/files_public/8cf1ac11d91374102d2263cf4b5cf72b/DBK_1128.pdf [Accessed 12 January 2021].

Deane-Drummond, C., and Artinian Kaiser, R., 2018. *Theology and Ecology Across the Disciplines: On Care for Our Common Home*. London: Bloomsbury.

Deutsche Kommission Justitia et Pax, 1997. *Reconciliation – A Gift from God and a Source of New Life* [online]. Available from: www.justitia-et-pax.de/jp/publikationen/pdf/guf_083_e.pdf [Accessed 18 January 2020].

Dilmaghani, M., 2018. Which Is Greener: Secularity or Religiosity? Environmental Philanthropy Along Religiosity Spectrum. *Environmental Economics and Policy Studies*, 20, 477–502.

EKD, 1977. *Ratserklärung der EKD zur Energiediskussion* (Council Declaration of the EKD on the Energy Discussion), Quoted from Grevel, J., 2015. *Mit Gott im Grünen: Eine Praktische Theologie der Naturerfahrung*. Berlin: Vandenhoeck und Ruprecht.

EKD and DBK, 1985. *Verantwortung wahrnehmen für die Schöpfung* [Take Responsibility for the Creation] [online]. Available from: www.ekd.de/schoepfung_1985_inhalt.html [Accessed 12 January 2021].

Ganoczy, A., 1983. *Schöpfungslehre*. Düsseldorf: Patmos.

Gardner, G., 2002. *Invoking the Spirit: Religion and Spirituality in the Quest for a Sustainable World, Worldwatch Paper 164*. Washington, DC: Worldwatch Institute.

Grevel, J., 2015. *Mit Gott im Grünen: Eine Praktische Theologie der Naturerfahrung*. Berlin: Vandenhoeck und Ruprecht.

Grim, J., 2011. The Roles of Religions in Activating an Ecological Consciousness. *International Social Science Journal*, 62, 205–206, 255–269.

Hallmann, David G., 2002. *Report on the World Summit on Sustainable Development* (WSSD) [online]. Available from: www.oikoumene.org/resources/documents/report-on-the-world-summit-on-sustainable-development-wssd [Accessed 7 February 2021].

Huber, F., 2022. Environmentalism in the Religious Field: The Role of the Establishment for Competition in Switzerland. *In*: J. Koehrsen, J. Blanc, and F. Huber, eds. *Religious Environmental Activism: Emerging Conflicts and Tensions in Earth Stewardship*. London: Routledge.

John Paul II, 1983. *Sacrae disciplinae leges* [online]. Available from: http://w2.vatican.va/content/john-paul-ii/en/apost_constitutions/documents/hf_jp-ii_apc_25011983_sacrae-disciplinae-leges.html [Accessed 18 January 2020].

Kearns, L., 1996. Saving the Creation: Christian Environmentalism in the United States. *Sociology of Religion*, 57 (1), 55–70.

Kim, S., and Day, K., 2017. *A Companion to Public Theology*. Leiden: Brill.

Koehrsen, J., 2015. Does Religion Promote Environmental Sustainability? – Exploring the Role of Religion in Local Energy Transitions. *Social Compass*, 62 (3), 296–310.

Koehrsen, J., 2018. Religious Agency in Sustainability Transitions: Between Experimentation, Upscaling, and Regime Support. *Environmental Innovation and Societal Transitions*, 27, 4–15.

Koehrsen, J., Huber, F., Becci, I., and Blanc, J., 2019. How Is Religion Involved in Transformations Towards More Sustainable Societies? A Systematization. *Historia Religionum*, 11, 99–116.

Konisky, D., 2018. The Greening of Christianity? A Study of Environmental Attitudes Over Time. *Environmental Politics*, 27 (2), 267–291.

Kopiec, P., 2016. From the Sustainable Development to the Sustainable Community: The World Council of Churches' Contribution to the Concept of Sustainability. *Studia Oecumenica*, 16, 293–304.

Mascia, M., Morandini, S., Numico, S., and Osti, G. eds., 2007. *Responsibilità per il Creato in Europa. L'impegno delle Conference Episcopali. Indagina europea sull'impegno per la responsibilità per il Creato delle Conferenze Episcopali aderenti al CCEE*. Padova: Fondatione Lanza.

McCright, A.; Clements, J., and Xiao, C., 2014. An Examination of the "Greening of Christianity" Thesis Among Americans, 1993–2010. *The Journal for the Scientific Study of Religion*, 53 (2), 1–14.

McGreevy, J., and Hamlin, C., 2006. The Greening of America, Catholicstyle, 1930–1950. *Environmental History*, 11 (3), 464–499.

Meadows, D., and Maedows, D., 1972. *The Limits to Growth. A Report for the Club of Rome's Project on the Predicament of Mankind*. New York: Universe Books.

Moltmann, J., 1985. *Gott in der Schöpfung. Ökologische Schöpfungslehre*. München: Kaiser.

Monnot, C., 2022. The Slow Greening of Established Churches in Switzerland: Tensions Between Local Parishes and Church Head Organizations. *In*: J. Koehrsen, J. Blanc, and F. Huber, eds. *Religious Environmental Activism: Emerging Conflicts and Tensions in Earth Stewardship*. London: Routledge.

Münk, H., 1987. Umweltkrise -Folge und Erbe des Christentums? Historisch-systematische Überlegungen zu einer umstrittenen These im Vorfeld ökologischer Ethik. *Jahrbuch für Christliche Sozialwissenschaften*, 28, 133–206.

Nothwehr, Dawn M., 2019. Weaving the Brown Thread. Toward Integral Ecology. *In*: D. O'Hara, M. Eaton, and M. Ross, eds. *Integral Ecology for a More Sustainable World. Dialogues with Laudato si'*. Lanham: Lexington Books, 111–126.

Ostheimer, J., and Blanc, J., 2021. Challenging the Levels: The Catholic Church as a Multi-Level Actor in the Transition to a Climate-Compatible Society. *Sustainability*, 13 (7), 3770.

Palmer, M., 2013. The Quiet Revolutionaries: Why the Faith-Based Environmental Movement Has Become the Largest Civil Society Movement in the Conservation World. *Biodiversity*, 14 (3), 180–182.

Pannenberg, W., 2000. *Beiträge zur Systematischen Theologie 2: Natur und Mensch – Und die Zukunft der Schöpfung*. Göttingen: Vandenhoeck & Ruprecht.

Paul VI, 1967. *Populorum Progressio* [online]. Available from: www.vatican.va/content/paul-vi/de/encyclicals/documents/hf_p-vi_enc_26031967_populorum.html [Accessed 18 January 2020].

Peppard, C., 2015. Pope Francis and the Fourth Era of the Catholic Church's Engagement with Science. *Bulletin of the Atomic Scientists*, 71 (5), 31–39.

Philipp, T., 2009. *Grünzonen einer Lerngemeinschaft*. München: Oekom.

Pope Francis, 2015. *Laudato Si'* [online]. Available from: www.vatican.va/content/francesco/de/encyclicals/documents/papa-francesco_20150524_enciclica-laudato-si.html [Accessed 18 January 2020].

Rappel, S., 1996. "*Macht euch die Erde untertan*". *Die ökologische Krise als Folge des Christentums*. Paderborn: Schöningh.

Rasmussen, L., 1995. The Integrity of Creation: What Can It Mean for Christian Ethics? *The Annual of the Society of Christian Ethics*, 15, 161–175.

Rosenberger, M., 2001. *Was dem Leben dient. Schöpfungsethische Weichenstellungen imkonziliaren Prozess der Jahre 1987–1989*. Stuttgart: Kohlhammer.

Roth, S., 2000. The Cosmic Vision of Hildegard of Bingen. *The Ecologist*, 30 (1), 40–42.

Sachs, W., 2017. The Sustainable Development Goals and Laudato si': Varieties of Post-Development? *Third World Quarterly*, 38 (12), 2573–2587.

Santos, Renan William dos, 2019. Direitos da natureza e deveres religiosos: tensões entre a ecologia católica e movimentos ambientalistas seculars. *Religião & Sociedade*, 39 (2), 78–99.

Taylor, B., 2016. The Greening of Religion Hypothesis (Part One): From Lynn White, Jr and Claims That Religions Can Promote Environmentally Destructive Attitudes and Behaviors to Assertions They Are Becoming Environmentally Friendly. *Journal for the Study of Religion, Nature and Culture*, 10 (3), 268–305.

Taylor, B., Van Wieren, G., and Zaleha, B., 2016. The Greening of Religion Hypothesis (Part Two): Assessing the Data from Lynn White, Jr., to Pope Francis. *Journal for the Study of Religion, Nature and Culture*, 10 (3), 306–378.

Teilhard de Chardin, P., 1958. *Le Phénomène Humain. Œuvres vol. 1*. Paris: Édition du Seuil.

Vogt, M., 2009. Beredtes Schweigen. Ein Kommentar zu ökologischen Aspekten der Sozialenzyklika "Caritas in vertitate". *AmosInternational*, 3, 27–35.

Vogt, M., 2013. *Prinzip Nachhaltigkeit. Ein Entwurf aus theologisch-ethischer Perspektive*. München: Oekom.

Vogt, M., 2019. *Christliche Umweltethik. Grundlagen und zentrale Herausforderungen*. Freiburg: Herder.

Vogt, M., and Numico, S., 2007. *Salvaguardia del creato e sviluppo sostenibile: orizzonti per le chiese in Europa*. Padua: Fondazione Lanza.

White, L., 1967. The Historical Roots of our Ecologic Crisis. *Science*, 155 (3767), 1203–1207.

World Commission on Environment and Development, 1987. *Our Common Future (Brundtland-Report)* [online]. Available from: https://sswm.info/sites/default/files/reference_attachments/UN%20WCED%201987%20Brundtland%20Report.pdf [Accessed 01 June 2020].

7 The Dissenting Voices

Perception of Climate Change and the Church's Responsibility in Nigeria

George C. Nche

Introduction

For many decades, there have been increasing concerns within Christianity about the need to address anthropogenic climate change. This was catalyzed partially by the indictments of Lynn White (1967), who blamed Christian values for the current state of global climate (Nche et al. 2016), and partially by growing environmentalist interest in the potential of faith communities following the failure of several UN climate change conferences and other environmentalist efforts in securing political will and public engagement to fuel robust action on climate change (Nche 2020a, 2020c; Wilkinson 2012; Rudd 2015). Nasr (1967) notes that the ecological crisis is essentially a crisis of values and that, religions being the principal sources of values in any culture, they are concerned in the decisions humans make about the environment. Also, studies (e.g., Corner 2013; Coward and Hurka 1993; Mastaler 2014; Millais 2006; Nche 2012; Posas 2007; Stults 2006; Tucker and Grim 2001) have suggested that the involvement of religion in addressing the global ecological crisis may yield the desired results. This is because organized religions have significantly shaped human cultural and ethical values around the world (Kaplan 2010). Faith communities, therefore, have spiritual resources and a unique ability to construct moral frameworks that can encourage human beings to protect the Earth (Bomberg and Hague 2018; Morrison et al. 2015; Tarakeshwar et al. 2001; Tucker 2003). In this light, some church leaders and organizations have established climate change initiatives and called for church-led climate action across the globe (see Pope Francis 2015; The South African Council of Churches 2009; Interfaith Power and Light 2015; The Cape Town Commitment 2011; World Council of Churches 2014; The National Religious Coalition on Creation Care 2019; Nche 2020c).

However, there are some dissenting voices among Christians with regard to anthropogenic climate change and the call for a church-based climate action. These are individuals and organizations who, informed by their religious beliefs and political ideologies, have argued that climate change is

DOI: 10.4324/9781003017967-9

not anthropogenic and that the church should have no business with it. For instance, polls showed that fewer than 50% of all US Protestants and Catholics believe the Earth is warming as a result of human actions (Pew Research Center 2015a, 2015b; Bloomfield 2019). Also, a survey of 1,000 Protestant pastors found that 41% strongly disagreed with the statement: "I believe global warming is real and manmade" (Hickman 2011). Furthermore, Jones et al. (2014) report that almost half of all American Christians attribute the severity of recent climate change induced natural disasters to biblical end times. Among white evangelicals, that number jumped to 77%. In Nigeria, research (Nche 2020a, 2020b) reports on some church leaders who believe that climate change is both natural and divinely orchestrated. The same trend was found among some Christian clerics in northern Nigeria (Shehu and Molyneux-Hodgson 2014). This not only shows how polarizing the issues of climate change and environmental stewardship are, but also explains why robust church-based climate action is still in short supply (see Nche 2020c).

In the present study, I examine the dissenting voices with respect to the cause of climate change and the Church's responsibility toward the phenomenon in Nigeria. Previous studies on Christian climate change skepticism and anti-environmentalism have focused mostly on evangelicals and other conservative Christians in the West, especially in the United States (e.g., Zaleha and Szasz 2015; Ronan 2017; Heikkinen 2018; Hickman 2011; Carr 2010; Routhe 2013). Little attention has been given to elements of climate change skepticism among Christians in Nigeria and its implications on church-based environmentalism in the country. My contribution focuses on the tensions among church leaders of different Christian denominations with respect to beliefs about the cause of climate change and the Church's responsibility toward it. The study is structured as follows: The first section discusses global religious environmentalism and dissenting voices, situating the present study within the broader field of research on climate change skepticism and anti-environmentalism. The second section discusses the methods employed in the recruitment of participants and the collection and analysis of the data. The third section presents the findings of the study. The fourth and final section discusses the findings and their implications for research, policy, and church-led climate action in Nigeria and around the world.

Global Religious Environmentalism: The Dissenting Voices

In 2015, Pope Francis through his encyclical *Laudato Si'* (Praise to you) invited the world to care for the Earth as our common home. This encyclical particularly laid out the attitudes that humans should have toward climate change. It speaks eloquently of stewardship for God's creation, care for the poor and those already affected by the exacerbatory impacts of climate change such as droughts, floods, heat waves, hurricanes, and other extreme weather conditions (Hayhoe 2015). The same themes of stewardship and care for the poor have, in recent years, been raised by the World Evangelical Alliance, who, in 2010, produced the Cape Town Commitment, at a conference which brought together 4,200 evangelical leaders from 198 countries, to discuss, among other issues, the challenge of climate change and the need for Christian action (see The Cape Town Commitment 2011). Also on 30 March 2015, 17 Anglican Bishops from all six continents made a climate change declaration under the leadership of Thabo Makgoba, the Archbishop of Cape Town and Primate of Southern Africa. This declaration commits the bishops to specific actions such as the promotion of energy conservation measures in church buildings, more renewable energy, nurturing biodiversity on church land, and supporting sustainability in water, food, agriculture, and land use. Furthermore, it commits to reviewing churches' investment practices, including a call for divestment, and closer ecumenical and interfaith co-operation (Interfaith Power and Light 2015). The Church of England also has an explicit environmental program which is premised on the statement conspicuously displayed on its website: "We believe that responding to climate change is an essential part of our responsibility to safeguard God's creation. Our environmental campaign exists to enable the whole church to address – in faith, practice and mission – the issue of climate change" (The Church of England n.d.).

However, amidst this growing need for robust global church-based environmentalism, there are voices which oppose the core assumptions of climate change science and environmentalism. These voices often come from religious conservatives who, convinced by their religious beliefs, specifically question the scientific basis of anthropogenic climate change and the call for faith-based action. For instance, in a study of Evangelicals in Ohio, Routhe (2013) found that 46% of the participants opposed anthropogenic climate change, with 19% specifically offering various forms of religious-based reasoning to justify this belief. Shao and Mccarthy (2020) also found that evangelical Protestants are less likely to believe in the existence and seriousness of global warming than others. Polls showed that fewer than 50% of all US Protestants and Catholics believe the Earth is warming as a result of human actions (Pew Research Center 2015a, 2015b; Bloomfield 2019). Again, a survey of 1,000 Protestant pastors found that 41% strongly disagreed with the statement: "I believe global warming is real and manmade" (Hickman 2011). Furthermore, Gander (2019) reports that over a third of evangelical Christians say there is "no solid evidence" that climate change is happening, and only 28% of white evangelicals accept that it is driven by human activity. Nche (2020a) and Shehu and Molyneux-Hodgson (2014) found some Christian leaders, among Nigerian samples, who reported that climate change is both natural and divinely ordained. In fact, as recently as 2018, polls have shown that Christian climate deniers outnumber Christian climate believers (Bloomfield 2019; Heikkinen 2018).

Some skeptical groups such as the Southern Baptists, the Cornwall Alliance, and other conservatives have gone beyond mere disagreements with scientific facts to champion stiff opposition to pro-environmental efforts and initiatives within and outside their ranks (see Zaleha and Szasz 2015; Ronan 2017). Ronan (2017) for instance, narrates how the National Association of Evangelicals (NAE) withheld its official support when the Evangelical Climate Initiative's (ECI) "Climate Change: An Evangelical Call to Action" was launched in 2006 with the goal of recognizing "human-induced climate change as a serious Christian issue" (see ECI 2006). This was because the NAE believed that the "call to action" did not represent the views of all evangelicals, many of whom rejected the idea of anthropogenic climate change. Since NAE represents a broad spectrum of evangelicals in the United States, it felt it should not take a position on an issue as controversial as climate change (Nazworth 2012; Ronan 2017). Conservative evangelical leaders reacted by singling out 22 politically influential evangelicals (such as the likes of Richard Land, head of the Southern Baptist Convention's Ethics & Religious Liberty Commission) and pressured signatories to withdraw their support for the ECI call to action (Zaleha and Szasz 2015; Ronan 2017; Colson et al. 2006). At the vanguard of this opposition to church-based environmentalism was the Cornwall Alliance (which was then named the Interfaith Stewardship Alliance). Founded by E. Calvin Beisner, the Cornwall Alliance is a prominent group of religious thinkers in the United States, which urges followers to resist the rising tide of environmentalism, which it refers to as the "Green Dragon" (Hickman 2011). Through several documents (e.g., A Call to Truth, Prudence, and Protection of the Poor: An Evangelical Response to Global Warming; Resisting the Green Dragon: Dominion not Death; etc.) and platforms (e.g., the U.S. Senate Environment and Public Works Committee; the Energy and Environment Subcommittee of the U.S. House of Representatives Committee on Energy and Commerce; the White House Council on Environmental Policy; etc.) the Cornwall Alliance has sought to promote disbelief in manmade climate change, deflate the core assumptions of environmentalists, and advocate a free-market approach to care for the environment (see Ronan 2017).

Their opposition has been attributed to conservative religious beliefs and orientations. Zaleha and Szasz (2015), for instance argue that the conflation of stewardship theology to neo-pagan nature worship, and end times induced environmental apathy, are responsible for evangelical protestants' opposition to climate change science and church-based environmentalism. This was also buttressed by Lucas Johnston, a professor of religion and environment at Wake Forest University. Cited by Bergman (2018), he says that "there is a longstanding antipathy toward environmental sentiments in Christians, and especially evangelical circles, because they have, for centuries, been imagined as pernicious and dangerous, and possibly bordering on paganism." Cheney (2019), however, holds a contrasting view. According to him, many American evangelicals do not believe in climate change for reasons that, "rather than being fully religious, have more to do with politics" (Cheney 2019). The moral orientation traceable to Jerry Falwell's "Moral Majority" movement in the 1980s, which encouraged tenacious adherence to traditional values in American politics is for Cheney what explains the current opposition to both religious and secular environmentalism in the United States. This traditional moral orientation predisposed evangelicals to oppose issues like gay rights, abortion, the teaching of evolution in schools and now, anthropogenic climate change. While these views appear to be opposing to each other, they are, in fact, like two sides of a coin which mutually reinforce each other. This is because evangelicals see climate change as a hoax for theological/religious reasons, and it is these conservative theological/religious orientations that informs their political conservatism. Moreover, studies have shown that there is a correlation between religious conservatism and political conservatism in the United States (see Sherkat and Ellison 2007; Taylor et al. 2016b; Ronan 2017). Be that as it may, tensions surrounding environmentalism within and beyond Christendom show that Christians do not all think alike about the environment, helping to explain the reason why robust church-based climate action is still in short supply. In the present study, I examined opposing views with respect to the cause of climate change and the church's response toward the phenomenon in Nigeria.

Methods

This study is part of a larger project that investigated climate change awareness/knowledge, role perception, and action among church leaders in Nigeria. In this sense, the methods section from the larger project will be adapted here. The project involved 30 church leaders from three denominations: Catholic, Anglican, and Pentecostal churches in Nigeria (i.e., 10 Catholics, 10 Anglicans, and 10 Pentecostals). The choice to focus on church leaders was informed by the fact that they occupy a central place in the global campaign for efficient faith-based climate action. In fact, the clergy and religious leaders have been found to be particularly influential in shaping the environmental worldviews of their congregations (Djupe and Hunt 2009; Mastaler 2014; Simkins 2008). Also, in recent times, there has been increasing emphasis on the role of religious leaders in the fight against climate change in Nigeria (Nche et al. 2017; Thomas-Odia 2018; Aliyu 2018; Nche 2020a, 2020b). Drawing participants from across three denominations (Catholic, Anglican, and Pentecostal churches) was necessary to allow for more robust information on the phenomenon under study as well as to evaluate diversity in thought patterns, which suits a work of this nature. The age of participants ranged from 26–52 with a mean of 40.6. All 26 of the participants were male. Efforts to interview female priests/pastors were unsuccessful as some of the female leaders approached declined and referred me to other persons. Participants were spread across five geopolitical zones in

Nigeria. These zones alongside their states include South – South (Akwa Ibom, Bayelsa, Cross River, Delta, Edo, and Rivers states), South East (Abia, Anambra, Ebonyi, Enugu, and Imo states), South West (Ekiti, Lagos, Ogun, Ondo, Osun and Oyo states), North Central (Abuja [the Federal Capital Territory], Benue, Kogi, Kwara, Nassarawa, Niger, and Plateau states), and North West (Jigawa, Kaduna, Kano, Katsina, Kebbi, Sokoto and Zamfara states). The North East (Adamawa, Bauchi, Borno, Gombe, Taraba, and Yobe states) was excluded on the basis of insecurity occasioned by Boko Haram insurgency.

The states in these five geographical zones were arranged alphabetically in clusters and the third states were systematically selected from each of the clustered arrangements. In this sense, the selected states included Cross River, Ebonyi, Kano, Kogi, and Ogun states. Then, the capital cities in these selected states were purposely selected for the study. This is because capital cities and urban areas have more educated church leaders and more populous churches than the rural areas. So, due to the need to interview persons who have great potential in influencing far-reaching climate action, it became necessary to recruit participants from cities. This, however, did not apply to Kano state where most of the local government areas are dominated by Muslims, except for the Sabon-garri area where there is a large presence of foreigners and Christians. Hence, I randomly selected churches in the Christian dominated Sabon-garri area in Kano state. The chosen capital cities in the other four states accordingly included Abakaliki (Ebonyi state), Abeokuta (Ogun state), Calabar (Cross river state), and Lokoja (Kogi state). Finally, the specific locations of churches within the selected cities were randomly selected using availability sampling technique.

The main data collection technique for this study involved in-depth semi-structured face-to-face interviews conducted over a period of 8 months (February to September 2018). The only exception was in the case of six church leaders in Abeokuta, Ogun state, who were interviewed over the phone. This became necessary as I could no longer embark on further travels due to financial constraints. Yet, there was not much discernable deficiency as all the interview questions were exhaustively attended to and recorded throughout the phone conversations. Although the interview protocol had questions addressing the broad objectives of the larger project, namely awareness/knowledge of climate change and perceptions of churches' role in addressing the phenomenon, the present study focused on participants' perceptions of the causes of climate change and the church's responsibility toward the phenomenon. The interviews were digitally recorded, each one lasting between 46 and 57 minutes. All interviews were conducted in the English language.

The data were analyzed following an adapted form of Colaizzi's guide (as cited in Sanders 2003; Speziale and Carpenter 2007; Shosha 2010) for descriptive data analysis. Relying on this guide, the following steps were taken in the process of the data analysis: After carefully transcribing each interview audio recording, I painstakingly verified each of the transcripts by listening to each interview audio recording to ensure accuracy. Each transcript was read and reread in order to obtain a general sense about the whole content and for each transcript, I extracted significant statements that pertain to the phenomenon under study. These statements were written on a separate sheet noting their pages and lines numbers; meanings were formulated from these significant statements in line with the major objective of the study; and the formulated meanings were sorted into categories and clusters of sub-themes. The findings of the study were then integrated into an exhaustive description of the phenomenon under study.

Findings

The Reality of Anthropogenic Climate Change: Real, But Not Anthropogenic

A majority of the church leaders reported that climate change is real and largely caused by human activities. For instance, Steven of the Anglican Church, while relying on his experience with the changes in weather, believes that climate change is real. In his words:

> Well, a blind man should believe when the weather changes and you know, when you have a mindset that by February or let's say November's ending or December, the weather should be cold, but suddenly you are still sweating in December at the 2nd week, you should know that something is wrong. So, I believe it.
>
> (Steven, Anglican Church)

In a similar vein, Godwin of the Anglican Church relied on his own experiences with the unpredictability of rainfall patterns and natural disasters (e.g., flood) in the country to admit the following:

> I don't have doubt about the reality of climate change because of the present happenings, such as . . . the natural disasters that are becoming common. We used to hear some of those things happening outside Nigeria, but they are already with us, which are signs that climate change is real. Such as the rain, in the past, for instance in Kogi State here, you know when the rainy season has come to stay and you know this time are for dry season, but now those things are no longer well defined again. This is one of the effects of climate change; you can't actually predict what will happen next. During dry season, you see rain start [to] coming down, which [it is] supposed not to be. So, to me it is a sign that climate change is real.
>
> (Godwin, Anglican Church)

With respect to the cause of climate change, Steven and Godwin also believe that it is largely caused by human activities such as the "emission of gasses, chemicals into the air." Furthermore, Michael of Christ Embassy strongly believes that climate change is real and willingly admits that it is largely anthropogenic. In his words:

> Definitely, climate change is real . . . (and) human activities contribute much to climate change (through) . . . the release of gas into the atmosphere. It affects the ozone layer in such a way that . . . there is not much penetration because of the gas disposition in the air. Again, . . . most times that people cut trees to make firewood, actually, the discharge of [those] gasses or the smoke also contribute. And you know, when there are more trees, they release . . . what we call . . . transpiration, whereby the release of vapour . . . increases humidity which also increases rain. But the more trees you cut down it also affects the humidity range.
>
> (Michael, Christ Embassy)

The same view was shared by Andrew, Lawrence, Helen, and many others (see Table 7.1).

However, there are some church leaders who believed that climate change is real but argued that it is not anthropogenic. In fact, Collins, a Catholic Seminarian, alongside Ayodele, a Catholic Catechist, express doubt or uncertainty about the scientific basis of the anthropogenic climate change. According to Collins, "if we don't know the future and cannot tell what tomorrow will look like" then there is no basis to conclude that human activities are the major cause of climate change. Nature and God, for these leaders, are the best options. For instance, when asked whether he believes climate change is real, Jonathan of Winners Chapel, says:

> Actually, why would I doubt it? If I can believe that [an] earthquake is happening, landslide[s] happen, why won't I believe in climate change even when I am feeling the effects.
>
> (Jonathan, Winners Chapel)

But when asked to say what he thinks causes climate change, Jonathan notes that "Actually, climate change . . . is a natural disaster . . . and because it is a natural disaster, that means, it is bound to occur" (Jonathan, Winners Chapel). Humphrey, a Catholic, also believes that climate change is real but shares the same view as Jonathan about the natural cause of the phenomenon. According to him, climate change refers to "natural seasonal variations . . . due to certain natural phenomenon . . ." (Humphrey, Catholic Church). When asked whether he believes that climate change is naturally caused, Humphrey retorts "that is what I believe, it is natural, it is natural" (Humphrey, Catholic Church).

Table 7.1 Summary of responses on the reality and causes of climate change

S/N	*Themes*	*Views/positions*	*Catholics (%)*	*Anglican (%)*	*Pentecostal (%)*	*Total No. of Respondents (%)*
1	**Reality of Climate Change**	Climate change is real	10 (100%)	10 (100%)	10 (100%)	30 (100%)
2	**Causes of Climate Change**	Human activities	7 (70%)	4 (40%)	4 (40%)	15 (50%)
		Nature	1 (10%)	3 (30%)	1 (10%)	5 (16.7%)
		Theological fatalism (Divinely caused)	1 (10%)	1 (10%)	1 (10%)	3 (10%)
		Uncertainty	1 (10%)	–	–	1 (3.3%)
		Not knowledgeable	–	2 (20%)	4 (40%)	6 (20%)

Anthony, a Catholic, and Henry of the Charismatic Renewal Ministry (CRM), on the other hand, believe in the reality of climate change but have theological fatalistic views (i.e., the belief that every event is divinely designed/predestined to happen) about the cause of the phenomenon. For instance, when asked what the cause of climate change is, Anthony says:

> If I could say, in my own opinion, . . . we can't attribute the [climate] change to a particular individual or a particular thing . . . it was meant to be so, because even if you go scripturally, they say, there is [a] time for everything. Therefore, there is always cause for change.
>
> (Anthony, Catholic Church)

For Henry, the only condition for believing in the reality of climate change is when it is connected with God. In his words:

> I believe in climate change as long as it has to do with God. You understand? Because we believe that those laws are rules in the natural circle . . ., most of the rules are natural. They are not issues that are created by human beings.
>
> (Henry, CRM)

When asked to state what he thinks is the cause of climate change, Henry says:

> That is what I am saying. For us, we believe that the climate [is] . . . created by God, and God is behind whatever changes. So, we look at them as natural things that actually happen. Those who are experts in the field will have some scientific reasons for all that, but for us in the realm of God, God is in charge; he dictates how the climate should be.
>
> (Henry, CRM)

In summary, these quotes show that while a majority of the church leaders believe that climate change is real and anthropogenic, some believe that, although real, it is not anthropogenic. Climate change, according to these dissenting voices, is rather caused by God and nature.

The Church's Responsibility: Soul-Winning and *Parousia*, Not Climate Action

A majority of church leaders across the three denominations accepted that the church/churches have some roles to play in addressing climate change. Some of the suggested roles include raising climate change awareness, charity for disaster victims and prayer (see Table 7.2).

With respect to "raising climate change awareness," Mathew of the Anglican Church, for instance, admits that the church should, despite its

Table 7.2 Summary of responses on the Church's responsibility

S/N	*Responsibility*	*Catholics (%)*	*Anglican (%)*	*Pentecostals (%)*	*Total (%)*
1	Raising climate change awareness	9 (90%)	5 (50%)	4 (40%)	18 (60%)
1a	Focus on soul winning/ salvation	–	–	1 (10%)	1 (3.3%)
1b	End-time/ second coming awareness creation	1 (10%)	–	1 (10%)	2 (6.6%)
2	Charity for disaster victims	5 (50%)	3 (30%)	3 (30%)	11 (36.7%)
3	Prayer	–	1 (10%)	3 (30%)	4 (13.3%)

spiritual focus, pay some attention to social issues, which includes climate change. He advocates that the church should sensitize their members on environmentally friendly behaviors as a way of contributing to the campaign against climate change. In his words:

> We have spiritual and social roles. I strongly believe that there is a part the church must play. What of the deforestation we have faced? Tearing down every bush, how many have we planted? We pulled down trees without reason, people get up to tear down the forest, level them for industrial developments, without actually finding a place to replace them. What of some of the animals – all these creatures contribute to man's well-being. Some of the chemicals we have developed to handle our grasses, have we actually sat down to ask ourselves the consequences? . . . We cannot run away from them. These are [the] realities around. The church can actually go a long way in sensitizing their members on what, [or] how to actually keep some of these trees around us. These trees can actually help us check the level of carbon emitted in the air . . . Today we live in the street [and] you barely see one tree standing, even when the government plants on the main road, we are too careless [towards] them, we even cut them down to make firewood. So, we are not conscious of our environment, we are not conscious of preserving the nature that God gave us. So, the church should actually sensitize their members on some of these things.
>
> (Mathew, Anglican Church)

Talking about the second role "charity for disaster victims" Anthony, a Catholic, believes that the Church also has the responsibility to help victims of climate change induced disasters as a show of God's love. According to him:

> It is the Church's responsibility to help every community who experience such [disasters]. Yeah, we said that if you can, before you say love God, you need to love . . . your neighbor first. So, it is [the] Church's responsibility to help others who are experiencing or who have [a] problem in their own locality, at least, with such help, they will overcome the problems or the feeling of rejection. So, it is the Church's role and responsibility to help them.
>
> (Anthony, Catholic Church)

On the third role which is "prayer" Henry, of CRM, having perceived climate change as a divinely caused phenomenon, has this to say:

> Yes, [to] the extent that I know about, we try to ask people to pray. Be very prayerful! Because we look at the effects [of] climate change, and we look at it as a disaster, and we believe that God has a way of handling things. So pray, so that we are not exposed to some of these disaster things.
>
> (Henry, CRM)

However, Michael (Christ Embassy), Anthony (Catholic), and Jumoke (Redeemed Christian Church of God) have different views with respect to the responsibility of the church toward climate change. According to Michael, the Church has a primary purpose and focus which is to reconcile humans with God and, as such, should not have any business with climate change. In his words:

> The Church is consecrated for something which is reconciling men to God, which is the primary reason of the church. The Church . . . you see, is not a building but the people that come in there to worship. And you know in every organization there is a focus, and when you derail from your focus, you miss your point and you don't get your objectives. So . . . the major focus is reconciling and making people know there is a life after here, and so reconciling them back to their maker and also telling them that we are not here by accident. . .. So, this is the major focus. So I don't think that . . . the Church is saddled with such responsibility [as] to educate their members on climate change, because it is not the major focus.
>
> (Michael, Christ Embassy)

In fact, for Michael,

> [The] Bible does not permit that. When a preacher goes to the pulpit, he is led by a divine institution, not by a secular world. [There is no] mandate to preach such things on the pulpit. Why? Because the pulpit is meant for reconciliation not for any other thing. You can't discuss politics on the pulpit. No!
>
> (Michael, Christ Embassy)

When asked if he believes that God is interested in the preservation of the environment, Michael answered affirmatively but feels it is rather the responsibility of the government to ensure that people are environmentally aware and friendly:

> because church has a lot of responsibility especially when it comes to building people spiritually, . . . the major thing that consumes the Church . . . is bringing people from the world in reconciliation with what Christ has done, the debt that has been paid, and making people aware that this debt has been paid, you don't need to let that down, you understand?
>
> (Michael, Christ Embassy)

While Jumoke, for her part, accepts that the church should educate members, her opinion differed with regard to the content of the education. Having related climate change to signs of the end times, she suggests that "the only way the Church can address the issue of climate change is to make it known to people that the end is near . . . just to let people know that the coming of Jesus is at hand." Anthony shares the same view on the content of the education the Church should give its members about climate change. According to him:

> The Church has a little role to play in addressing climate change in the sense that we need to educate our members that the world is coming to an end, that they should see what is happening just as a change which God has . . . predestined. So, with that, we can draw a conclusion. So, the church's role here should be at least to educate its members that what they are seeing, the changes around them are signs of the end times.

Interestingly, however, church leaders who did not support educating congregations about climate change, did not oppose the view that the church should support and cater to climate change disaster victims. With respect to the need for charity toward climate change victims, Michael, for instance, says:

> Yeah, [the] Church can come in there because if I have a member that has been displaced, I don't expect him to be in church, when he doesn't

> have a place [to] live . . . this is what the Church is all about. The Church now has to come in, rendering assistance in one way, or to give that person a better habitation where [possible], because psychologically, if a person does not have a house, he might not be able to participate in most things that we do. So indirectly or directly, it affects the Church.
>
> (Michael, Christ Embassy)

Anthony meanwhile has this to say:

> It is the Church's responsibility to help every community who experience such [distasters]. Yeah, we said that if you can, before you say love God, you need to love . . . your neighbor first. So, it is [the] Church's responsibility to help others who are experiencing or who have [a] problem in their own locality, at least, with such help, they will overcome the problems or the feeling of rejection. So, it is the Church's role and responsibility to help them.

In summary, these quotes show that despite an overwhelming consensus among Church leaders to support active church-based climate action, some participants insist that the Church should have no business with the phenomenon. This highlights the tensions and conflicts among the Church leaders with respect to environmental stewardship in Nigeria.

Discussion and Conclusion

This study examined dissenting voices with respect to the reality of anthropogenic climate change and the Church's responsibility toward the phenomenon in Nigeria. The study found that a minority of participants reported that climate change is occurring but is not of anthropogenic origin. While 16.7% of these participants believed that climate change is caused by nature, 10% believed it is divinely caused. A majority (50%), however, agreed that climate change existed and is caused by humans. This finding contradicts those of previous studies on conservative evangelicals who were found to doubt the reality of climate change (see Routhe 2013; Pew Research Center 2015a, 2015b; Bloomfield 2019; Hickman 2011), while showing that a minority do share the same opinion as these conservative evangelicals in doubting anthropogenic causation argument (Gander 2019; Hickman 2011; Pew Research Center 2015a, 2015b; Bloomfield 2019; Carr 2010). This shows that Christians do not all think alike about the environment. It also shows how contentious the issues around the causes of climate change are. Yet, while the cause of climate change has been one of the central issues in climate change debates, a majority of scientists argue that global warming is largely anthropogenic (see Romm 2016; Nche 2020a).

However, it is possible that the reason that some Church leaders accepted that climate change is happening but doubted anthropogenic causation is due to a lack of adequate scientific knowledge. This is likely because some of them relied solely on personal experiences with the weather when forming a view on climate change. This, by the way, reinforces Lujala et al.'s (2015) finding that personal experience with the impacts of climate change is a key factor in determining people's perceptions of the phenomenon. These participants might not have had enough access to climate change information, which is generally poor in Nigeria (see Nche 2020a). This is unlike white evangelicals in the United States who, despite ample access to the relevant scientific information, still deny the reality of anthropogenic climate change. In fact, at the vanguard of evangelical climate change skepticism in the United States are climate experts, one of whom is G. Cornelis van Kooten, an expert reviewer for the Intergovernmental Panel on Climate Change's (IPCC) Fourth Assessment Report (AR4) (Ronan 2017).

Hence, the availability of adequate access to climate change information in Nigeria, especially for Church leaders, might make a difference with respect to their climate beliefs. Besides, studies (e.g., Morrison et al. 2015; Stehr and Grundmann 2012; Ojomo et al. 2015) have shown that climate change awareness/knowledge is the first major step toward climate change mitigation and adaptation. It is therefore important for the Nigerian government and other stakeholders to provide adequate sources of scientific information in order to build awareness/knowledge among faith leaders in Nigeria. Beyond mere provision of information sources, efforts could be channeled toward educating Church leaders on how to educate their congregations. This, however, is not to deny the potency of religious influence on climate change beliefs which, in some cases, holds sway despite adequate climate change knowledge as seen in the case of white evangelicals. Be that as it may, disbelief in anthropogenic climate change not only discourages climate action but could also fuel a strong church-based anti-environmentalism movement in Nigeria. The activity of the Cornwall Alliance is a typical case (see Ronan 2017; Zaleha and Szasz 2015).

Moreover, the study found that a couple of participants expressed dissenting views with respect to the church's responsibility toward climate change. This is unlike the majority who supported Church-led climate action and suggested raising awareness, charity for disaster victims, and prayer as roles churches can play in addressing climate change. One (i.e., a Pentecostal) of these dissenting participants insisted that salvation of souls or soul-winning, not climate change, should be the focus of the Church, which corresponds to Carr's (2010) finding among evangelicals in the United States. Indeed, a majority of his participants said that in their opinion, churches have foci more important than climate change. Reporting the views of some of these participants, Carr writes:

> Julie and Margery, for example, feel that environmental issues, and global warming in particular, have no place in the Church at all. Julie says she would "shudder" if her church ever publicized an environmental position because that would take the focus off of Jesus. In the long

> run, she says, environmental issues are temporary, but faith in Christ offers eternal salvation, therefore salvation should be the one and only emphasis of the Church.
>
> (Carr 2010, p. 165)

The rest (i.e., one Catholic and an Anglican) of these dissenting participants said that the Church, instead of creating climate change awareness, as suggested by the majority, should be preoccupied with creating awareness on the imminence of the end-times. This also reinforces findings made by previous studies among conservative Christians in the United States. Ronan (2017), and Zaleha and Szasz (2015) for instance, have shown how conservative evangelicals have built strong opposition against church-based environmentalism due to end-time beliefs and other reasons. This also confirms the conclusion that end-times or eschatological beliefs instill apathy or discourage pro-environmental behaviors/climate actions among conservative Christians (see Routhe 2013; Carr 2010). These beliefs, according to Nche (2020b), discourage pro-environmental behaviors in the sense that people who hold them see climate change as a sign or fulfillment of end-time prophecies. To these people, there is therefore no basis to worry or address it; they rather focus on improving their spiritual lives in preparation for the second coming (*parousia*) of Christ.

In all, these findings suggest that religious beliefs and values are strong schemas that do not only determine perceptions of climate change among conservative Christians but also how far they could go in addressing it or building anti-environmentalism positions. Both the participant who felt that climate change is completely outside the foci of the Church and the ones who felt the Church should rather preach the imminence of the second coming, were all influenced by their religious/theological convictions. This reinforces the call for a more thorough understanding of the latter in order to decide how best to communicate climate change to conservative Christians, especially their leaders. In fact, it has been observed that the reason many people do not feel concerned about climate change is significantly connected to how climate change is being communicated (Greco 2019). Church/Christian leaders occupy central positions and have the moral authority to play a decisive role in swaying public policy toward (or away from) action to mitigate climate change (Zaleha and Szasz 2015). In Nigeria, the influence of Church leaders is even more pronounced and far-reaching. In fact, for Nigerians to be aware and committed to climate action, church leaders, according to Nche (2020a), are indispensable as about 86 million persons congregate under these church leaders on a weekly basis in Nigeria (see Diamant 2019). Hence, special attention needs to be given to convincing and bringing these leaders on board in the fight against climate change. And convincing these leaders will require strategic communication and engagement. This, according to climate change communication experts, would involve treating conversations on climate change as dialogues, locating common values, and avoiding too much reliance on science (see Bloomfield 2019; Handricks 2017; Greco 2019; Corner et al. 2018).

For instance, in the present study, it was found that despite the dissenting views expressed by some participants about the church's responsibility toward climate change, they all agreed that the church owes climate disaster victims care and charity. The feeling of sympathy/empathy for climate disaster victims constitutes a common value across those who supported church-based climate action and those who opposed it. As such, in order to convince these dissenting participants, one should communicate climate change to them from the point of view of the devastations the phenomenon has brought upon the poor and vulnerable. This strategy may go a long way to bringing these dissenting participants on board in the global campaign against climate change.

Finally, these findings have some implications for research. First, they show that the skepticism both toward the causes of climate change and the church's responsibility toward the phenomenon cut across all denominations – Catholic, Anglican, and Pentecostal. This suggests that religious conservatism is not an exclusive feature of any Christian denomination. There are Catholic conservatives, just like there are Anglican and Pentecostal conservatives. The same applies to religious liberalism. This reinforces Wuthnow's (1988, 1996) categorization of Christians in the United States – a categorization which has been validated by Taylor et al. (2016a, 2016b), Zaleha and Szasz (2015) and many more. According to Wuthnow, to understand differences within American Christianity one needs to look beyond denominational labels and to think, instead, of three different categories of Christians: one theologically conservative, a second quite liberal, and between them a broad swathe of amorphous and ill-defined moderates. The present findings have validated this categorization and shown that it applies to Nigeria Christianity. Hence, future studies on Christians, especially within Nigeria, should be conducted in light of this categorization. Again, while climate change believers outnumbered skeptics in the present study, the findings do not necessarily reflect the general reality among Christians in Nigeria, especially given the minute nature of the current sample. This of course highlights a major limitation of qualitative studies generally. It therefore follows to say that a larger quantitative study/survey on climate change perceptions and the Church's role in Nigeria could reveal an opposite trend, arguably making a quantitative survey imperative. Another limitation of this study derives from its focus on educated church leaders in urban areas. This could explain why the study found fewer participants who opposed church-based climate action. It is possible that a focus on rural areas may reveal that climate skepticism is a broader phenomenon among Christians in the country. Nonetheless, the mere presence of climate skeptics among Church leaders in Nigeria as the present study has shown does not bode well for church-based environmental activism in the country.

References

Aliyu, A., 2018. NCF Seeks Multifaith Approach to Climate Change, Environmental Challenge. *DailyTrust* [online]. Available from: www.dailytrust.com.ng/ncf-seeks-multifaith-approach-to-climate-change-environmental-challenge.html [Accessed 28 September 2018].

Bergmann, M.M., 2018. What Would Jesus Do? Talking with Evangelicals About Climate Change. *The Guardian* [online]. Available from: www.theguardian.com [Accessed 7 April 2020].

Bloomfield, E.F., 2019. Understanding Christians' Climate Views Can Lead to Better Conversations About the Environment. *The Conversation* [online]. Available from: https://theconversation.com [Accessed 20 January 2020].

Bomberg, E., and Hague, A., 2018. Faith-Based Climate Action in Christian Congregations: Mobilisation and Spiritual Resources. *Local Environment*, 23 (5), 582–596.

Carr, W.A., 2010. *The Faithful Skeptics: Conservative Christian Religious Beliefs and Perceptions of Climate Change*. Missoula: University of Montana.

Cheney, J., 2019. Why Many Evangelicals Believe Climate Change Is a Hoax. *Religion Unplugged* [online]. Available from: https://religionunplugged.com [Accessed 8 April 2020].

Colson, C.W., Dobson, J.C., Hagee, J.C. et al., 2006. A Letter to the National Association of Evangelicals on the Issue of Global Warming. *Interfaith Stewardship Alliance* [online]. Available from: www.pbs.org/moyers/moyersonamerica/green/nae_response.pdf [Accessed 13 April 2020].

Corner, A., 2013. Christianity and Climate Change: The Relationship Between God and Green. *The Guardian Newspaper* [online]. Available from: www.theguardian.com [Accessed 25 March 2020].

Corner, A., Shaw, C., and Clarke, J., 2018. *Principles for Effective Communication and Public Engagement on Climate Change: A Handbook for IPCC Authors*. Oxford: Climate Outreach.

Coward, H.G., and Hurka, T., eds., 1993. *Ethics and Climate Change: The Greenhouse Effect*. Waterloo: Wilfred Laurier Press.

Diamant, J., 2019. The Countries with the 10 Largest Christian Populations and the 10 Largest Muslim Populations. *Pew Research Center: Religion and Public Life* [online]. Available from: www.pewresearch.org/fact-tank/2019/04/01/the-countries-with-the-10-largest-christian-populations-and-the-10-largest-muslim-populations/ [Accessed 4 July 2020].

Djupe, P.A., and Hunt, P.K., 2009. Beyond the Lynn White Thesis: Congregational Effects on Environmental Concern. *Journal for the Scientific Study of Religion*, 48 (4), 670–686.

Evangelical Climate Initiative (ECI), 2006. *Climate Change: An Evangelical Call to Action* [online]. Available from: www.npr.org [Accessed 22 March 2020].

Gander, K., 2019. What Evangelical Christians Really Think About Climate Change. *Grist* [online]. Available from: https://grist.org [Accessed 4 April 2020].

Greco, C.J., 2019. What's Wrong with the Way We Communicate Climate Change? *Environmental and Energy Study Institute* [online]. Available from: www.eesi.org/articles/view/whats-wrong-with-the-way-we-communicate-climate-change [Accessed 10 March 2020].

Handricks, R., 2017. Communicating Climate Change: Focus on the Framing, not Just the Facts. *The Conversation* [online]. Available from: https://theconversation.com [Accessed 20 January 2020].

Hayhoe, K., 2015. Science and Faith Can Solve Climate Change Together: Moral Conviction, Backed by Facts, Could Finally Inspire Global Action. *Scientific American* [online]. Available from: www.scientificamerican.com [Accessed 25 January 2020].

Heikkinen, N., 2018. *Climate Wire: Christianity Is Not Getting Greener: U.S. Christians' Concerns About the Environment Have not Shifted Much in the Past Two Decades* [online]. Available from: www.scientificamerican.com/article/christianity-is-not-getting-greener1/ [Accessed 23 February 2020].

Hickman, L., 2011. The US Evangelicals Who Believe Environmentalism Is a 'Native Evil': The Cornwall Alliance, a Prominent Group of Religious Thinkers in the US, Explains Why It Urges Followers to 'Resist the Green Dragon'. *The Guardian* [online]. Available from: www.theguardian.com/environment/blog/2011/may/05/evangelical-christian-environmentalism-green-dragon [Accessed 23 February 2020].

Interfaith Power and Light, 2015. *17 Anglican Bishops from All Six Continents Have Call for Urgent Prayer and Actions on the "Unprecedented" Climate Crisis* [online]. Available from: www.interfaithpowerandlight.org [Accessed 13 November 2017].

Jones, R.P., Cox, D., and Navarro-Rivera, J., 2014. *Believers, Sympathizers, & Skeptics: Why Americans Are Conflicted About Climate Change, Environmental Policy, and Science. Findings from the PRRI/AAR Religion, Values, and Climate Change Survey*. Washington, DC: Public Religion Research Institute.

Kaplan, M.S., 2010. Will Religions Guide Us on Our Dangerous Journey? *In*: K.D. Moore and M.P. Nelson, eds. *Moral Ground: Ethical Action for a Planet in Peril*. San Antonio: Trinity University Press.

Lujala, P., Lein, H., and Rod, J.K., 2015. Climate Change, Natural Hazards, and Risk Perception: The Role of Proximity and Personal Experience. *Local Environment: The International Journal of Justice and Sustainability*, 20 (4), 489–509.

Mastaler, J.S., 2014. The Role of Christian Ethics, Religious Leaders, and People of Faith at a Time of Ecological and Climate Crisis. *The New Theology Review*, 26 (2), 43–48.

Millais, C., ed., 2006. *Common Belief: Australia's Faith Communities on Climate Change*. Sydney: Climate Institute.

Morrison, M., Duncan, R., and Parton, K., 2015. Religion Does Matter for Climate Change Attitudes and Behavior. *PLoS One*, 10 (8), 1–16.

Nasr, S.H., 1967. *Man and Nature: The Spiritual Crisis in Modern Man*. Chicago: Kazi Publishers.

Nazworth, N., 2012. Evangelicals and Climate Change: Global Warming Activists. Pt. 2. *The Christian Post* [online]. Available from: www.christianpost.com/news/evangelicals-and-climate-change-globalwarming-activists-pt-2-75939/ [Accessed 13 April 2020].

Nche, G.C., 2012. Challenges of Climate Change: The Role of Christian Religious Leaders. *Journal of Educational and Social Research*, 2 (10), 73–80.

Nche, G.C., 2020a. Beyond Spiritual Focus: Climate Change Awareness, Role Perception and Action Among Church Leaders in Nigeria. *Weather Climate & Society*, 12 (1), 149–169.

Nche, G.C., 2020b. The Religion-Environment (Climate Change) Connection: Evidence from Nigeria. *Worldviews: Global Religions, Culture and Ecology*, 24 (1), 81–115.

Nche, G.C., 2020c. The Church Climate Action: Identifying the Barriers and the Bridges. *Transformation: Transformation: An International Journal of Holistic Mission Studies*, 37 (3), 222–241.

Nche, G.C., Achunike, C.H., Okoli, A., and Diara, B.C., 2017. Challenges of Climate Change and the Culpability of Churches: Towards an Effective Church Climate Change Action in Nigeria. *Missionalia*, 45 (2), 168–187.

Nche, G.C., Okwuosa, N.L., and Nwaoga, T.C., 2016. Revisiting the Concept of Inculturation in a Modern Africa: A Reflection on Salient Issues. *HTS Teologiese Studies/Theological Studies*, 72 (1), 3015.

Ojomo, E., Elliot, M., Amjad, U., and Bartram, J., 2015. Climate Change Preparedness: A Knowledge and Attitudes Study in Southern Nigeria. *Environments*, 2 (4), 435–448.

Pew Research Center, 2015a. *Religion and Science: Religion and Views on Climate and Energy Issues* [online]. Available from: www.pewresearch.org [Accessed 4 April 2020].

Pew Research Center, 2015b. *Catholics Divided Over Global Warming: Partisan Differences Mirror Those Among General Public* [online]. Available from: www.pewforum.org [Accessed 4 April 2020].

Pope Francis, 2015. *Laudato Si': Encyclical Letter of the Holy Father Francis on Care for Our Common Home*. Rome: Vatican Press.

Posas, P.J., 2007. Roles of Religion and Ethics in Addressing Climate Change. *Ethics in Science and Environmental Politics*, 7 (1), 31–49.

Romm, J., 2016. *Climate Change: What Everyone Needs to Know*. New York: Oxford University Press.

Ronan, M., 2017. Religion and the Environment: Twenty-First Century American Evangelicalism and the Anthropocene. *Humanities*, 6 (92), 1–15.

Routhe, A.S., 2013. *Religion's Influence on Environmental Concern: U.S. Evangelicals' Construction of Climate Change Perception*. Tennessee: University of Tennessee.

Rudd, K., 2015. Paris Can't Be Another Copenhagen. *New York Times* [online]. Available from: www.nytimes.com/2015/05/26/opinion/kevin-rudd-paris-cant-be-anothercopenhagen.html [Accessed 10 February 2020].

Sanders, C., 2003. Application of Colaizzi's Method: Interpretation of an Auditable Decision Trail by a Novice Researcher. *Contemporary Nurse Journal*, 14 (3), 292–302.

Shao, W., and McCarthy, A.F., 2020. Understanding Evangelical Protestant Identity, Religiosity, Extreme Weather, and American Public Perceptions of Global Warming, 2006–2016. *Geographical Review*, 110 (4), 1–20.

Shehu, M.U., and Molyneux-Hodgson, S., 2014. Faith Communities and Environmental Degradation in Northeast Nigeria. *International Journal of Environmental Sustainability*, 10 (1), 27–40.

Sherkat, D.E., and Ellison, C.G., 2007. Structuring the Religion-Environment Connection: Identifying Religious Influences on Environmental Concern and Activism. *Journal for the Scientific Study of Religion*, 46 (1), 71–85.

Shosha, G.A., 2010. Employment of \zi's Strategy in Descriptive Phenomenology: A Reflection of a Researcher. *European Scientific Journal*, 8 (27), 31–43.

Simkins, R.A., 2008. Religion and Environment: An Introduction: The Legacy of Lynn White, Jr. *Journal of Religion & Society*, Supplement Series 3, 1–4.

Speziale, H.J., and Carpenter, D.R., 2007. *Qualitative Research in Nursing: Advancing the Humanistic Imperative*. Lippincott: Williams and Wilkins.

Stehr, N., and Grundmann, R., 2012. How Does Knowledge Relate to Political Action? *The European Journal of Social Science Research*, 25 (1), 2944.

Stults, M., 2006. Religious Groups Becoming a Factor in Climate Policy Debate. *Climate Institute* [online]. Available from: http://web. archive. org/web/20070407060340/www. climate. org/topics/localaction/religio n-climate-change. shtml [Accessed 4 April 2022].

Tarakeshwar, N., Swank, A.B., Pargament, K.I., and Mahoney, A., 2001. The Sanctification of Nature and Theological Conservatism: A Study of Opposing Religious Correlates of Environmentalism. *Review of Religious Research*, 42 (4), 387–404.

Taylor, B., Van Wieren, G., and Zaleha, B.D., 2016a. The Greening of Religion Hypothesis (Part Two): Assessing the Data from Lynn White, Jr, to Pope Francis. *Journal for the Study of Religion, Nature and Culture*, 10 (3), 306–378.

Taylor, B., Van Wieren, G., and Zaleha, B.D., 2016b. Lynn White Jr. and the Greening-of Religion Hypothesis. *Conservation Biology*, 30 (5), 1000–1009.

The Cape Town Commitment, 2011. *A Confession of Faith and a Call to Action* [online]. Available from: www.lausanne.org [Accessed 26 January 2020].

The Church of England, n.d. *The Environment and Climate Change* [online]. Available from: www.churchofengland.org/environment [Accessed 12 April 2020].

The National Religious Coalition on Creation Care (NRCCC), 2019. *Religious Declaration of Unprecedented Human Emergency* [online]. Available from: http://nrccc.org [Accessed 27 January 2020].

The South African Council of Churches, 2009. *Climate Change – a Challenge to the Churches in South Africa*. Stellenbosch: SUN Media.

Thomas-Odia, I., 2018. Interfaith Groups Create Awareness on Climate Change. *Guardian* [online] Available from: https://guardian.ng/property/interfaith-groups-createawareness- on-climate-change/ [Accessed 28 December 2018].

Tucker, M.E., 2003. *Worldly Wonder: Religions Enter Their Ecological Phase*. Chicago: Open Court.

Tucker, M.E., and Grim, J.A., 2001. The Emerging Alliance of World Religions and Ecology. *Daedalus: Journal of the American Academy of Arts and Sciences*, 130 (4), 1–22.

White, L., 1967. The Historical Roots of Our Ecologic Crisis. *Science*, 155 (3767), 1203–1207.

Wilkinson, K.K., 2012. *Between God and Green: How Evangelicals Are Cultivating a Middle Ground on Climate Change*. Oxford: Oxford University Press.

World Council of Churches (WCC), 2014. *Interfaith Statement on Climate Change: Climate, Faith and Hope: Faith Traditions Together for a Common Future* [online]. Available from: www.oikoumene.org [Accessed 27 January 2020].

Wuthnow, R., 1988. *The Restructuring of American Religion: Society and Faith Since World War II*. Princeton, NJ: Princeton University Press.

Wuthnow, R., 1996. Restructuring of American Religion: Further Evidence. *Sociological Inquiry*, 66 (3), 303–329.

Zaleha, B.D., and Szasz, A., 2015. Why Conservative Christians Don't Believe in Climate Change. *Bulletin of the Atomic Scientists*, 71 (5) 19–30.

Appendix

Table 7.3 Socio-demographic information of participants

S/N	Pseudonyms	Age	Sex	Highest Ed. Qualification	Weekly church attendance	Position in the church	Denomination	Geo-political Zone
1	Philip	51	M	Diploma	7 days	Parish Catechist	Roman Catholic	South-East, Nigeria
2	Clement	34	M	B.A	7 days	Parish Priest	Roman Catholic	South-East, Nigeria
3	Humphrey	48	M	Diploma	7 days	Parish Catechist	Roman Catholic	North-West, Nigeria
4	Collins	33	M	B.A	7 days	Parish Priest	Roman Catholic	North West, Nigeria
5	Dominic	47	M	B.A	7 days	Parish Priest	Roman Catholic	North Central, Nigeria
6	David	52	M	M.Sc.	7 days	Chairman, Vocational Committee	Roman Catholic	North Central, Nigeria
7	Ushie	35	M	OND	7 days	Parish Catechist	Roman Catholic	South-South, Nigeria
8	Anthony	25	M	B.A	7 days	Seminarian	Roman Catholic	South-South, Nigeria
9	Fredrick	42	M	PhD	7 days	Parish Priest	Roman Catholic	South West, Nigeria
10	Ayodele	38	M	B.Sc	7 days	Parish Catechist	Roman Catholic	South West, Nigeria
11	Solomon	51	M	Diploma	7 days	Parish Catechist	Anglican Church	South-East, Nigeria
12	Matthew	40	M	B.A	7 days	Parish Priest	Anglican Church	South-East, Nigeria
13	Steven	39	M	Diploma	7 days	Parish Catechist	Anglican Church	North-West, Nigeria
14	Andrew	42	M	B.A	7 days	Parish Priest	Anglican Church	North West, Nigeria
15	Lawrence	39	M	B.A	7 days	Parish Priest	Anglican Church	North Central, Nigeria
16	Godwin	40	M	M.Sc.	7 days	Chairman, Vocational Committee	Anglican Church	North Central, Nigeria
17	Helen	47	F	PhD	4 days	Parish Catechist	Anglican Church	South-South, Nigeria
18	Okon	42	M	B.A	7 days	Archdeacon	Anglican Church	South-South, Nigeria
19	Yomi	40	M	B.A	7 days	The Vicar	Anglican Church	South West, Nigeria
20	Simi	45	F	B.Sc	7 days	Women leader	Anglican Church	South West, Nigeria

(Continued)

Table 7.3 (Continued)

S/N	Pseudonyms	Age	Sex	Highest Ed. Qualification	Weekly church attendance	Position in the church	Denomination	Geo-political Zone
21	Peter	58	M	Ph.D.	4 days	Senior Pastor	Pentecostal	South East, Nigeria
22	Moses	54	M	B.L. B.Sc.	3 days	A deacon and Church Financial Secretary	Pentecostal	South East, Nigeria
23	Michael	40	M	B.Sc.	3 days	A cell leader	Pentecostal	North West, Nigeria
24	Frank	29	M	B.Ed.	6 days	Pastor	Pentecostal	North West, Nigeria
25	Angela	26	F	B.Sc.	4 days	A cell leader, Choir Member	Pentecostal	North Central, Nigeria
26	Henry	47	M	B.L.	5 days	Church Administrator	Pentecostal	North Central, Nigeria
27	Daniel	37	M	B.Sc.	7 days	Assistant Resident Pastor	Pentecostal	South-South Nigeria
28	Jonathan	27	M	B.Sc.	7 days	Secretary to the Pastor	Pentecostal	South-South, Nigeria
29	Jumoke	30	F	B.Sc	4 days	Church interpreter, Youth President	Pentecostal	South West, Nigeria
30	Adekunle	40	M	B.Sc	4 days	Area Pastor	Pentecostal	South West, Nigeria

Note: Names used in the table are pseudonyms.

Part III

Interreligious Tensions

8 Environmentalism in the Religious Field

The Role of the Establishment for Competition in Switzerland

Fabian Huber

Introduction[1]

A number of religious groups and initiatives address issues of environmentalism, as is emphasized by both religious activists and scientists (Gottlieb 2006; Shibley and Wiggins 1997; Veldman et al. 2014). References to the religious leadership, such as the encyclical "Laudato Si'" usually serve as evidence for environmental commitment. In addition, quantitative studies show the environmental commitment of individual believers (Carlisle and Clark 2018; Taylor et al. 2016). Often, these studies simply assume that congregations take over tendencies from the leadership. Congregations and regional umbrella organizations then serve as a kind of hinge that transmits ecological awareness to the followers. Vaidyanathan et al. (2018) describe how local religious communities can shape the ecological commitment of their members, but admit that this influence is only very limited. Yet studies on the environmental commitment of religious groups at a local or regional level are rare. To fill this gap, this chapter focuses on congregations and regional umbrella organizations. To do so, I compare the engagement of congregations from different faith backgrounds in Switzerland in terms of their environmental commitment. I argue that differences in these groups' commitment are mainly attributable to the degree of establishment and therefore a distinction between *Established* and *Non-established*. This distinction goes back to the work of Weber (2001 [1921]) and Bourdieu (1971a, 1971b). According to them, actors compete with each other in a religious field. On the one hand, there are those who have power and a privileged position – the Established; on the other, there are those with less power, who seek to improve their own position – the Non-established. This constellation leads to struggles between the two, and thus tensions exist.

Drawing upon the Established/Non-established distinction, this chapter addresses two questions: (1) To what extent is the position of power within the religious field decisive for environmental commitment? To address this question, I explore whether the Established are more committed to environmental issues than the others, and if the degree of establishment serves as

DOI: 10.4324/9781003017967-11

an explanation for the commitment. (2) To what extent does environmental commitment play a role in the competition within the religious field? I will investigate how struggles arise from environmental engagement and how they lead to tensions. Here the focus is on interreligious tensions.

The first section describes the theoretical framework, including the religious field according to Bourdieu and Weber, the Established and others within this field as well as the development of different forms of establishment, distinguishing particularly between *legal establishment* and *de facto establishment*. It also explains why Switzerland constitutes a valid case study for this approach. The second section addresses the potential of religious communities in environmentalism, formulating questions that cluster around environmental commitment within religious communities. The chapter then moves through a presentation of the data and methods, followed by the results. These describe the environmental engagement of the religious communities; how the Established/Non-established distinction explains differences in this engagement; and how tensions arise between the different positions. The chapter then ends by highlighting the importance of the Established/Non-established distinction when explaining differences in environmental engagement among religious actors, but also by reflecting on the limitations of this approach.

Theoretical Background

In order to explain the different levels of environmental commitment of religious communities and the resulting tensions, I have designed a theoretical model of the relationships among religious actors. First, I will outline the religious field according to Bourdieu. Then I will go into the distinction between the Established and the Non-established within this field. After explanations on various forms of establishment, I will conclude with a description of the religious field in Switzerland.

The Religious Field

Weber's sociology of religion has not only strongly influenced Bourdieu's understanding of religion, but also his field concept (Bourdieu 1987, p. 33). According to Bourdieu (1993a, p. 108), a social field is a structured space, which reflects the status of the power relations of the actors and institutions involved and/or the distribution of the specific capital. The actors fight for the monopoly of the legitimate symbolic violence (or specific authority) in the field, that is, for the preservation or reassignment of power. Bourdieu used various metaphors to characterize a field.[2] At one point, he describes them as "play spaces," as autonomous spheres, in which one plays according to the field's specific rules (1984, p. 8). He also speaks of "battlefields," in which the fight is waged for the preservation or change of the balance of power (1982, p. 24).

In his consideration of the religious field, Bourdieu refers in particular to Weber, who describes how various types of religious actors compete with one another ([1921] 2001): on one side, the priests; on the other, the prophets and magicians. Priests represent the tradition, the orthodox religious message, while prophets proclaim a new religious message, and magicians solve individual problems. Given that magicians have no permanent followers, they do not play a major role in the competition for the favor of the laity. Rather, this competition takes place between the priests and prophets. Bourdieu further develops Weber's approach by perceiving it as a market,[3] in which religious actors offer services to the laity and in return are legitimized by the laity (Egger 2011, p. 269). In his essay "Interpretation of Religion According to Max Weber," Bourdieu (1971a, p. 5) emphasizes that first one has to elaborate the structures of the religious field to understand the interactions between the positions. Competition for religious power owes its peculiarity to its exertion of great influence on people. Religious power influences lay people's practice and worldview by imposing and imprinting a religious habitus on them (Bourdieu 1971b, p. 320).[4] The fundamental struggle is between those who hold a socially recognized and institutionalized capital of religious authority and those who do not hold this capital. Therefore, "in general, Bourdieu sees this opposition occurring between the established agents and the new arrivals in fields" (Swartz 1996, p. 80). In order to maintain their position, the established agents are willing to prevent the entry of new salvation enterprises into the market. The new arrivals question the monopoly by their very existence (Bourdieu 1971b, pp. 319–320).[5] Since this approach of the religious field focuses on struggles and competition, it is particularly conducive for the investigation of inter-religious tensions.

Established and Non-established in the Religious Field

As is apparent from these explanations, within the religious field there are mainly two types of actors: those who occupy a privileged position and defend it and those who enter the field and strive for a privileged position. I refer to these two forms as Established and Non-established. The term "Established" itself refers to Norbert Elias and John L. Scotson's study, "The Established and the Outsiders" (1990). According to them, the Established are a coherent group that is able to maintain a surplus of power over another group or individual. This enables them to gain access to socially significant positions and resources. In order to secure their power superiority, the established group secures its status socially by excluding members of other groups. The central questions are how this constellation came about, how it remains stable, and why the outsiders accept their role. As with Bourdieu, the understanding of demarcation mechanisms and the explanation of power differences are elementary.[6] For the other groups, I simply use the neutral term "Non-established." This makes it possible to unite these groups well with one term without overemphasizing certain characteristics. Such a restriction would result from the use of the term "Outsider" according to Elias and Scotson (1990), for example, since the stigmatization would move into the foreground. Also, the term "Newcomer," as used by Stolz and Monnot (2018) for a study in Switzerland, ultimately proves to be too narrow, as the temporal dimension is weighted too heavily.[7] Hence, this article will use as open a term as possible.

External Factors: Two Forms of Establishment

Various external factors influence the positions within a field. Therefore, the privileged position of the established religious groups depends on the support they receive from outside the religious field. In this sense, establishment refers to the preferential treatment of a religious group, its members, and institutions. There are two different forms: *legal establishment* (Whelan 1990) and *de facto establishment* (Beyer 2013).

With the *legal establishment* this preferential treatment is afforded by the state and can include various rights (e.g., collecting of taxes, religious teaching in schools), but also obligations (e.g., financial disclosure, democratic governance). The dependency of the fields is also evident for Bourdieu.[8] Autonomy of the religious field does not mean absolute independence, especially from the political field (Dianteill 2003, p. 537). For a privileged position within a field, the relationship with state authorities is important (Bourdieu and Wacquant 1992, p. 90). Political actors have an interest in influencing who is powerful in the religious field because religion legitimizes the established order and shapes the religious habitus of the laity.[9] In the end, the interplay between the religious field and the political field determines the structure of the religious field (Bourdieu 1971b, pp. 328–330).

In the case of the *de facto establishment*, the society grants the preferential treatment. The established groups enjoy the recognition of the population because they have a long tradition or constitute the majority. Here the central role of the relationship between the religious actors and the laity becomes apparent. According to Bourdieu (1971a, p. 14), the religious message has to meet the worldly interest of the lay people as a prerequisite for its success. It must therefore be adapted to the interests of the laity (Weber 2001, p. 206; Bourdieu 1971a, p. 9). The message that can best satisfy the religious demand of the laity is also the one that has the most power (Bourdieu 1971b, p. 314).

Switzerland as a Case Study

Switzerland is well suited to study the Established/Non-established distinction, as Stolz and Monnot (2018) have shown. There are two established religious actors – the Roman Catholic Church and the Reformed Church – while all other groups are non-established. There are several reasons for this. First, the cantonal church-state system (*Landeskirchentum*): the Roman Catholic and Reformed Churches are recognized as corporate bodies under public law in (almost) every single canton.[10] The others, on the other hand, are (mostly) not (see Stolz and Monnot 2018, p. 112).[11] This recognition bestows various rights to a religious organization, such as pastoral care, religious education, and tax sovereignty (tax collection among the members). There are also duties such as democratic constitution, and disclosure of finances (Cattacin et al. 2003). Second, the long historical religious tradition of the two actors: the two established traditions originated before 1800, with non-established Christians and Jewish communities only settling later.[12] More recently, other religions came, most of them only since 1975 (Stolz and Monnot 2018, p. 102). Third, their majority position: over 60% of the Swiss population are either Roman Catholics or Reformed. All other religious communities account for only about 12% (5% of them Muslim) of the population (Bundesamt für Statistik 2018).[13] Finally, their "Public service": the two Christian denominations also have strong support in society as a whole. The population sees these Churches as providers of a "public service," especially in the areas of welfare and help for the needy.[14]

In their results, Stolz and Monnot (2018, pp. 105ff.) show that the established groups have far more resources (financial, buildings, staff). The question now arises as to whether and how these differences between Established and the others become manifest in the environmental engagement of religious actors.

Environmentalism in the Religious Field

Academic debates strongly emphasize the potential of religion for environmentalism. These potentials include the economic and material resources of religious communities, their moral function as role models, their political and public influence, and the large number of their members (Gardner 2003; Gottlieb 2006; Veldman et al. 2014). According to Koehrsen (2015, 2018), religious communities can engage in environmental issues on three different levels: (1) *Public campaigning*, relating to the public dimension of religion (cf., Casanova 1994). Thus, religious communities have a wide reach among the population and influence on public opinion, enabling them to take a stand on environmental issues. They can express themselves in the media, take part in public discussions, and lobby (Gardner 2006). (2) *Materialization*, relating to the concrete "material" implementation of environmentally friendly projects (cf., Gottlieb 2006; Harper 2011). This can involve the introduction of renewable energies (e.g., the installation of solar panels) or the conversion to a more efficient use of resources (e.g., reduced heating output in buildings, purchasing and consuming regional, sustainable products, or recycling). (3) *Dissemination of values*: By actively promoting environmentally friendly values and worldviews, religious communities could encourage their members to adopt more environmentally friendly lifestyles

(Gardner 2003, 2006; Gottlieb 2008; Rolston 2006; Tucker 2006; Bergmann 2009). Religious actors can accomplish this, for example, by emphasizing the value and protection of the environment in sermons and religious instruction (Djupe and Hunt 2009; Shibley and Wiggins 1997).

Drawing upon the theoretical framework presented here, I argue that the established groups can mostly exploit the aforementioned potentials and will tend to be more environmentally committed, as only they have the necessary resources and the support of politics and the population. By contrast, the Non-established lack the resources and the support. Thus, they hardly have any opportunity to exhaust the potentials.

Against this backdrop, I assume that established groups are more committed to environmentalism than the others. Consequently, two questions arise: Is there any evidence of a greater environmental commitment on the part of the established groups? Is the status of being established an essential reason for this commitment?

In this chapter, however, the question is not only about how religious communities are involved in environmental issues. Rather, one must also analyze how environmental commitment can shape the religious field. The increasing importance of the topic also has an effect on religions. As a consequence, environmentally friendly reinterpretations have taken place within various religious traditions, which is often described as a religious "greening process" (Kanagy and Willits 1993; Shibley and Wiggins 1997; Taylor 2010). Environmental protection and climate change are major challenges today. Civil society actors are strongly demanding more commitment, as illustrated by the popular "Fridays for Future" movement. In the "Worry Barometer 2018," 23% of the population stated that environmental protection was one of Switzerland's five most important problems (Golder et al. 2018).[15] This drives the religious greening process forward. As the Established in Switzerland represent the majority of the population, they must meet the needs of different lay groups, and bear a certain responsibility to society. Therefore, they must also take up these demands for more environmental protection. This looks different for the Non-established. Both society and its own members will have fewer expectations for their environmental engagement.

Environmental commitment may also affect the competition within the religious field. In order to maintain or improve their position in the field, religious communities need recognition from the laity and society. The criterion of recognition leads to symbolic capital.[16] Bourdieu (1984, p. 3) describes symbolic capital – commonly referred to as prestige, reputation, and so on – as a perceived and legitimately recognized form of the other capitals. This includes all forms of credit for social recognition or appreciation that can establish themselves within a social field (Bourdieu 1987, p. 160), leading to the following proposition: Environmental commitment serves as a form of symbolic capital for the competition in the religious field. Again, this raises further questions. Do religious actors use environmentalism in competition to strengthen their own position? Do tensions arise between the religious actors as a result?

In sum, drawing upon the theoretical framework, I argue that the Established have the resources to engage in environmental issues and, at the same time, face the pressure to do so. In addition, this commitment may also strengthen their position of power in the religious field. By contrast, the Non-established lack both the resources and the social pressure to get involved. However, if they perceive environmental engagement as a symbolic capital, they may use it to improve their position in the field.

Data and Methods

My data stems from the SNSF project "Urban Green Religions? Religion in Low Carbon Transitions in Two Western European Cities," which has been carried out between 2018 and 2020 at the Centre for Religion, Economy, and Politics (ZRWP), University of Basel. Together with my colleagues, we conducted 43 semi-structured interviews with representatives of religious communities and umbrella organizations.[17] Of these interviews, 12 took place in Switzerland and 31 in Germany, plus the representative of Jehovah's Witnesses who spoke for both countries.

In this chapter, I will focus on the interviews concerning Switzerland. In the sample, the following groups represent the Established: the Roman Catholic Church (2 interviews), the Reformed Church (3), and a joint environmental umbrella organization of these two (1). The following are the Non-established: Islamic communities (2), Evangelical Churches[18] (2), a Buddhist association (1), a Jewish community (1), and Jehovah's Witnesses (1).

The method of semi-structured interviews has a couple of advantages: the adherence to an interview structure ensures comparability. At the same time, this form offers a certain openness, which allows reacting spontaneously to our interlocutors (Mayring 2002, pp. 66f.). Over the course of the 60 to 90 minutes of interviews, the conversation partners told us how and to what extent communities deal (or do not deal) with environmental issues and for what reasons. We transcribed the interviews and imported them into the MaxQDA program. The analysis was carried out partly based on fixed theoretical codes (e.g., the three functions of the environmental commitment of religious communities), and partly openly according to the coding approach of "grounded theory" (Strauss and Corbin 1998).

The field access and the planning of the interviews already led to initial results concerning the Established/Non-established distinction. It was easier to find interview partners from the established communities. For the other communities, it was sometimes quite difficult and we had to make several attempts. Some requests (Hinduism, Baha'i) were even unsuccessful. Furthermore, the Established also had experts like architects who explicitly addressed environmental issues within the community. Among the non-established groups, this was the case only with the Jehovah's Witnesses, but the latter was responsible for Germany, Liechtenstein, Luxembourg, Austria, and Switzerland.

In addition to the interviews, we collected and analyzed documents of the religious communities concerning environmental issues. These include books, brochures, press releases, or the website of religious communities, but also debates in the media.

Results

Environmental Activities of Religious Communities?

The research revealed how the religious communities engage with environmental issues. The following description of religious environmental commitment follows the distinction between Established and Non-established as well as the three levels of public campaigning, materialization, and value dissemination (see Table 8.1).

The established communities are involved in various ways in *public campaigning*. Above all, the umbrella organizations are active in this domain, especially the association "Oeku Kirche und Umwelt." An advisory body for ecological issues of the Reformed and Roman Catholic Churches,[19] Oeku produces a number of publications, such as the environmental handbook "Let there be green" (*es werde grün*) and guidelines on energy saving.[20] To a certain extent, the established groups also engage in lobbying. For example, on the occasion of the climate conference in Katowice (COP24):

Table 8.1 Religious environmentalism in the religious field

	Established	*Non-established*
Public Campaigning	Oeku (environmental umbrella organization) Publications Lobbying (COP24) "Churches for Climate"	AKU (environmental umbrella organization) Publications
Materialization	"Green Rooster" Energy Funds Heating, Isolation, Renewable Energies, Palm oil free cleaning products, Fair Trade Coffee, Regional Products, Recycling	Fair Trade Coffee Regional Products Recycling
Value Dissemination	Training Staff Sermons Education Camps	Holy Scriptures Sensitize Members

in a joint letter, the Swiss Bishops' Conference, the Christ Catholic Church of Switzerland, and the Federation of Swiss Protestant Churches appealed to the Federal Council to advocate a fair climate policy.[21] Recently, various actions have also taken place on the local level. Particularly in the context of the "Fridays for Future," the Christian churches have provided the climate strikers with rooms, prayed for them, and organized panel discussions. They even created a working group "Churches for Climate." On 12 March, at 5 to 12, the clock of St. Peter's in Zurich, the largest church clock in Europe, stood still to show that time is running out fast and created visibility for climate change.

By contrast, there are only a few activities in the domain of public campaigning among the Non-established. With "AKU" (*Arbeitsgemeinschaft Klima, Energie, Umwelt* [Working Group for Climate, Energy, Environment of the Evangelical Alliance]), the Evangelical Churches also founded a working group that explicitly deals with environmental issues. However, AKU was mainly active in the years 2011/2012, and has seen relatively little activity since then. We found almost nothing among the other non-established religions. Only the Muslim "VIOZ" (*Vereinigung der Islamischen Organisationen in Zürich* [Association of Islamic Organizations in Zurich]) published an environmental brochure, "Environmental Protection and Sustainability in Islam" (*Umweltschutz und Nachhaltigkeit im Islam*). The data also does not reveal any activities in lobbying or participation in discussions on environmental issues.

On the level of *materialization*, there are considerable differences between the Established and the others. Christian communities can be certified with the "Green Rooster" (*Grüner Güggel*). For this, environmental auditors accompany the communities and measurable targets are set (e.g., saving thermal energy or reducing waste).[22] Furthermore, the interview partners point out various projects in which they are involved, such as the optimization of the heating, lighting, and electrical systems; the lowering of the heating temperature and the improvement of building; investments in renewable energies; and the use of palm oil-free cleaning products. Some non-established groups also referred to achievements in this area. For example, they reported that they recycle or offer Fairtrade coffee and regional products. Interestingly, the Established ones did not address these points. However, when asked, they said that these were self-evident facts.

Regarding the *dissemination of values*, interestingly, the Non-established talked much more about this dimension of religious environmentalism than the Established in the interviews.[23] For example, the representatives of the Muslim and Jewish communities emphasized that the holy scriptures already include environmental protection. If believers live their life according to these, they are automatically environmentally friendly.[24] The Evangelicals stated that they try to sensitize their members to environmental issues, but in the end, it is up to each individual how they want to live. By contrast, the established religious groups train their staff in environmental issues, and sometimes give sermons on environmental topics. Interviewees from the established groups often emphasized that the environment is important in child and youth education. As such, scouts or youth camps address environmental protection as a topic. For example, the Catholic Church is co-organizer of the "fair camp."[25]

In total, the Established and the Non-established show strong differences in their environmental engagement on all three levels. In each area, the Established are more involved in environmental issues. Now the question arises as to what extent the status of being established leads to these differences.

Established/Non-Established Distinction as Explanation

In the interviews, various indicators suggest that the environmental commitment depends mainly on the Established/Non-established distinction. The position occupied by one group in the religious field gives rise to various differences that influence environmental commitment. There are two different factors: on the one hand, those related to structural conditions and resources; on the other hand, those caused by social pressure.

Structural Aspects and Resources

A first point lies already in the fact of recognition as corporate bodies under public law (see later). In order to be recognized, a religious group must have a democratic constitution and a democratic internal organization (Cattacin et al. 2003, p. 17). In order to fulfill these prerequisites, the Established have built up a democratic "secular" hierarchy in addition to the "clerical" one. Therefore, the Established have a dual structure: on the one hand, there are the people who deal with theological questions; on the other hand, there is the whole administration. Within the "secular" structure different areas developed. Nowadays, there are particular experts, such as architects who deal with environmental issues.[26] In addition, they also have specific organizations, such as Oeku, for environmental issues. By contrast, the Non-established have no comparable administrative infrastructure and, therefore, no experts on environmental issues. Thus, the "spiritual" leadership itself is responsible for worldly issues such as the environment. Apart from AKU, there is no association that dedicates itself explicitly to the topic. Since the Established own their buildings, they have to deal with maintenance issues. For the others, these questions arise less frequently, as the norm is the opposite, and only a few groups own buildings. One representative of an Islamic community reported on the reconstruction of a mosque. However, environmental aspects did not play a role; it simply had to be as inexpensive as possible. This alludes to the importance of financial resources (economic capital), which have a major impact on environmental commitment. The following two statements illustrate this. A representative of the Roman Catholic Church describes the creation of an energy fund as follows:

> So, the whole thing actually started with us in such a way that we had a kind of surplus – surplus earnings – and this money, that was about 1.4 million at that time, where we then thought we would make a fund out of it.
>
> (Interview 1, conducted on 7/23/2018)

The financial experience of a representative of a Muslim community is completely different:

> Environmental protection is such an issue for the wealthy. If you don't have any financial difficulties, if you know, yes tomorrow I have enough to eat, tomorrow I can pay the electricity in the mosque, tomorrow I can also pay the Imam, that won't be a problem, then you can say: Okay, now we can also dedicate ourselves to this topic. But if you're always in this financial emergency, as is the case with 99 percent of the mosques . . . you just don't have time to deal with this topic properly now.
>
> (Interview 2, conducted on 7/5/2018)

With the Established, it seems almost as if there was too much money, which they had to spend. As such, they started an energy fund. The Non-established, on the other hand, do not even have enough money for operating costs; investments in environmental technologies and activities seem almost utopian.

Social Pressure

The assessment of the expectations of one's own members leads to more motivations. As the Established have more members, they have to cover the interests of different lay groups. They assume that their members (or at least part of it) care about the environment. As a representative of the Reformed Church said:

> The audience that is engaged in the church, I don't have to "preach" that anymore. I no longer have to "preach" about ecological topics. Actually, they all have a sensitivity for questions of sustainability.
>
> (Interview 3, conducted on 9/27/2018)

Furthermore, the Established see themselves as representatives of the whole society. Moreover, they find that environmental issues are becoming increasingly important in society. Therefore, it seems logical for them to deal with environmental issues. The Non-established assume that other issues are more important to their members. For some, discrimination is an issue. Often the members are migrants, some of them from war zones. In the absence of corresponding demands for environmental engagement from the laity, the religious leadership likewise does not see the need to establish this engagement in the given religious group.

Differences in the availability of resources combined with dissimilar societal expectations also give rise to different priorities. The Established are able to cover various areas, including the environment. The Non-established, on the other hand, have to limit themselves and set priorities. The topics of discrimination, integration, pastoral care, and poverty in the world are perceived as more important. In the end, it is a matter of fulfilling the core task. As a Buddhist puts it:

> It is not the task of Buddhist communities to educate people about environmental protection, this is not the task. The task of Buddhism is the training of the spirit. I mean a sports or football club does not have the task to enlighten about environmental protection either.
>
> (Interview 4, conducted on 6/22/2018)

In summary, the Established/Non-established distinction helps to explain the differences in environmental commitment. While the Established dispose over sufficient resources and face societal pressure for environmental engagement, the others lack the resources, have to adjust their priorities accordingly, and concentrate on the essential congregational tasks.

Tensions – Environmentalism as Symbolic Capital

Now, I will address the question of the extent to which environmental commitment plays a role in competition in the religious field and thus leads to interreligious tensions. The previous results (concerning the expectations) suggest that the Established see environmentalism as important for their legitimacy and therefore constitutes symbolic capital for them. A representative of the Roman Catholic Church believes that it is important to cultivate environmental concern:

> For example, we try to take up such topics in our media, in the parish gazette and in the website . . . for the whole church, which is also represented in the general population, so I rather see the task to reach people who do not belong to the core.
>
> (Interview 5, conducted on 10/11/2018)

This shows that environmental issues can be an advantage in competition. It allows for reaching new people and thus strengthening their own position. The non-established groups (except for the Buddhists) stated that, though the environment is topical, there were more pressing obligations. Other non-established communities see possible symbolic capital, and that an involvement in such issues could increase the prestige of one's own community. A Muslim representative said he would like to see more engagement on environmental issues, because:

> That would be one of those signs. One would simply get away from the everyday stuff that Muslims are accused of. That they are terrorists, that one has a bad picture of Islam, of the Muslims. Where one could simply show: Wow, Muslims also think for our environment.
>
> (Interview 2, conducted on 7/5/2018)

In this sense, environmentalism can serve to improve the image of one's own community in society and position in the religious field.

Both the Established and Non-established see environmental commitment as a symbolic capital. In the sense of competition, there are tensions between the communities concerning the environment. It serves the Established in defending their position. In addition, they even see it as an opportunity to generate even more power in the field. Even the Non-established see the potential in environmentalism to improve their own position and thus the opportunity to rise in the field. However, the importance of environmentalism in competition is low and tensions between religious actors in this regard are only subliminal.

There are two explanations that these interreligious tensions remain latent. First, the emphasis on the importance of environmental issues in the interviews probably resulted from the interview situation and social desirability. This effect was evident among both, Established and Non-established. Although they regard environmentalism as a sort of symbolic capital, other forms of symbolic capital appear to be more important. This is especially the case for social issues such as help for the needy, justice, or integration.[27] Even the engagement of the Established remains modest. For them, too, the environment is not a priority. This becomes evident in the case of Oeku: The main environmental association of the two established churches has struggled for financial means and jobs. To date, it has not been possible to achieve the desired size. Since 2019, the engagement at the local level has been growing strongly. This is especially due to "Fridays for Future." The activities now initiated, however, also show how religious actors take the interests of lay people into account. Environmentalism is gaining in importance as a social issue and thus also in the religious field. A second explanation for the low level of tensions can be found in Elias and Scotson (1990, pp. 27f.), who claim that the tensions between the Established and the Outsiders (Non-established) remain silent when the power gap is too wide. In such cases, the Established simply ignore the other groups. This is also the case for environmental commitment. Since the Non-established lack resources and thus opportunities for increased commitment, they do not pose a threat to the Established in this respect. Environmental issues cannot call into question the privileged position of the Established.

Conclusion

Research on religion and ecology emphasizes that religious traditions shape attitudes toward the environment through theology, both in a negative sense as environmental degradation (White 1967) and in a positive sense as environmental protection (Gottlieb 2006; Kearns 1996; Shibley and Wiggins 1997; Veldman et al. 2014). On this basis, the studies suggest differences between religions. Thus, Lorentzen and Leavitt-Alcantara (2006) state that certain religious traditions such as Buddhism, Neo-Paganism, or some indigenous religions are comparatively "green." However, the studies often focus on theological or ethical aspects and neglect the social dimension of religion and the importance of interactions (Djupe and Hunt 2009; Vaidyanathan et al. 2018). Especially the institutional context and the position in the field are decisive. This study shows that the distinction between Established and Non-established helps to explain the different environmental commitment of religious groups. The Established were more committed to the environment than the others. Their commitment depends essentially on their position within the religious field. The available resources and the expectations were decisive in this respect.

However, the Established/Non-established distinction is just an analytical instrument. Even if it seems to be quite clear in Switzerland, the differences between the groups are gradual. As such, this study also reveals differences among the Non-established: For example, Evangelical and Jewish communities have more resources than other Non-established. In this respect, they occupy an intermediate position. Concerning environmental issues, the Evangelicals are closest to the Established within the Non-established group. Compared to the other Non-established, the Evangelicals are the most committed. Their integration into the society helps to explain their elevated commitment. Contrasting other non-established groups, their members have usually no migration background. As a result, they are more integrated in structural, legal, and political terms (Stolz and Huber 2014). Depending on the canton, there will be differences within the established group. For example, in some cantons the Catholic Church is dominant. Accordingly, the Reformed Church can almost become a non-established group and not be very involved in the environmentalism due to lack of resources.

Tensions between religious communities due to environmental issues remain at a low level. There are indications that environmental commitment can serve as symbolic capital and thus bring advantages in competition within the religious field. However, since all communities (including the Established) show relatively little commitment in this area and since the power gap between the Established and the Non-established is too wide, the potential tensions stay subtle. This could change in the future due to rising environmental awareness among the population, as illustrated by the climate strikes.

Finally, Bourdieu's field approach allows investigating intrareligious as well as religious-societal tensions. Possible tensions between religious and other actors could rise when religious actors are active in other fields (Glaab 2022). Bourdieu also refers to tensions that can arise within a community. Here, a special form of struggle occurs between orthodoxy and heresy. This is a conflict for power within the Church, hinging on the denial or defense of the ecclesiastical hierarchy. This becomes apparent, as different ecological ethics compete within the same faith traditions (Kearns 1996; Wardekker et al. 2009). In the end, the environmental commitment of a religious community is therefore a result of the interaction of internal struggles and external constraints (Bourdieu 1971b, pp. 322–325).

Notes

1 I am thankful for the support of the "Urban Green Religions" team, Jens Koehrsen, Julia Blanc, Vera Schaffer, Anabel Da Pra, and Vanessa Heiniger.

2 This suggests that the "field" is more of a metaphor than a clearly defined term (Fröhlich and Rehbein 2014, p. 100).

3 Bourdieu often uses the terms "field" and "market" synonymously (Fröhlich and Rehbein 2014, p. 100).

4 Influencing the habitus is of enormous importance and, according to Bourdieu, occupies a central position. Habitus forms are systems of permanent and transferable dispositions. They are structured structures that are designed to function as structuring structures. Habitus serves as a schema of the perception of thought and action, producing individual and collective practices (Bourdieu 1972, pp. 175, 188, 1980, pp. 88, 92).

5 This already marks the transition from doxa to orthodoxy and heterodoxy (Bourdieu 2009, p. 332; see also Glaab 2022).

6 The two approaches work well together: Both deal with the interaction and competition between different groups. In an interview with Hessischer Rundfunk in 1983, Bourdieu himself stressed his closeness to Elias: "Among all living sociologists . . . stands Elias, actually closest to my approach. It has shown the existence of such structures of objective relations – what I call the 'field' – shining in his writings is depicted" (see Fröhlich and Rehbein 2014, p. 36.) Bourdieu and Elias kept in touch, exchanged letters, and met in person several times (see Fröhlich 2014, p. 42; Hasselbusch 2014, pp. 9, 244–270). However, Elias understands power not as a quality one possesses, but as a relationship. Therefore, he speaks of "balance of power" (Elias 2014, p. 76).

7 However, the term "Newcomer" could also be linked to Bourdieu. In his comments on the fields (some Properties of fields), Bourdieu uses the term "new": "nouveaux entrants" in French (2002, p. 115), "new players" in English (1993b, p. 74) "Neulinge" in German (1993a, p. 109).

8 This has to do with the field of power. The Field of Power denotes a space between the fields in which the Established within the field fight to improve the position of their own field (Bourdieu 1998, p. 51).

9 Since the Established in the religious field want to remain autonomous, tensions also arise between the fields.

10 The cantons of Geneva and Neuchâtel are exceptions in this respect, since this form of recognition does not exist due to the greater separation between church and state. Here, however, the national churches are recognized as playing a role in public life (see Cattacin et al. 2003, pp. 65f., 75f.)

11 There are some exceptions. The Jewish groups are recognized corporate bodies under public law in the cantons of Basel Stadt, St. Gallen, Fribourg, and Bern, and as playing a role in public life in the cantons of Vaud and Zurich. Another exception is the Christian Catholics, which I will not cover in this chapter due to lack of data (Cattacin et al. 2003).

12 The first Jewish settlement activities in Switzerland date back to the end of the 12th century. Nevertheless, the majority of the communities came into being much later. (Knoch-Mund 2016). Most of the congregations originated after 1800, many even after 1975 (Stolz and Monnot 2018, p. 102).

13 www.bfs.admin.ch/bfs/de/home/statistiken/bevoelkerung/sprachen-religionen/religionen.assetdetail.7726966.html

14 Even many people for whom the Church is not personally important say that the Church is important for society and especially for the underprivileged (Stolz et al. 2014, p. 276).

15 This put the topic in fourth place.

16 For Bourdieu, symbolic capital is, after the economic, cultural, and social capital, the fourth basic form of capital and usually occurs in conjunction with one of the other three types of capital. For the three types, see Bourdieu 1983. In the course of time, Bourdieu increasingly developed other types of capital, such as the "intellectual," "political," "legal," and, of course, "religious" capital.

17 In the second phase of the project, We conducted 24 interviews with non-religious actors.

18 Evangelicalism here means a "conversion Christianity" characterized by Biblicism, individual conversion, the central position of Jesus Christ, emphasis on mission, and interdenominational dimension (Huber and Stolz 2017, pp. 275–276). For the history and streams of Evangelicalism in Switzerland, see Favre 2006.

19 Following simply called Oeku. About 600 parishes, church organizations, and individuals are members of the association, which was founded in 1986. Oeku is recognized by the Swiss Bishops' Conference (Schweizer Bischofskonferenz: SBK) and the Federation of Swiss Protestant Churches (Schweizerischer Evangelischer Kirchenbund: SEK) (www.oeku.ch/de/ueber_uns.php). For more on Oeku, see Becci and Monnot 2016, pp. 99–102.

20 In this respect, Oeku is also active in the areas of value dissemination and, indirectly, materialization. The same applies to AKU (see later).

21 https://christkatholisch.ch/wp-content/uploads/181024_Brief_Bundesrat_Weltklimakonferenz-2018.pdf.

22 To date, however, there are only 17 certified parishes and church institutions in Switzerland (see: www.oeku.ch/de/documents/RegisterGrunerGuggel-StandFeb2019.pdf).

23 However, this has to be taken with caution: The Non-established may have reported so much here because they had less to show in the other areas.

24 The interview with the representative of the Jewish community actually dealt almost exclusively with this point.

25 www.faires-lager.ch.

26 This was already evident in the field access. It was easier to find interviews with representatives of the established communities. For the other communities it was sometimes quite difficult and we had to make several attempts. Furthermore, the Established also had individuals who explicitly addressed environmental issues within the community.

27 In addition, the "tendential dominance of the economic field" shows itself in the religious field (Bourdieu 1984, p. 3). Thus, tensions over economic capital come to the fore.

References

Becci, I., and Monnot, C., 2016. Spiritualité et religion: nouveaux carburants vers la transition énergétique? *Histoire, monde et cultures religieuses*, 40 (4), 93–109.

Bergmann, S., 2009. Climate Change Changes Religion. *Studia Theologica – Nordic Journal of Theology*, 63 (2), 98–118.

Beyer, P., 2013. Deprivileging Religion in a Post-Westphalian State: Shadow Establishment, Organization, Spirituality and Freedom in Canada. *In:* W.F. Sullivan and L.G. Beaman, eds. *Varieties of Religious Establishment*. Farnham: Ashgate, 75–92.

Bourdieu, P., 1971a. Une interprétation de la théorie de la religion selon Max Weber. *European Journal of Sociology*, 12 (1), 3–21.

Bourdieu, P., 1971b. Genèse et structure du champ religieux. *Revue française de sociologie*, 12 (3), 295–334.

Bourdieu, P., 1972. *Esquisse d'une théorie de la pratique. Précédé de trois études d'ethnologie kabyle*. Genève: Droz.

Bourdieu, P., 1980. *Le sense pratique*. Paris: Minuit.

Bourdieu, P., 1982. *Leçon sur la Leçon*. Paris: Minuit.

Bourdieu, P., 1983. Ökonomische Kapital, kulturelles Kapital, soziales Kapital. *In:* R. Kreckel, ed. *Soziale Ungleichheiten. Soziale Welt, Sonderband 2*. Göttinge: Schwartz, 183–198.

Bourdieu, P., 1984. Espace social et genèse des classes. *Actes de la recherche en sciences sociales*, 52 (53), 3–14.

Bourdieu, P., 1987. *Choses Dites*. Paris: Minuit.

Bourdieu, P., 1993a. *Soziologische Fragen*. Frankfurt am Main: Suhrkamp.

Bourdieu, P., 1993b. *Sociology in Question*. London: Sage.

Bourdieu, P., 1998. *Praktische Vernunft. Zur Theorie des Handelns*. Frankfurt am Main: Suhrkamp.

Bourdieu, P., 2002. *Questions de Sociologique*. Paris: Minuit.

Bourdieu, P., 2009. *Entwurf einer Theorie der Praxis auf der ethnologischen Grundlage der kabylischen Gesellschaft*. Frankfurt am Main: Suhrkamp.

Bourdieu, P., and Wacquant, L., 1992. *Réponses. Pour une anthropologie réflexive*. Paris: Editions du Seuil.

Bundesamt für Statistik, 2018. *Religion 2015–2017* [online]. Available from: bfs.admin.ch/bfs/de/home/statistiken/kataloge-datenbanken/grafiken.assetdetail.7726985.html [Accessed 15 August 2022].

Carlisle, J.E., and Clark, A.K., 2018. Green for God: Religion and Environmentalism by Cohort and Time. *Environment and Behavior*, 50 (2), 213–241.

Casanova, J., 1994. *Public Religions in the Modern World*. Chicago: University of Chicago Press.

Cattacin, S., Famos, C.R., Duttwiler, M., and Mahning, H., 2003. *Staat und Religion in der Schweiz – Anerkennungskämpfe, Anerkennungsformen*. Bern: Eidgenössische Kommission gegen Rassismus.

Dianteill, E., 2003. Pierre Bourdieu and the Sociology of Religion: A Central and Peripheral Concern. *Theory and Society*, 32 (5–6), 529–549.

Djupe, Paul A., and Hunt, P.K., 2009. Beyond the Lynn White Thesis: Congregational Effects on Environmental Concern. *Journal for the Scientific Study of Religion*, 48 (4), 670–686.

Egger, S., 2011. Pierre Bourdieus Religionssoziologie. Eine werkbiographische Skizze. *In:* F. Schultheis and S. Egger, eds. *Pierre Bourdieu. Religion*. Berlin: Suhrkamp, 257–278.

Elias, N., 2014. *Was ist Soziologie?* Weinheim: Juventa.

Elias, N., and Scotson, J.L., 1990. *Etablierte und Außenseiter*. Frankfurt am Main: Suhrkamp.

Favre, O., 2006. *Les Eglises évangéliques de Suisse. Origines et identités*. Geneva: Labor et Fides.

Fröhlich, G., and Rehbein, B., 2014. *Bourdieu Handbuch. Leben – Werk – Wirkung*. Stuttgart: J.B. Metzler.

Gardner, G.T., 2003. Engaging Religion in the Quest for a Sustainable World. *In*: Worldwatch Institute, ed. *State of the World, 2003: A Worldwatch Institute Report on Progress Toward a Sustainable Society*. New York: W.W. Norton & Company, 152–175.

Gardner, G.T., 2006. *Inspiring Progress: Religions' Contributions to Sustainable Development*. New York: W.W. Norton & Company.

Glaab, K., 2022. The Green, the Secular, and the Religious. The Legitimacy of Religious Environmentalism in Global Climate Politics. *In:* J. Koehrsen, J. Blanc, and F. Huber, eds. *Religious Environmental Activism: Emerging Conflicts and Tensions in Earth Stewardship*. London: Routledge.

Golder, F., Jans, C., Venetz, A., Wattenhofer, K., Bohn, D., and Herzog, N., 2018. *Schlussbericht Credit Suisse Sorgenbarometer 2018*. Bern: gfs.bern.

Gottlieb, R.S., 2006. Introduction: Religion and Ecology – What Is the Connection and Why Does It Matter? *In*: Gottlieb, ed. *The Oxford Handbook of Religion and Ecology*. Oxford: Oxford University Press, 3–24.

Gottlieb, R.S., 2008. You Gonna Be Here Long? Religion and Sustainability. *Worldviews: Global Religions, Culture, and Ecology*, 12 (2), 163–178.

Harper, F., 2011. Greening Faith: Turning Belief into Action for the Earth. *Zygon®*, 46 (4), 957–971.

Hasselbusch, I., 2014. *Norbert Elias und Pierre Bourdieu im Vergleich. Eine Untersuchung zu Theorieentwicklung, Begrifflichkeit und Rezeption*. Dissertation, Pädagogische Hochschule Karlsruhe.

Huber, F., and Stolz, J., 2017. Das evangelikale Milieu. *In:* F. Elwert, M. Radermacher, and J. Schlamelcher, eds. *Handbuch Evangelikalsimus*. Bielefeld: Transcript, 275–287.

Kanagy, C.L., and Willits, F.K., 1993. A "Greening" of Religion? Some Evidence from a Pennsylvania Sample. *Social Science Quarterly*, 74 (3), 674–683.

Kearns, L., 1996. Saving the Creation: Christian Environmentalism in the United States. *Sociology of Religion*, 57 (1), 55–70.

Knoch-Mund, G., 2016. Judentum. *In:* Historisches Lexikon der Schweiz [online]. Available from: https://hls-dhs-dss.ch/de/articles/011376/2016-02-01/ [Accessed 30 March 2022].

Koehrsen, J., 2015. Does Religion Promote Environmental Sustainability? – Exploring the Role of Religion in Local Energy Transitions. *Social Compass*, 62 (3), 296–310.

Koehrsen, J., 2018. Religious Agency in Sustainability Transitions: Between Experimentation, Upscaling, and Regime Support. *Environmental Innovation and Societal Transitions*, 27, 4–15.

Lorentzen, L.A., and Leavitt-Alcantara, S. 2006. Religion and Environmental Struggles in Latin America. *In*: R.S. Gottlieb, ed. *The Oxford Handbook of Religion and Ecology*. Oxford: Oxford University Press, 510–534.

Mayring, P., 2002. *Einführung in die qualitative Sozialforschung. Eine Anleitung zu qualitativem Denken*. Weinheim: Beltz.

Rolston, H., 2006. Caring for Nature: What Science and Economics Can't Teach Us but Religion Can. *Environmental Values*, 15 (3), 307–313.

Shibley, M.A., and Wiggins, J.L., 1997. The Greening of Mainline American Religion: A Sociological Analysis of the Environmental Ethics of the National Religious Partnership for the Environment. *Social Compass*, 44 (3), 333–348.

Stolz, J., and Huber, F., 2014. Wie kann man die Integration religiöser Gemeinschaften in die Gesellschaft erklären? *In:* E. Arens, M. Baumann, and A. Liedhegener, eds. *Integration durch Religion? Geschichtliche Befunde, gesellschaftliche Analysen, rechtliche Perspektiven*. Zürich: TVZ, 21–40.

Stolz, J., Könemann, J., Schneuwly Purdie, M., Engelberger, T., and Krüggeler, M., 2014. *Religion und Spiritualität in der Ich-Gesellschaft. Vier Gestalten des (Un-) Glaubens*. Zürich: Theologischer Verlag Zürich.

Stolz, J., and Monnot, C., 2018. The Established and the Newcomers. A Weberian-Bourdieusian View of Congregations in the Swiss Religious Field. *In:* C. Monnot and J. Stolz, eds. *Congregations in Europe*. Cham: Springer, 91–116.

Strauss, A.L., and Corbin, J.M., 1998. *Basics of Qualitative Research. Techniques and Procedures for Developing Grounded Theory*. Thousand Oaks: SAGE.

Swartz, D., 1996. Bridging the Study of Culture and Religion: Pierre Bourdieu's Political Economy of Symbolic Power. *Sociology of Religion*, 57 (1), 71–85.

Taylor, B., 2010. *Dark Green Religion: Nature Spirituality and the Planetary Future*. Berkeley: University of California Press.

Taylor, B., van Wieren, G., and Zaleha, B.D., 2016. Lynn White Jr. and the Greening-of-Religion Hypothesis. *Conservation Biology*, 30 (5), 1000–1009.

Tucker, M.E., 2006. Religion and Ecology: Survey of the Field. *In*: R.S. Gottlieb, ed. *The Oxford Handbook of Religion and Ecology*. Oxford: Oxford University Press, 398–418.

Vaidyanathan, B., Khalsa, S., and Ecklund, E.H., 2018. Naturally Ambivalent: Religion's Role in Shaping Environmental Action. *Sociology of Religion: A Quarterly Review*, 79 (4), 472–494.

Veldman, R.G., Szasz, A., and Haluza-DeLay, R., 2014. *How the World's Religions Are Responding to Climate Change. Social Scientific Investigations*. London, New York: Routledge.

Wardekker, J.A., Petersen, A.C., and van der Sluijs, J.P. 2009. Ethics and Public Perception of Climate Change: Exploring the Christian Voices in the US Public Debate. *Global Environmental Change*, 19 (4), 512–521.

Weber, M., 2001. Religiöse Gemeinschaften. *In*: H.G. Kippenberg, ed. *Max Weber – Gesamtausgabe. Band I/22,2*. Tübingen: Mohr Siebeck Verlag.

Whelan, F.G., 1990. Church Establishments, Liberty & Competition in Religion. *Polity*, 23 (2), 155–185.

White, L., 1967. The Historical Roots of our Ecologic Crisis. *Science*, 155 (3767), 1203–1207.

9 "What does religion have to do with nature conservation?" Investigating the Tensions in an Interreligious Nature Conservation Project in Germany

Carrie B. Dohe

Introduction

Throughout the 20th century and increasingly since the 1970s, the idea has been spreading that religious traditions can shape our views of and behavior toward the nonhuman world. All major religions are said to be in essence nature-preservationist, when rightly understood, and could thus contribute to solving the global crisis of environmental degradation (summarized by Taylor 2016). Some scholars have contested these claims, arguing that religious environmental ethics stem less from ancient scriptures and traditions than from modern notions of nature, ecology, and environment (e.g., Harris 1991, 1997; Kalland 2005; Pedersen 1995; Tomalin 2009). Still, people of faith across the religious spectrum seek to retrieve, recover, reinterpret, or reconstruct their religious teachings and practices to make them more environmentally beneficial (Bauman et al. 2011; Gottlieb 1996; Tucker and Grim 2001).

This is also the case in Germany, home to several religious environmentalist projects in Christianity, Islam, Judaism, and Buddhism.[1] Yet aside from two sociological studies (Koehrsen 2015, 2018), religious environmentalism there remains empirically understudied (but see both Blanc and Gojowczyk, this volume). Interreligious environmentalism has yet to be investigated at all. This lack of scholarly attention is striking, because the German federal government has repeatedly funded ecumenical and interreligious events and publications throughout the 2000s.[2] Moreover, from the federal to the local level, Germany has demonstrated a strong interest in interreligious dialog as a way to maintain social cohesion and reduce conflict as its population becomes ever more diverse.[3] These dialog initiatives often focus on conflicts with and the integration of Muslims into a Christian-based, yet largely secularized society (Klinkhammer et al. 2011).

Might an interreligious initiative for the environment be a particularly effective way to increase social cohesion by focusing on the common ground (literally and figuratively) that all members of society share? A new project

DOI: 10.4324/9781003017967-12

in Germany called "Religions for Biological Diversity" has given me the opportunity to examine this very question. Funded by the Federal Agency for Nature Conservation (*Bundesamt für Naturschutz*, or BfN), supported by the Center for Societal Responsibility (*Zentrum Gesellschaftliche Verantwortung*, or ZGV) of the Protestant Church of Nassau und Hessen, and administered by the Abrahamic Forum in Germany, it is running for three years (2017–2019). Thus, unlike the one-off events prior to the project's founding, it provides a venue to examine the kinds of issues that arise in an ongoing interreligious collaboration focused on the natural world. "Religions for Biological Diversity" uses the previously mentioned claim that nature protection is a value shared by religions[4] to bring together very diverse and often mutually antagonistic actors from different religions. It seeks to build networks between these actors and nature conservation organizations, typically officially non-confessional, to increase the awareness and conservation of nature. To achieve its aims, it has developed four sub-projects: an interreligious network; the "greening" of properties owned by religious communities (following the models in Saan-Klein and Wachowiak 2008); religion and nature conservation dialog teams; and an interreligious week for nature conservation, using the Christian ecumenical Day of Creation at the beginning of September[5] as its point of departure. This interreligious week is the most ambitious part of the overall project, because it brings together multiple religious communities, nature conservation groups, city employees, politicians, the media, and the public, and has been the arena for most of the tensions I have witnessed. Does this project face similar tensions and conflicts as other interreligious initiatives, or, given the special focus on nature preservation and its location in Germany, do these tensions differ in significant ways? Using document analysis, participation observation, interviews, and events organization, I found tensions that go well beyond the issue of how best to protect nature. Indeed, the focus on nature conservation became yet another outlet for the expression of tensions concerning secularism, Islam, and the place of religion in German society today.

Before launching into an overview of the "Religions for Biological Diversity" project, I will contextualize it against the background of literature on interreligious dialog in Germany and on interreligious cooperation for the environment. From there, I will explain the range of methods employed to examine the project over the span of 2 years and multiple cities. This leads into a description of several encounters with the tensions just mentioned. Most of these tensions revolve around specific Muslim groups and Islam as a whole, both in the privacy of interviews and the public space of news reports and blogs. This is followed by an overview of tensions with Christianity, which were expressed only in private, and finally with "religion," which emerged in numerous interviews and encounters with environmentalists. In the final discussion, the divergences and overlaps of these tensions are examined.

Interreligious Dialog in Germany and Interreligious Cooperation for the Environment

Thus far, interreligious cooperation for the environment in Germany has not been studied, despite a jump in interest in interreligious dialog more generally after 9/11. Throughout the 2000s, a growing number of publications stressed "dialog" as a "sociopolitical task," which should accomplish three goals: (1) contribute to a peaceful co-existence between German natives and newcomers, (2) find solutions to problems of integration, and (3) propose specific ways to effect social participation (Klinkhammer 2008). From the federal to the local level, many interreligious initiatives focus particularly on dialog with Muslims, who constitute only about 5.5% of the population. According to a 2018 survey on views of religious plurality and democracy, a large majority of Germans (87%) see Buddhism, Hinduism, Judaism, and Christianity as enriching society; however, a majority consider Islam as threatening (51% of western and 58% of eastern German respondents respectively) (Pickel 2019). Indeed, the fear of Muslims and Islam, concerns about integrating Muslims into a Christian-based, secular society, and issues with mosque building and Muslim women's veiling practices have been the impulses for many interreligious initiatives (Klinkhammer 2008).

Hence, participation in interreligious initiatives is not necessarily motivated by religious interest per se; rather, participants often wish to improve social cohesion, reduce prejudices, increase social participation, and improve the status quo (Ohrt and Kalender 2018; Satilmis 2008). To achieve these aims, interreligious dialog ideally engenders both self-reflection and a deep understanding of diverging worldviews, guiding participants toward constructive ways to deal with these differences (Klinkhammer and Satilmis 2008). Dialog should happen at "eye-level," cultivating the democratic principles of mutual respect, debate, and communication among equals (Malik 2008; Satilmis 2008). Where power differentials exist, those people with more power relinquish some of it in order to promote cooperation (Johnston 2014).

Successful interreligious environmentalist dialog necessitates "worldview translation" (Johnston 2013, p. 150), because terminology favored by one group may alienate another (Marshall 2016), and scientific language is often incomprehensible to non-scientists (Fonseca et al. 2018). Beyond translation, an exchange is necessary: Environmentalists have scientific knowledge that religious actors often lack, whereas religious traditions can add to scientific approaches a missing focus on ethics as well as local ways of meaning-making and knowing (Chungprampree 2016). Miscommunication due to different worldviews have led some authors to formulate universal, abstract ethical codes (e.g. Küng and Kuschel 1993; Tucker 2008), yet others argue that such codes fail to speak to people at the grassroots level, who have cultural and emotional ties to specific religious traditions (Johnston 2013). At the same time, removing rituals and narratives from religion-specific contexts and combining them in new forms in interreligious and religious-secular settings can invoke in participants a feeling of unity in diversity and a collective feeling of belonging across the religious spectrum and the religion-secular divide (Frisk 2015).

Good communication skills belong to the moral competencies that successful interreligious environmentalist dialog projects demonstrate. An additional competency involves partnering with minority groups and the sharing of power (Del Vecchio 2018). However, minority group members must be willing to share their worldview (Patel and McKermott 2016). Conflicts between different populations can block effective cooperation (Tucker and Grim 2001), and the unequal places that actors occupy in the social hierarchy may hinder projects (Baugh 2017). Even when groups are willing to work together, projects may fail due to a lack of clear leadership and financial resources (Lysack 2014), whereas strong political support and leadership may sustain a project over the long term (Johnston 2013). In any case, initiatives need a long "courtship" for a good "marriage" to result (Ibid., p. 142; Miksch 2008).

Methods

Document analysis was used to examine "explicit and implicit discourses" of promotional materials, meeting minutes, press announcements, and emails (Davie and Wyatt 2014, p. 158). Because the field was not yet defined or researched, participant observation was used (Franke and Maske 2011) while attending dialog forums and planning meetings in different cities as well as the 2017 Religious Week of Nature Conservation in Darmstadt (hereafter Darmstadt Week). This field, shifting across localities and involving different actors, is the space where the discourse of interreligious environmentalism is "created, negotiated and elaborated upon" (Garsten 2010, p. 59). Finally, action research was used in my role as activist-organizer of the 2018 Interreligious[6] Week of Nature Conservation in Cologne and Environs, Sept. 2–9, 2018 (hereafter Cologne Week) to develop practical knowledge about how a project for interreligious environmentalism is developed. Action research "seeks to bring together action and reflection, theory and practice, in participation with others, and the pursuit of practical solutions to issues of pressing concern to people, and more generally the flourishing of individual persons and their communities" (Reason and Bradbury 2008, p. 4).

The 2018 Cologne Week consisted of 13 events in eight days, focusing on garbage reduction and bee protection. Baha'i, Buddhist, Catholic, Muslim, Protestant, and Sikh organizations participated. The Baha'i and Sikh communities also held open-door events, where the public could become acquainted with the respective religion as well as the environmentalist projects of the worldwide communities to which the Cologne groups belong. Individuals from other religious communities operated behind the scenes.

Friends of the Earth Cologne, the city's garbage service, a bee protection program, and an organization that redistributes unsaleable but still edible groceries also participated.

After the week ended, I conducted semi-structured interviews with the project's leaders to tease out the intentions and attitudes behind the project's promotional materials (Lamnek 2010, p. 456). I also held similar interviews with 18 of my collaborators in Cologne (in German; quoted material that follows was translated by me). All interviews were coded and analyzed thematically. Not all collaborators had participated regularly in planning meetings, if at all, since the organization of individual events often took place within the hosting religious communities.

Those interviewed in Cologne were born in Germany and raised Christian (mainly Lutheran Protestantism or Catholicism), except for a Cologne-born Sikh representative and a Turkey-born Sunni Muslim. Thirteen interviewees represent religious communities: Protestantism (two), Catholicism (three), Baha'i (two), Buddhism (three), Brahma Kumaris (one), Sikh (one), and Islam (one). Five represent environmentalist or nature conservationist groups. Eight of the 18 interviewees are women. All interviewees completed secondary education; five, an additional occupational training; thirteen, a degree of higher education. Fourteen are regularly employed; two are full-time students. Two-thirds of the interviewees are over 50, mainly married with grown children.

First Tensions: The "Religions for Biological Diversity" Project Begins

The "Religions for Biological Diversity" project started in February 2015 with a dialog forum in Bonn. Ninety people attended the two-day event, including myself. The forum began with lectures that focused on abstract, universal ethical principles related to environmentalism, balanced by the religion-specific contributions of representatives of Alevism, Baha'i, Buddhism, Christianity, Hinduism, Judaism, Islam, Sikhism, and Yazidism[7] (Interkultureller Rat 2015).[8] Four subprojects were proposed: an interreligious network, the "greening" of religious communities' properties, religion and nature conservation dialog teams, and an interreligious week for nature conservation. A *Joint Declaration on Religions for Biological Diversity* was also discussed. This document states that the goal of both the German National Strategy for Biological Diversity and the UN Decade on Biodiversity is to protect the integrity of nature; indeed, this has always been an essential component of the ethics of the world's religions as transmitted through stories, rules, and modes of behavior (Abrahamisches Forum in Deuschland e.V. 2015). Overall, the public greeted the forum with enthusiasm. Yet the project was delayed due to internal politics and funding issues, finally getting off the ground in early 2017 with a highly reduced budget.

Given that the project seeks to connect religious communities with nature conservation groups, the latter of which are usually officially non-confessional, the *Joint Declaration* had been formulated so as to appeal to as many disparate groups as possible, both religious and secular. Yet the sought-after concordance was marred by signs of religio-political discord. One unnamed group (the Alevi, I later learned from an employee of this community) at first refused to sign the *Declaration*, fearing that doing so would have negative repercussions for Alevis in Turkey, where they regularly face persecution. We were also told that a Jewish representative was prohibited from attending the interreligious conference. In the background was the tension generated by the Israel-Gaza conflict the year before. Finally, Irfan Ortac of the Yazidi Community of Hesse stated that the paradisiacal Earth that the Divine had granted to Adam had become a living hell for the Yazidi, who were facing genocide at the hands of the Islamic State. All three of these conflicts involve Islam in some form. This would presage what was to come.

NourEnergy and Islamophobia in Darmstadt

Fast forward two and a half years to a meeting after the 2017 Darmstadt Week, where the collaborators discussed the Week's successes and possible areas of improvement. I mentioned my attendance at an event by the Muslim environmentalist organization NourEnergy e.V.[9] NourEnergy was founded in 2010 in Darmstadt, Hesse, with a focus on solar energy, and has expanded its expertise to rainwater management, energy efficiency, and permaculture. In 2013, NourEnergy put solar panels on two mosques, including the Emir Sultan Mosque in Darmstadt. After I made a positive comment about the group, a Jewish representative replied tensely that NourEnergy was "problematic." She did not elaborate on her cryptic statement, nor did anyone ask for explanation.

The source of her tension was probably a news article published in January 2017. NourEnergy had won a prize for its social engagement from the renewable energy supplier Entega, which has its headquarters in Darmstadt. Yet on January 27, 2017, the Hessian radio program HR-Info reported that NourEnergy had been stripped of its prize. This news item was immediately reported in numerous other outlets. Following headlines such as "Company Withholds Prize: Proximity to Extremists Suspected," these articles explained that NourEnergy had held joint events with the Bilal Center in Darmstadt. The Bilal Center has environmental and fair-trade projects, works against Islamic radicalization, and gives lectures on racism and anti-Semitism. Yet according to an unnamed speaker for the Hessian Office for the Protection of the Constitution (*Landesamt für Verfassungsschutz*, or LfV),[10] the center had been classified as "extremist." Suddenly, NourEnergy appeared to the broader public not as an environmentalist organization making a positive contribution to society, but rather as a double-dealer, showing an accommodating face to the majority society and another to an Islamic fundamentalist base, the aims of which run contrary to those of the society in which it lives.[11]

This story continued to haunt the Darmstadt Week throughout 2018. The Week was held as well in Cologne and Osnabrück, and in all three cities, it ended on September 9. Two days later, the Abrahamic Forum sent out a press announcement, filled with good news: over 50 events were held throughout Germany, at which people from across the religious spectrum acted peacefully together to conserve nature. A "high point" in the Week was the accolade awarded by the UN Decade on Biodiversity in Germany to the Turkish-Islamic Community of Darmstadt. Its Emir Sultan Mosque had partnered with a landscape architect from Nature and Biodiversity Conservation of Germany (NABU). Following an interreligious tree-planting event during the 2017 Darmstadt Week, the landscape architect planted indigenous bushes and flowers as homes for native birds and insects. On the roof, he erected nest boxes for bats and a tower for a falcon. At the awards ceremony, the Turkish general consul of Frankfurt, the president of the Central Council of Muslims in Hesse, and a councilwoman of Darmstadt spoke (Abrahamisches Forum in Deutschland e.V. 2018).

Three days later, a representative of ZGV sent an email to the Darmstadt Week mailing list about a news report on the Week. Claudia Kabel, a journalist for the *Frankfurter Rundschau*, reported on an anonymously written "Open Letter to the Organizers of the Religious Nature Conservation Week Darmstadt from 2 to 9 September 2018," sent both to her paper and the *Darmstadter Echo*, but not to the Week's organizers. The anonymous letter writer had accused the Muslim organizations participating in the Week of a "lack of differentiation via-a-vis Islamist-fundamentalist ideas with links to terrorist communities" and mentioned the withdrawal of the Entega prize from NourEnergy. The problem was no longer the joint events with the Bilal Center; rather, NourEnergy allegedly had "intensive connections to the militant Muslim Brotherhood." Furthermore, the Al-Rahma Mosque, which had cooperated with the Bilal Center on an event to greenify mosques, had supposedly been deemed by the LfV 3 years previously to be "influenced by Islamic fundamentalism and thus classified as anti-constitutional." The Christian Matthäus Congregation had partnered with the Association of Islamic Information & Services (IIS) in Frankfurt, which, "according to media reports and blog entries," was also being monitored by the LfV. Only at the end of the article did Kabel mention the 50 events of the interreligious weeks for nature conservation that had taken place across the country (Kabel 2018).

Kabel did not question whether the Al-Rahma Mosque was still being monitored, nor did she indicate whether the blog entries were written by credible scholars or by self-named experts with only a layperson's knowledge of Islam.[12] I found no mention of the participating Darmstadt Muslim groups in the LfV Reports for the years 2014–2017. Kabel did talk to Jürgen

Micksch, the director of the Abrahamic Forum, who explained that he had always worked with groups being monitored by the LfV and even served on a committee to have them removed from the report. After all, the organizations that are monitored by this office often change over time. He lamented the "rumors that are constantly being spread" that insinuated that most Muslims are extremists.

The ZGV representative included in his email part of his response to the *Frankfurter Rundschau*, where he had pointed out that the accusation regarding NourEnergy's links to the Muslim Brotherhood were "absolutely inadmissible" because wrong and already long clarified. Rather, the "intensive connections" were, in his opinion, an exaggeration of the journalist herself. He also indicated that the vague term "contact guilt" (*Kontaktschuld*) had been used in the context of the letter. He then quoted the anonymous letter, which suggested that the same Muslim groups promoting environmentalism in Germany were involved in Islamic fundamentalist regimes and the persecution of Jews: "Is the installation of a solar system on the roof sufficient for the Protestant and Catholic Church as well as the Jewish community to join forces with organizations that elsewhere demand the introduction of Sharia law and prefer to drive Jews into the sea?" The ZGV representative lamented the question's "lack of differentiation." He then underscored to the Darmstadt group his belief that an interreligious project for nature conservation could counteract the kinds of anti-democratic tendencies in contemporary society reflected in the anonymous letter.

DITIB and Concerns Over National Sovereignty and Integration in Cologne

Tensions concerning another Muslim association have emerged in Cologne and center around DITIB (*Diyanet İşleri Türk İslam Birliği*). Founded in 1984 in Cologne, DITIB is the largest organization of Muslims (specifically of Turkish descent) in Germany and has over 900 mosque communities there. It has close ties to the Office for Religious Affairs (*Diyanet*) in Ankara, which educates and pays imams to work for 4-year stints in Germany (Kiefer 2012). A DITIB employee claims this is of benefit to Germany, yet a one-time Catholic turned atheist environmentalist I interviewed questioned whether the imams are "indoctrinated" to preach against "certain populations" (without clarifying which ones).

DITIB is mentioned in the 2019 Report for the Federal Office for the Protection of the Constitution (*Bundesamt für Verfassungsschutz*, or BfV) in the section on "espionage and other intelligence service activities," along with the Turkish umbrella group Union of International Democrats (UID). According to the BfV, both groups "behave emphatically modest in front of the public and attempt to highlight the autonomous and independent character of their organizations and to downplay the connections and dependencies with Turkey" (Bundesamt für Verfassungsschutz 2019,

p. 310). However, apparently only UID is being monitored because of its alleged ties to Turkish intelligence services: After the aforementioned quote, DITIB is no longer mentioned, nor is it listed in the register's annex, which names all the groups that the BfV considers anti-constitutional. Still, despite DITIB's absence from the BfV Report, DITIB is regularly accused of being the "long arm of [Turkish president Recep Tayyip] Erdoğan" in Germany (Keller 2018).

Critique against DITIB has been manifold. For example, at a neighborhood meeting for a city greening project, I discussed the upcoming interreligious week for nature conservation. Opposition was expressed to DITIB's involvement by two members of the meeting. One person, a left-leaning Catholic nature conservationist, was incensed over Turkey's invasion of Syria on January 20 and its attack on the Kurdish enclave Afrin. Several weeks later, a Catholic adult educator participating in the Cologne Week, who also has several years' experience of working with DITIB, called me to read out loud an article in the local newspaper. It reported that DITIB was organizing a youth trip to visit President Erdoğan in Turkey. Rather than referring to him as "president" (Turkish, *cumhurbaşkan*), however, DITIB had called him "our supreme military commander" (*Başkomutan*). Volker Beck, a former Green Party representative and religious studies instructor at the University of Bochum, accused DITIB of using religion as a facade for the promotion of "nationalist state propaganda" (*Kölner Stadt-Anzeiger* 2018a). Finally, the atheist environmentalist reported that DITIB has been accused of harassing people of Turkish descent within Germany for a lack of loyalty toward Turkey. Because of these kinds of reproaches, he and the Catholic educator[13] both expressed concern that DITIB might instrumentalize environmentalist projects to improve its public image.

The fears around DITIB as Erdoğan's agent were reinforced by the official opening of DITIB's central mosque on 29 September 2018. Erdoğan had been invited to speak; no German politicians were present. It was also reported that Cologne's mayor Henriette Reker cancelled her appearance after being told only the evening before that she could speak at the event (Kölner Stadt-Anzeiger 2018b). My DITIB interlocutor claimed the previous chairman had invited the German president Frank-Walter Steinmeier, but failed to pursue the matter. DITIB had also refrained right away from extending an invitation to Reker to speak at the mosque because, from DITIB's perspective, she had created a media scene about the event. It was also reported that the entire event was held exclusively in Turkish. With no German translation, it was impossible for non-Turkish-speaking people to participate. Karsten Fiedler, the editor-in-chief of the local newspaper, reported that many Cologne politicians had expected the DITIB Central Mosque to become a center of interreligious dialog[14] and a sign of integration. Indeed, my DITIB interlocutor argued that Erdogan opened the mosque "not as a sign of isolation . . . [but] of cohesion and coexistence," yet Fiedler instead argued that the event had turned the mosque into a symbol of seclusion. DITIB, he claimed, had damaged "the integration and the relationship between German Turks and [German] Germans" (Fiedler 2018).

I talked to two Catholic social workers whose youth center had participated in the 2018 Cologne Week and is near DITIB's central mosque. Both expressed their dismay at the opening event, not only because German politicians but also non-Muslim religious leaders were not invited. One felt that Erdoğan had split Islam in Germany. The consecration of the mosque appeared to be more of a political event than a religious one, which was problematic for the many Muslims of Turkish descent in Germany who do not support the Turkish president. The other asserted that DITIB's exclusionary actions handicap her ability to convince the young clients of her center to accept and tolerate other religions and cultures. She said, "Then I understand . . . that some people say: 'I'm afraid of Islam' . . . because we have another [way of] thinking . . . about freedom and about acceptance of the other." It is necessary to stay in dialog, she argued, if DITIB wants a peaceful co-existence in Germany.

When I asked my DITIB interlocutor about the lack of German translation at the mosque's consecration, he did not answer directly; instead, he spoke about the Balkans, where Turks lived for 500 years among other ethnic and linguistic groups. Every people spoke their own language and practiced their own religion. "That's how they learned to live together," he argued. However, Germany has long cultivated the self-image of being a people of "poets and philosophers." Yet Turks are not "mountain people;" rather, they, too, "produced an advanced civilization," which Germans have apparently forgotten. In line with DITIB's reported stance on promoting a Turkish-linguistic Islam (Kiefer 2012), he argued that Turkish immigrants simply wish to cultivate their own language and religion. He agreed with me that Turkish native speakers should learn German for their daily affairs. But modern states have constitutions, which is the "basis for living together." As long as people uphold the constitution, he queried, why should it be problematic if they speak other mother tongues and practice other religious and cultural traditions? "That isn't a problem for me," he said, "but abundance."

As with NourEnergy, the story of DITIB has a sequel. DITIB has participated in both the 2018 and 2019 Cologne Weeks. For the 2019 Week, I was invited to the Friday prayer service to speak at a podium discussion in connection to the planned "greening" of the mosque's open plaza. When I arrived, I was surprised to hear the imam speaking about Islam and nature conservation not only in Turkish, but also in German. My DITIB interlocutor told me that Friday prayer services have been held in both languages for some years. Furthermore, although both the Turkish and German press had been invited to the greening event, only the Turkish press showed up. I had not attended the consecration of the mosque the year before and cannot therefore verify what happened. However, both the example of NourEnergy and of DITIB suggest that one must be careful about relying on media reports concerning Islam in Germany.

Islamic Fundamentalism and Fears of Funding Loss in Cologne

Cologne had been chosen to be the sponsored city in 2019, therefore receiving BfN funding. After the events planning had come to a close in May, the Week's brochure was sent to BfN to be reviewed. There, two events were struck from the brochure, because the hosting community, Islamic Community (*Islamische Gemeinschaft*) Millî Görüş (IGMG), had been found listed in the BfV 2018 Report in the section on "Islamism/Islamic extremism." The BfV Report states that IGMG is a member of the larger umbrella organization Millî Görüş ("national vision"), which was inspired by Necmettin Erbakan (1926–2011). Once the Prime Minister of Turkey (1996–1997), he was also a critic of western influence in Turkey, and of state secularism. Millî Görüş promulgates the political-religious ideology that regimes based on "divine revelation" should replace those invented by humans. More specifically, the West is seen as a regime promoting violence, injustice, and domination, it should therefore be replaced by a "just regime" based exclusively on Islamic principles. Through this ideology, Millî Görüş unites smaller associations by varying degrees, which nevertheless operate independently. IGMG still has connections to the main movement in Turkey, but has been dissolving them over the years. References to extremism have also decreased in the group across Germany, and the group's current head is trying to create an independent profile with a focus on religious education (Bundesamt für Verfassungsschutz 2019).

I discussed the removal of the events from the Cologne brochure with a collaborator who once worked for the Federal Ministry for the Environment, Nature Conservation and Nuclear Safety (BMU). This ministry funds BfN, and the collaborator played a leading role in creating the "Religions for Biological Diversity" project. She explained that, although this project is small, its funding ultimately belongs to the federal budget, which is debated by the German Parliament. Were it suggested in any way that Islamic fundamentalists had participated in the "Religions for Biological Diversity" project, the project could be stripped of its funding.

Stereotyping of Muslims in Cologne

These examples involving specific groups point to concerns with Islam more generally in the non-Muslim German population. These concerns were expressed in the interviews I held with my collaborators, and they centered around the issues of gender and sexuality as well as secularism. In this regard, the concerns expressed often intersected with similar concerns about Christianity and religion as a whole. For example, one Baha'i member insisted that the Koran advocates gender equality, but found that the behavior of Muslim men spoke otherwise. Moreover, two women described negative emotional reactions aroused by the veiling of Muslim women.[15] One of these, a nature conservationist, had referred to two headscarf-wearing Muslim environmentalists she had just met at an event as appearing "conservative," in part because they had, in accordance with German social rules, used the formal form of "you" (Sie) to an adult person not yet of acquaintance. In the environmentalist circles that the nature conservationist operated in, it was more common for people to use the informal form of "you" (du). Another critic of the headscarf was the previously mentioned Catholic social worker. She also mentioned Muslim families of her acquaintance where the daughters were oppressed due to their gender. Her outspokenness about gender inequality in Islam derived in part from witnessing teenage boys in her hometown's youth center radicalize and turn Salafist. "I reach my limit," she said, "when I realize that there is a current or a direction that is very misogynistic," denying women human dignity and equal rights. Finally, a Protestant adult educator mentioned a frightening attack on women on New Year's Eve 2015/2016 at Cologne's main train station, where over 600 women departing the station were groped and robbed by groups of young men, many of whom appeared to be Arabic or of North African descent (Michel et al. 2016; Diehl et al. 2016). He said this mass attack increased the "anti-Muslim affect" in Cologne.

Critique of Christian Sexual Morality in Cologne

Christianity also received criticism for its sexual morality. For example, one ecumenical environmentalist mentioned antiquated prohibitions on sex before marriage and religiously mixed marriages. The Catholic-turned-atheist was incensed about a woman in a state of emergency being denied "the pill" in a Catholic hospital, and another male environmentalist denounced the relative lack of women's right to be heard in the Christian ecumenical movement. He then added, "Of course, something comes to mind again that I heard about the religion of Islam, but I have to say, [it] speaks for Christianity almost as well: religion **is** in most cases always something **of** men, **for** men" (emphasis in original). Finally, scandals related to Catholic leaders' sexual abuse of children and attempts to cover up the crimes (BBC News 2019) were criticized by the Catholic adult educator. He admitted that this horrific news had probably led many people to think regarding the Church, "I don't want any more dealings with this business." Such scandals cause people to lose sight of the religion's positive aspects, he said, or to fear they will look bad should they continue to be involved with the Church.

Mistrust of Religion and Support for Secularism in Cologne

Finally, religion as such came under fire during the interviews I conducted with collaborators in Cologne. For example, the atheist environmentalist said that the mixing of religion and state runs contrary to the norms of German society, while also admitting that its secularization remains unfinished.

"We live more or less in laicism here," he said, "[although] the French are in fact much further than we are, because here we still have the church tax. So one can also say, 'we are not really that laicist.'" However, he continued, separation of church and state is the ideal. This is not the case in Turkey, which makes DITIB's mixing of the two problematic. The Catholic adult educator explained that a historical phase of the church, in which one was a member by birth, was coming to an end. A new phase based on free association had begun, but this demands reflection about one's faith. German Christians, however, "have not yet developed their own freedom and their own competence to reflect why [and] what to believe," leaving them feeling uncertain. This uncertainty is presumably compounded by pastoral alienation from contemporary society, a problem mentioned by other interviewees. These issues would then contribute to a growing indifference to the Church, as reflected in people cancelling their church membership and in falling church attendance rates (Gabriel 2015). As the ecumenical environmentalist quipped, German Christians typically "go only three times per year to church: Easter, Christmas . . . and for a funeral."

Secularism as a principle governing state and society was touted most strongly by nature conservationists – but not the ones that came to the planning meetings. As the organizer for the Cologne Week, I was regularly faced with many environmentalists' skepticism around religion and even the outright refusal to partner with the Week. Some groups wished to remain neutral vis-à-vis any would-be "ideology"; for example, a community center refused to let us rent its space for the opening event, because it does not rent to any groups representing either religion or politics. Another time, I met with two men from Friends of the Earth Germany, one a Christian, who told me privately that he tends to keep quiet about his faith when around other members of the organization. The second man spoke disdainfully of religious groups that hold to "anti-democratic principles," without naming which ones he had in mind. For the 2019 Week, I had to justify why the events should be listed in the online calendar of the city of Cologne. The employee responsible for the calendar had called to tell me that his supervisor, who oversees environmental management in Cologne, had "certain reservations." "Yeah, yeah," I replied, "'What does religion have to do with nature conservation?'" "Exactly," he said. This response did not surprise me: After all, this was also reportedly the mayor's first reaction when a Buddhist representative of the Council of Religions presented the upcoming 2018 Week at the Council's spring meeting. In the end, the Council voted not to support the Week.

Two nature conservationists I interviewed from Friends of the Earth reported that some of their colleagues consider themselves to be "absolutely irreligious," immediately rejecting anything that has to do with religion. One of the two conservationists explained that her colleagues consider religion as something that they have "overcome." For example, they reproach the Church as a force that "basically prevented progress." They point to "how long it has taken for the church and the state to really be separate – they are still not properly separated – but how long has it taken that the church does not have that much influence." Moreover, "[i]t makes you feel particularly progressive when it's no longer as in some Muslim countries," by which her colleagues apparently meant "age-old traditions, which do not really fit into modern times any more" but "continue to be sustained in the name of religion." An environmentalist from another group also described his colleagues' reactions as deeply negative: they recoiled from the very word, he said. I told him this matched my own observation when I talked to nature conservationists about the Week at an event for bee protection, one of whom literally jerked back when I mentioned the word "religion." This corporeal rejection, he explained, stemmed from the association of religion with a set path that one may not leave; whoever is not on that path will be targeted for missionizing. Should an environmentalist group cooperate with a religious community, he continued, the once-independent organization would fear being taken over by religious advocates and their members "**indoctrinated**" (emphasis in the original).

Conclusion: Divergences and Overlaps in the Tensions around Islam, Christianity, and Religion

I began this chapter by explaining how the promotional materials for the "Religions for Biological Diversity" project assert that nature conservation is a shared value across religions. The project uses this claim to bring together a wide spectrum of actors, not just those in different religious communities, but also employees in the public sector, secular environmentalist organizations, and even the general public. It seeks to balance universal ethical statements with religion-specific ones on nature conservation (cf., Johnston 2013), in order to create the image of a collective "us" across religions and the religion-secular divide. This ideal is turned into lived experience at dialog forums (Frisk 2015).

Yet the tensions in a largely secularized Germany around Christianity and Islam counteract this attempt. Indeed, the focus on nature conservation has in some ways even brought them more to the fore, as the resistance to and even outright rejection of religion by many nature conservationists indicate. Moreover, the criticism reveals a tendency in many people – even those interested in and open to interreligious dialogue – to view religions as stereotyped wholes, rather than distinguishing among individuals and themes.

As with many of the interreligious initiatives in Germany, concerns around Islam (Klinkhammer et al. 2011) have dominated the "Religions for Biological Diversity" project. This is so, despite the fact that nine different religions are represented on the advisory council. The tensions around Islam in Germany span the spectrum from justifiable fear of radicalism and terrorism to blanket stereotypes and to outright bigotry. Despite the project's attempt to form common ground between different social and religious groups in

Germany, the tensions around Islam expressed in private and in newspapers and blogs correspond to those that drive non-environmentalist interreligious dialog initiatives in Germany. These do not concern how best to preserve nature, but how to maintain social cohesion, integrate social newcomers, effect social participation, and improve the status quo (Klinkhammer 2008; Ohrt and Kalender 2018; Satilmis 2008).

No doubt there are Islamic fundamentalists in Germany who have committed acts of terror, as well as groups that promote a radical form of Islam and oppose the German constitution. Yet over 90% of German Muslims support democracy and find it a proper form of government (Pickel 2019). Hence, one can speak of the reality of Muslim life in Germany, as well as of a "reality" that in part exists only in the collective German imagination (Rohe 2016, p. 13).

Moreover, it is not only the justifiable fear of Islamic fundamentalism but also Muslims' successful integration that are grounds for increased social tensions. Muslims have moved from the margins of society as ethnically designated (usually Turkish) "guest workers" in the 1970s–1990s (Kiefer 2012; Spielhaus 2013), to citizens and actors in all segments of German society today, with their own organizations established in accordance with German law. Hence, they are now visible in a way they never were before. Having moved into the public sphere, they have begun adding their voices to the discussion about how that space should be arranged. Metaphorically speaking, as more people have found a seat at the collective dining table, they not only expect their fair share of the meal, but have even suggested an expansion of items on the menu (El-Mafaalani 2018). This may be exemplified by the DITIB interviewee defending his right to the institutional cultivation of his native language and customs as well as NourEnergy promoting a scientifically sound form of environmentalism that accords with Islamic principles.

Yet this is not just a story about Islamophobia and the fear of those deemed foreign, as the critiques of Christianity and the resistance to religion as such make clear. Rather, it entails a struggle to define the foundation of German culture today as a secular state. The resistance expressed by many nature conservationists to supporting the Week indicates the view that religion should be unseen and unheard, private not public, because its faith elements are seen as incompatible with a rationalized modern society. This fear, already expressed regarding Christianity, becomes stronger vis-a-vis Islam. The specter of Islamic fundamentalism and the mixing of religion and politics conjure the fear of a regressive move to a society dominated by religious ideology, where the free choice of the individual is constrained by undemocratic social controls and indoctrination.

Mistrust of the other is the underlying element of the tensions described in this chapter. And trust is an issue that is sometimes only implied and other times explicitly addressed in the literature reviewed previously. To what degree this project can speak to this literature in the future depends not only on whether the project grows and stabilizes, but on what the conditions are for organizing, planning, and collaborating in each city. For example, self-reflection and a better understanding of others (Klinkhammer and Satilmis 2008) has been generated even at one-off events, but the development of mutual respect and debate among equals (Malik 2008; Satilmis 2008) will probably only be realized in the context of regular planning meetings that participants commit to attending. So far, this has not been the case in Cologne. The lack of regular face-to-face encounters means also that "worldview translation" (Johnston 2013, p. 150) has only taken place at a few events (such as the aforementioned open-door events of the Sikh and Baha'i communities). At other events, nature conservationists have been able to make their terminology available to religious actors (cf. Chungprampree 2016; Fonseca et al. 2018) interested in learning about practical nature conservation, yet the terminology favored by religious actors often alienates nature conservationists (cf., Marshall 2016); even the word *religion* has engendered strong negative reactions.[16] It may be possible to investigate these issues should the "Religions for Biological Diversity" project win another 3 years of funding.

Given the tensions that I have spelled out here in this chapter, may one surmise that the unity within diversity being promoted by the "Religions for Biological Diversity" project is an unachievable goal? In answer, I point to the over 50 events that took place across Germany during the second year of this project. Although many of these events were religion-specific, being held by the members of a given religion within that religion's quarters, these were open to the public. Furthermore, many other events were hosted by interreligious groups. When Muslims are involved in such events, they are surely perceived as an imminent danger by those who believe that any good deed by a Muslim is a façade masking the much more sinister plot to realize an Islamic fundamentalist takeover of Germany. That holds as well for those defenders of secularism who see religion as such (in Germany, meaning first Christianity and then Islam) as regressive, antimodern, and undemocratic. However, one could argue that the dogmatism of both the extreme anti-Muslim and the extreme secular positions pose as much of a threat to the principles of an open, pluralistic, democratic society as does any form of religious fundamentalism, Islamic or otherwise.

Notes

Thanks to Jiska Gojowczyk and Sofiah Jamil for their comments on earlier versions of this chapter.

1 Examples include Jews Go Green; NourEnergy and Hima e.V., both Muslim; the German Buddhist Union Committee on Buddhism and Environmentalism; and the church-based Green Rooster environment management program.

2 In chronological order: Orth 2002; Saan-Klein and Wachowiak 2008; Miksch et al. 2015; Singh and Steinau-Clark 2016. The Protestant Academy of Loccum held an interreligious conference called "The Contribution of the Religions for a Sustainable Way of Life" in 2003, partly funded by the Federal Ministry of

the Environment, and another in 2013, called "Act Ecologically, even in God's Name!"

3 Twenty percent of the German population has "a migration background," meaning that at least one parent has not had German citizenship since birth (El-Mafaalani 2018).

4 The project has produced postcards, one for each of the nine religions represented in the project, all of which carry the same message on the back side: "The protection of nature is a common task of the religions" (Abrahamic Forum, n.d., my translation).

5 Proposed in 1989, Creationtide begins on September 1, the beginning of the Orthodox ecclesiastical calendar, and ends on the Day of St. Francis, since 1979 the patron saint of those who "promote ecology" (John Paul II 1990). In 2009, the ecumenical Consortium of Christian Churches in Germany (Arbeitsgemeinschaft Christlicher Kirchen in Deutschland, or ACK) voted in the new holiday. Creation Day takes place on the first Friday of September; Creationtide concludes with the Harvest Thanksgiving on 4 October (Arbeitsgemeinschaft Christlicher Kirchen in Deutschland n.d.a, n.d.b).

6 The project's advisory board decided that "religious" is synonymous with "interreligious." (Interview notes from February 20, 2019). The Cologne group flatly rejected this proposition, arguing that "religious" in Germany implies "Christian" and thus voted to call the Week "interreligious."

7 In Germany, Alevism is recognized as a religion independent of but related to Islam (Deutsche Islam Konferenz 2009). The nine listed religions are those considered to have a public presence in Germany. Members of each sit on the project's advisory board (interview notes, February 20, 2019). The two dominant Christian confessions in Germany, Roman Catholicism and Lutheran Protestantism, are represented by a member of the ACK.

8 Both the lectures and the individual religious contributions may be found in the cited source.

9 E.V. stands for "eingetragener Verein," or "registered association." Most non-Christian religious communities in Germany are registered associations, as are, say, sports clubs and environmentalist groups.

10 One has to question why the speaker was not named, since one could justifiably assume that it is a speaker's job to communicate an agency's official position to the public.

11 This is the insinuation of the blogger Sigrid Herrmann-Marschall (2016, 2019), a self-styled independent "expert" on sects and Islamic fundamentalism. See footnote 12.

12 For example, the previously mentioned Herrmann-Marschall has been criticized by German scholars of Islam as having only a layperson's knowledge of Islam and of accusing people of Islamic fundamentalism without evidence. Yet German officials have been known to listen to her, leading to Muslim employees being suspended from their jobs before the soundness of her charges had been examined (Ramadan 2017).

13 In another conversation with the Catholic adult educator, I mentioned the prize-winning project concerning the Emir Sultan Mosque and NABU. The mosque's youth were so impressed by the falcon that it has since become the mascot for the mosque. I saw this enthusiasm as a normal reaction on the part of kids; my Catholic interlocutor, however, replied that the eagle is a symbol of rulership in Arab culture. True; but a falcon is not an eagle and the Emir Sultan Mosque is for people of Turkish descent, not Arab.

14 DITIB engages in several interreligious initiatives, such as the Christian-Muslim Peace Initiative of Germany. Furthermore, my contact at DITIB is responsible

for interreligious dialog. He is also a member of the Cologne Council of Religions and has been part of the "Religions for Biological Diversity" project since its first dialog forum in 2015.

15 This is hardly surprising, given the prevalent use of pictures of hijab-wearing women to symbolize the immigrant needing integration (Spielhaus 2013).

16 After finalizing this chapter, I attended the Project's Dialog Forum, "Celebrating Nature," on October 17–18, 2019 in Mainz. Apparently, the resistance to religious engagement in nature conservation is not particular to Cologne, because it was encountered by the organizers of these Weeks (and Days), whether in Osnabruck, Darmstadt, or Wetterau.

References

Abrahamisches Forum in Deutschland e.V., n.d. *Postcards for Nature Conservation.* Darmstadt.

Abrahamisches Forum in Deuschland, e.V., 2015. *Joint Declaration on Religions and Biological Diversity.* [pamphlet]. Darmstadt.

Abrahamisches Forum in Deutschland, e.V., 2018. *Pressemitteilung: Großes Interesse an Religiöser Naturschutzwoche.* [Received as email attachment. 13 September 2018].

Arbeitsgemeinschaft Christlicher Kirchen in Deutschland., n.d. a. *Geschichte und Anliegen* [online]. Available from: www.oekumene-ack.de/themen/glaubenspraxis/oekumenischer-tag-der-schoepfung/geschichte-und-anliegen/ [Accessed 20 May 2015].

Arbeitsgemeinschaft Christlicher Kirchen in Deutschland., n.d. b. *Was wir tun* [online]. Available from: www.oekumene-ack.de/themen/glaubenspraxis/oekumenischer-tag-der-schoepfung/geschichte-und-anliegen/ [Accessed 20 May 2015].

Baugh, A., 2017. *God and the Green Divide. Religious Environmentalism in Black and White.* Oakland: University of California.

Bauman, W., Bohannon II, R., and O'Brien, K., 2011. *Grounding Religion. A Field Guide to the Study of Religion and Ecology.* Abingdon, Oxon: Routledge.

BBC News, 2019. *Catholic Church Child Sexual Abuse Scandal* [online]. Available from: www.bbc.com/news/world-44209971 [Accessed 14 April 2019].

Bundesamt für Verfassungsschutz, 2019. *Verfassungsschutzbericht 2018.* Berlin: Bundesministerium des Innern, für Bau und Heimat.

Chungprampree, S., 2016. Spirituality and Ecology from a Buddhist Perspective: Engaged Buddhism Across Asia. *The Journal of Interreligious Studies*, 19, 35–40.

Davie, G., and Wyatt, D., 2014. Document Analysis. *In:* M. Stausberg and S. Engler, eds. *The Routledge Handbook of Research Methods in the Study of Religion.* Oxon: Routledge, 151–160.

Del Vecchio, K., 2018. Interreligious Environmentalism. Pragmatic Projects and Moral Competencies that Address Climate Change. *The Journal of Interreligious Studies*, 22, 46–63.

Deutsche Islam Konferenz, 2009. Gott, Muhammad und Ali: Alevitischer Religionsunterricht in Deutschland [online]. Available from: www.deutsche-islam-konferenz.de/DIK/DE/Magazin/IslamBildung/AlevitischerRU/alevitischer-unterricht-node.html [Accessed 30 April 2017].

Diehl, J., Kollenbroich, P., Stenzel, K., and Ternieden, H., 2016. Silvester-Übergriffe und die Folgen: Eine Nacht in Köln. *Spiegel* [online]. Available from:

www.spiegel.de/panorama/justiz/koeln-was-in-der-silvesternacht-am-bahnhof-geschah-a-1070625.html [Accessed 13 April 2019].

El-Mafaalani, A., 2018. *Das Integrationsparadox. Warum gelungene Integration zu mehr Konflikten führt*. Cologne: Kiepenheuer & Witsch.

Fiedler, C., 2018. *Kommentar zu Erdogan-Besuch in Köln Die Moschee ist nun ein Symbol der Abschottung* [online]. Available from: www.ksta.de/politik/kommentar-zu-erdogan-besuch-in-koeln-die-moschee-ist-nun-ein-symbol-der-abschottung-31372362 [Accessed 13 April 2019].

Fonseca, R., Villela, M., and Vogas, A., 2018. Fé no Clima: Faith Communities and Climate Change. *The Journal of Interreligious Studies*, 19, 14–23.

Franke, E., and Maske, V., 2011. Teilnehmende Beobachtung als Verfahren der Religionsforschung. *In:* S. Kurth and K. Lehmann, eds. *Religionen erforschen: Kulturwissenschaftliche Methoden in der Religionswissenschaft*. Wiesbaden: VS Verlag für Sozialwissenschaften, 105–134.

Frisk, K., 2015. Citizens of Planet Earth: The Intertwinement of Religion and Environmentalism in a Globalization Perspective. *Journal for the Study of Religion, Nature & Culture*, 9 (1), 68–86.

Gabriel, K., 2015. Christen in Deutschland: Zunehmend umorganisierte Randgruppe oder systemrelevanter Akteure? *In:* M. Rohe et al., eds. *Christentum und Islam in Deutschland. Grundlagen, Perspektiven und Erfahrungen des Zusammenlebens*. Freiburg: Herder, 14–38.

Garsten, C., 2010. Ethnography at the Interface: 'Corporate Social Responsibility' as an Anthropological Field of Enquiry. *In:* M. Melhuus, J. Mitchell, and H. Wulff, eds. *Ethnographic Practice in the Present*. New York: Berghahn Books.

Gottlieb, R., 1996. *This Sacred Earth. Religion, Nature, Environment*. London: Routledge.

Harris, I., 1991. How Environmentalist is Buddhism? *Religion*, 21 (2), 101–114.

Harris, I., 1997. Buddhism and the Discourse of Environmental Concern: Some Methodological Problems Considered. *In:* M. Tucker and D. Williams, eds., *Buddhism and Ecology: The Interconnection of Dharma and Deeds*. Cambridge, MA: Harvard University Press, 377–402.

Herrmann-Marschall, S., 2016. *Erneubarer Islamismus* [online]. Available from: https://vunv1863.wordpress.com/2016/10/12/erneuerbarer-islamismus/ [Accessed 8 August 2019].

Herrmann-Marschall, S., 2019. *Die Gegengesellschaften und ihre Trugbilder* [online]. Available from: https://vunv1863.wordpress.com/2018/11/06/die-gegengesellschaften-und-ihre-trugbilder/ [Accessed 8 August 2019].

Interkultureller Rat, 2015. *Religionen und Naturschutz – Gemeinsam für biologische Vielfalt* [online]. Available from: www.interkultureller-rat.de/wp-content/uploads/Natursch.Einlad-22-1-2015.pdf [Accessed 10 March 2015].

John Paul II, 1990. *Peace with God the Creator, Peace with All of Creation. World Day of Peace Sermon*, 1 January 1990 [online]. Available from: www.vatican.va/holy_father/john_paul_ii/messages/peace/documents/hf_jp-ii_mes_19891208_xxiii-world-day-for-peace_en.html [Accessed 26 March 2013].

Johnston, L., 2013. *Religion and Sustainability. Social Movements and the Politics of the Environment*. Abingdon, Oxon: Routledge.

Johnston, L., 2014. Sustainability as a Global Faith? The Religious Dimensions of Sustainability and Personal Risk. *Journal of the American Academy of Religion*, 82 (1), 47–69.

Kabel, C., 2018. *Mangelnde Abgrenzung zum Islamismus* [online]. Available from: www.fr.de/rhein-main/darmstadt/evangelische-kirche-org26909/mangelnde-abgrenzung-islamismus-10959724.html [Accessed 13 September 2018].

Kalland, A., 2005. The Religious Environmentalist Paradigm. *In:* B. Taylor, ed. *Encyclopedia of Religion and Nature*. London: Continuum International, 1367–1371.

Keller, C., 2018. Moscheeverband Ditib: Der lange Arm Erdogans in Deutschland. *Tagesspiegel* [online]. Available from: www.tagesspiegel.de/politik/moscheeverband-ditib-der-lange-arm-erdogans-in-deutschland/13961072.html [Accessed 18 June 2018].

Kiefer, M., 2012. Die DITIB in der Zuwanderungsgesellschaft: Garant oder Hindernis der Integration? *In:* T. Schneiders, ed. *Verhärtete Fronten: Der schwere Weg zu einer vernünftigen Islamkritik*. Wiesbaben: Springer, 209–216.

Klinkhammer, G., 2008. Der Dialog mit Muslimen. *In:* G. Klinkhammer and A. Satilmis, eds. *Interreligiöser Dialog auf dem Prüfstand. Kriterien und Standards für die interkulturelle und interreligiöse Kommunikation*. Berlin: Lit, 21–46.

Klinkhammer, G., Frese, H., Satilmis, A., and Siebert, T., 2011. *Interreligiöse und Interkulturelle Dialoge mit MuslimInnen in Deutschland. Eine Quantitative und Qualitative Studie*. Bremen: University of Bremen.

Klinkhammer, G., and Satilmis, A., 2008. Einleitung. *In:* G. Klinkhammer and A. Satilmis, eds. *Interreligiöser Dialog auf dem Prüfstand. Kriterien und Standards für die interkulturelle und interreligiöse Kommunikation*. Berlin: Lit, 1–15.

Koehrsen, J., 2015. Does Religion Promote Environmental Sustainability? Exploring the Role of Religion in Local Energy Transitions. *Social Compass*, 62 (3), 296–310.

Koehrsen, J., 2018. Eco-Spirituality in Environmental Action. Studying Dark Green Religion in the German Energy Transition. *Journal for the Study of Religion, Nature and Culture*, 12 (1), 34–54.

Kölner Stadt-Anzeiger, 2018a. *Besuch bei 'unserem obersten Heerführer'. Ditib plant Jugendreise zu Erdogan* [online]. Available from: www.ksta.de/politik/besuch-bei – unserem-obersten-heerfuehrer – ditib-plant-jugendreise-zu-erdogan-29852680# [Accessed 13 June 2018].

Kölner Stadt-Anzeiger, 2018b. *Moschee-Eröffnung: Oberbürgermeisterin verteidigt Absage* [online]. Available from: www.ksta.de/nrw/moschee-eroeffnung – oberbuergermeisterin-verteidigt-absage-31370224 [Accessed 13 April 2019].

Küng, H., and Kuschel, K., eds., 1993. *A Global Ethic. The Declaration of the Parliament of the World's Religions*. London: SCM Press, New York: Continuum.

Lamnek, S., 2010. *Qualitative Sozialforschung*. Basel: Beltz.

Lysack, M., 2014. Stepping Up to the Plate: Climate Change, Faith Communities and Effective Environmental Advocacy in Canada. *In:* R. Veldman, A. Szasz, and R. Haluza-DeLay, eds. *How the World's Religions are Responding to Climate Change: Social Scientific Investigations*. Abingdon, Oxon: Routledge, 157–173.

Malik, J., 2008. Interreligiöser Dialog – ein Integrationswerkzeug? *In:* G. Klinkhammer and A. Satilmis, eds. *Interreligiöser Dialog auf dem Prüfstand. Kriterien und Standards für die interkulturelle und interreligiöse Kommunikation*. Berlin: Lit, 141–163.

Marshall, G., 2016. Communicating with Religious Communities on Climate Change. Research Overview and Emergent Narratives. *The Journal of Interreligious Studies*, 19, 24–34.

Michel, A.M., Schönian, V., Thurm, F., and Steffen, T., 2016. Übergriffe an Silvester: Was geschah in Köln? *Die Zeit* [online]. Available from: www.zeit.de/gesellschaft/

zeitgeschehen/2016-01/koeln-silvester-sexuelle-uebergriffe-raub-faq [Accessed 13 April 2019].

Miksch, J., 2008. Was bringen interkulturelle und interreligiöse Dialoge? *In:* G. Klinkhammer and A. Satilmis, eds. *Interreligiöser Dialog auf dem Prüfstand. Kriterien und Standards für die interkulturelle und interreligiöse Kommunikation.* Berlin: Lit, 83–100.

Miksch, J., Khurshid, Y., Meisinger, H., and Mues, A., eds., 2015. *Religionen und Naturschutz. Gemeinsam für biologische Vielfalt.* Bonn: Bundesamt für Naturschutz.

Ohrt, A., and Kalender, M., 2018. Interreligious Practice in Hamburg: A Mapping of Motivations, Conditions, Potential Benefits and Limitations from a Participant's Perspective. *In:* J. Ipgrave, T. Knauth, A. Körs, D. Vieregge, and M. Lippe, eds. *Religion and Dialog in the City. Case Studies on Interreligious Encounter in Urban Community and Education.* Munster, New York: Waxmann, 55–83.

Orth, G., 2002. *Die Erde – lebensfreundlicher Ort für alle. Göttinger Religionsgespräch 2002 zur Umwelt- und Klimapolitik* Berlin: Lit.

Patel, G., and McKermott, M., 2016. Hindu Worldviews and the Religious-Environmental Movement. *The Journal of Interreligious Studies*, 19, 41–45.

Pedersen, P., 1995. Nature Religion and Cultural Identity: The Religious Environmentalist Paradigm. *In:* O. Bruun and A. Kalland, eds. *Asian Perceptions of Nature: A Critical Approach.* London: Routledge, 258–273.

Pickel, G., 2019. *Weltanschauliche Vielfalt und Demokratie. Wie sich religiöse Pluralität auf die politische Kultur auswirkt.* Gütersloh: Bertelsmann Stiftung.

Ramadan, D., 2017. *Islamismus mit Islam bekämpfen* [online]. Available from: www.sueddeutsche.de/politik/deradikalisierung-islamismus-mit-islam-bekaempfen-1.3616942 [Accessed 7 August 2019].

Reason, P., and Bradbury, H., 2008. *The SAGE Handbook of Action Research.* Participative Inquiry and Practice. Los Angeles: SAGE.

Rohe, M., 2016. *Der Islam in Deutschland. Eine Bestandsaufnahme.* Munich: C.H. Beck.

Saan-Klein, B., and Wachowiak, M., eds., 2008. Vielfalt als Gewinn. Kirchengemeinden und Biodiversität. *Forschungsstätte Evangelischer Studiengemeinschaft e.V.* Berlin: Digital Business and Printing.

Satilmis, A., 2008. Chancen und Grenzen interkultureller und interreligiöser Dialoge – Evaluationsergebnisse. *In:* G. Klinkhammer and A. Satilmis, eds. *Interreligiöser Dialog auf dem Prüfstand. Kriterien und Standards für die interkulturelle und interreligiöse Kommunikation.* Berlin: Lit, 101–139.

Singh, K., and Steinau-Clark, J., eds., 2016. *Voices from Religions on Sustainable Development.* Bonn: German Federal Ministry for Economic Cooperation and Development (BMZ).

Spielhaus, R., 2013. Vom Migranten zum Muslim und wieder zurück. Die Vermengung von Integrations- und Islamthemen in Medien, Politik und Forschung. *In:* D. Halm and H. Meyer, eds. *Islam und die deutsche Gesellschaft.* Wiesbaden: Springer, 169–194.

Taylor, B., 2016. The Greening of Religion Hypothesis (Part One): From Lynn White, Jr. and Claims That Religions Can Promote Environmentally Destructive Attitudes and Behaviors to Assertions They Are Becoming Environmentally Friendly. *Journal for the Study of Religion, Nature and Culture*, 10 (3), 268–305.

Tomalin, E., 2009. *Biodivinity and Biodiversity. The Limits to Religious Environmentalism.* Surrey: Ashgate.

Tucker, M., 2008. World Religions, the Earth Charter, and Sustainability. *Worldviews*, 12, 115–128.

Tucker, M., and Grim, J., 2001. Introduction. The Emerging Alliance of World Religions and Ecology. *Daedalus*, 130 (4), 1–22.

10 Finding Ubuntu in the Bible

How the Zion Christian Church in South Africa Relates to Concepts of Ecology in African Traditional Religions

Juliane Stork and Charel du Toit

Introduction

It was in 2019 that the environment began to play a role in the official theology of the Zion Christian Church (ZCC) in South Africa. The Church's leader, His Grace the Right Reverend Bishop Dr. Barnabas E. Lekganyane, spoke exclusively on the environment in public for the first time in October 2019 by giving a speech entitled "Ecological Sustainability as a Topic of Redemption Within the Theology of the Zion Christian Church." The speech was part of an international conference on the role of Southern African churches as actors for ecological sustainability. It was held at the Faculty of Theology and Religion at University of Pretoria which jointly hosted the conference alongside the German development organization Bread for the World and the Research Programme on Religious Communities and Sustainable Development at Humboldt-University Berlin.

Participatory observation of this occasion and qualitative content analysis of the speech will be used to unlock new insights into the ZCC's recent eco-theology and into how it relates to concepts of the environment prevalent in African Traditional Religions (ATRs).

The ZCC is the biggest church in South Africa, with a self-proclaimed[1] 15 million members (Lekganyane 2019c, p. 4). It has a strong influence on its members' everyday lives through strict behavioral rules such as abstinence from pork, gambling, and alcohol and also through the star badge that members are asked to wear visibly at all times. The ZCC was founded in 1910 by the current leader's grandfather, Engenas Lekganyane. As an African Independent Church, the ZCC is part of the fast-growing movement of African Initiated Christianity.

African Initiated Churches[2] (AICs) were founded in three phases in the 19th and 20th centuries (Öhlmann et al. 2020, p. 6). The ZCC is part of the second founding phase that aimed for a contextualization of the Christian teachings that had been brought to Africa by European missionaries. Gaining independence from European historic mission churches, which had

DOI: 10.4324/9781003017967-13

been emphasized during the first founding phase at the end of the 19th century, also meant gaining independence from the western way of teaching Christianity. Active contextualization thus characterized the second founding wave. AICs produced new forms of Christian worship, singing, praying, teaching, and reading the Bible. This contextualization continues today as the oral theology of AICs keeps on growing and changing in accordance with their members – some of whom experience the effects of climate change on a daily basis.

In the context of South Africa, AICs represent well over half of the population (Öhlmann et al. 2020, p. 11) and are thus significant agents for development and change. On a continent that will be affected by climate change faster than any other, (UNEP 2016, pp. 16–17), AICs will become even more important agents for epistemological, social, and behavioral human adaptation to climate change. In the past, Martinus Daneel described how AICs in Zimbabwe became involved with tree planting and advocacy against climate change (Daneel 2006). Although his environmental movement ceased to exist, recent findings from interviews conducted by the Research Program on Religious Communities and Sustainable Development at Humboldt-University Berlin suggest that AICs show an increasing interest in ecological concerns beyond the known cases in Zimbabwe (Stork and Öhlmann 2019). Although AIC leaders ranked the ecology-related SDGs (numbered 13, 14, and 15) among the last when asked to name the most important ones, qualitative interviews showed that church leaders often cited ecological protection when asked for their visions of a good life. Especially in East Africa, which has a large rural population dependent on subsistence farming, AIC leaders deemed ecology important where it was connected to increasing poverty of their own congregants caused by climate change (Stork and Öhlmann 2019, p. 2). According to the interviews, many churches already developed and implemented their own programs against plastic pollution, against deforestation, or to increase resilience against climate change (Stork and Öhlmann 2019, p. 4). So far, no such programs have been implemented by the ZCC though.

Regarding eco-theology in ZCC, similarities between the perceptions of the environment in the ZCC and in African Traditional Religions (ATRs) stand out. Religious practices and convictions on the environment that are prevalent in ATRs seem to be reflected in the religious practices in AICs – although altered through a Christian prism. The ZCC usually denies close ties to ATRs and actively rejects traditional beliefs and practices on an official level (Lekganyane 2019c, p. 5). At the same time, AICs often emphasize the "African-ness" of their traditions, reflecting the contextualization of Christian dogmas in a context of African worldviews (Ngada 1999, pp. 1–2). This tension can be considered as part of AICs' bigger identity challenges that Kwame Bediako describes as "fundamental questions of African existence." He regards African Christianity's main task as resolving the dilemma of a continent "uncertain of its identity, poised between the impact of the West and the pull of its indigenous tradition" (Bediako 1997, p. 5). It is this identity conflict that is mirrored in the reinterpretation of ecological concepts by Bishop Lekganyane.

A key hypothesis of this chapter is that the ZCC's leadership has reacted to this tension using an entirely new interpretation of traditional categories in a way that is neither the equivalent of ATRs, nor goes along the lines of the teachings of historic mission churches. The void that opens up when traditional categories are banned gives rise to the possibility of the church filling it with convincing reinterpretations. We exemplify how this is done in the qualitative content analysis of a speech by Bishop Lekganyane.

To analyze the tension between the ecological concepts underpinning ATRs and the Christian historic mission churches further, this chapter first presents a condensed literature review on ecology in both ATRs and AICs. While literature is available on both religious strands individually, the relationship between ATRs and AICs in the domain of ecology has not yet been explored. In the second part of the chapter, the theoretical background and resulting hypothesis are presented. Both are derived mainly from the descriptions of Zionist theology in Southern Africa by Christoffer Grundmann, and literature on the ZCC by Retief Müller and David Chidester, all of whom acknowledge that the ZCC creates a safe worship space for its members to ward off threats from traditional spirits. Through a content analysis of Bishop Lekganyane's speech on ecology, this theoretical concept is developed further in the fourth section. The final section proceeds to a discussion of the speech analysis and of how the ZCC reinterprets traditional ecological concepts and creates a safe liminal space against the spiritual and existential threats from climate change in traditional categories.

Looking at the broader field of religion and environment, recent decades have seen increasing research on green religious perspectives and voices against climate change. In the article "The Greening of Religions Hypothesis," Taylor describes this trend in detail before arguing that it has become a normative trend among researchers to claim that religions have the power to change environmental ethics and consequently behaviors (Taylor et al. 2016, p. 296). He criticizes followers of this "religious environmentalist paradigm" for favoring "alternative religions" (Taylor et al. 2016, p. 296) and claiming "that indigenous societies often have unique, religion-embedded 'Traditional Ecological Knowledge,' which promotes environmentally sustainable societies," (Taylor et al. 2016, p. 288) – despite a lack of empirical evidence. Our research, however, seems to fall in line with those who have claimed that religions are increasingly interested in ecology, such as Chaplin 2016. In response to Taylor's criticism, however, it puts the greening of religions hypothesis under closer scrutiny by *empirically* analyzing exactly how the new ecological argument is presented in the ZCC.

African Traditional Religions and African Initiated Churches

This chapter follows Öhlmann, Frost, and Gräb in their definition of AICs as churches that were founded by Africans in colonial and post-colonial Africa (2020, pp. 4–5). According to them, AICs can be ordered in three founding phases that each have distinct characteristics. The ZCC belongs to the second founding phase that:

> is characterised by indigenisation and hybridisation. Elements of African Traditional Religions and local cultures were incorporated into Christian religious practice and belief. These churches are characterised by a worldview that assumes spiritual forces to be intertwined with the social and material world, synthesising African spiritual worldviews with Christian beliefs. Spiritual forces, whether good or evil, are considered to have an influence on people's lives, well- being, and social relations.
>
> (Öhlmann et al. 2020, p. 6)

In line with this observation, the ZCC incorporates characteristics of ATRs into its theology and assumes the interference of good and evil spirits within its member's lives. At the same time, the ZCC's current leader, Lekganyane, rejects the participation of ZCC members in ATR rituals (see later). Research has not yet reviewed this tension and this chapter will contribute to filling this research gap.

ATRs are equally as manifold as AICs and differ with regions and times. With good reason, Mbiti, probably the most famous scholar of African Religion and Philosophy, refrains from defining ATRs, and instead describes shared topics of African Religions as, for example, spirits, practices, and concepts of time (Mbiti 1997). Chidester, in *Religions of South Africa*, describes ATRs as the "beliefs and practices of the indigenous inhabitants of Southern Africa" (Chidester 1992, p. 1). However, he uncovers that the term "traditional" was in the past often misunderstood to imply that traditions were passed on unchanged from one generation to the next. Instead, he argues, traditions are rather picked up by younger generations "as an open set of cultural resources and strategies" (ibid.). Retief Müller (2015) challenges the term ATR altogether as a category that links religions, which are still alive, to an essentialized past. Additionally, he refers to Talal Asad's (1997) insights on the invention of religion by Europeans, and points out that ATRs, understood as "religions" that existed before white missionaries spread Christianity, are constructs created by Africa's white "explorers" (Müller 2015, pp. 183–184). Müller instead argues for the use of the term "African indigenous religions" as it describes the phenomenon as rooted in a current culture instead of defining it through mention of its historic origin. In opposition to Müller's opinion, Bahr points out that the basis of African indigenous religions is indeed *tradition*, because ATRs rely on oral dissemination of customs as there is no foundation in sacred texts to instruct followers on right behaviors or beliefs (Bahr 2005, p. 38). We adhere to the term ATRs as most African scholars also use it as a description (Bediako 1997, p. 192; Mbiti 1997, p. 6; Olúpọ̀nà 2014, p. 2). It is explicitly not meant to imply any hint to "backwardness" (Müller 2015, p. 184). Probably the most precise definition of ATRs can be found in the report on Christianity and Islam in Africa by the North American Pew Research Center. For their extensive quantitative studies, the research team defined the four most distinct features of ATRs to be the use of traditional medicines, the use of sacred objects, participation in ancestral rites, and participation in traditional puberty rituals (Lugo et al. 2010, p. 34).

For this research, we want to add the discursive dimension of defining ATRs in the post-colonial debate. The term ATRs is often used here to positively emphasize an "original" African religiosity that is independent from and has existed earlier than any white missionaries' attempts of proselytization.[3]

When eco-theologies are discussed in AICs, the relationship between ATRs and AICs becomes a crucial focal point to understand their background further.

This relationship has been subject to debates for a long time. Already in 1985, Ngada et al. ascribe in their book *Speaking for Ourselves*, an important role to the Western Christian missionaries in defining the essential opposition between ATRs and Christian teachings that still prevails today. They hold that this break line between ATRs and Christianity, as the missionaries taught it, led to an alienation of Africans from their culture and an uprooting of African Christians that can be observed in the ongoing enslavement of black people by white people (Ngada 1985, pp. 17–18).

In his speech on the history and theology of the ZCC, Bishop Lekganyane argued in a similar vein to Ngada et al.:

> It is clear that in its own way, the founding of the Zion Christian Church marked a move from the views previously held, that Christianity and African cultural values are absolutely mutually exclusive. Indeed, previous views were that a convert has to absolutely change as a person in order to become properly Christian. This was accompanied by encouragement for abandoning one's traditions. My grandfather, the founder of the church, knew that there was nothing inherently unchristian about being an African.
>
> (Lekganyane 2019c, p. 5)

In his speech Lekganyane himself proclaimed that being Christian and being (traditionally) African are not mutually exclusive. Nevertheless, at the same time, the ZCC rejects traditional religious practices by members. Bishop

Lekganyane stated in the same speech at the Humboldt-University of Berlin in July 2019 that:

> I have observed with disappointment misconceptions of the tolerance of cultures and customs that the ZCC practices, being confused with promoting ancestral worship and other customary practices forbidden in the Bible. As a church, we respect people's identities, cultures and customs. However, we do not ourselves practice or encourage the practicing of aspects of traditions related to ancestral worship and healing through traditional medicines. While those who themselves believe in traditional healing and ancestral worship are welcome to attend our church services, as is everyone else, members of the church are not allowed to participate in ceremonies of ancestral worship, traditional initiation schools and to use traditional medicines.
>
> (Lekganyane 2019c, p. 5)

In this part of his speech Bishop Lekganyane pointed out that, although his church is tolerant toward traditional beliefs of non-members, *members* of his church are not allowed to follow practices that are part of ATRs. The ZCC does not ask their members to become "born-again" believers with entirely new traits like some Pentecostal churches do (Müller 2015, p. 182). Instead, Bishop Lekganyane stated here that everyone's background was respected, but that ancestral worship, traditional healing, and traditional initiation schools (Lekganyane 2019c, p. 5) would be rejected in the context of the ZCC practices. These are interestingly three out of the four features that define ATRs according to the Pew Research Centre (see previous). To adhere to what Bishop Lekganyane said, believers must rid themselves of ATR practices when following the ZCC.

The ZCC seems entangled in this tension between either opposing or complementing the positions of ATRs and historic Christian teachings. We chose the theoretical lens of a safe liminal space to explain this tension further.

Scientific literature on ATRs, AICs, and ecology has increased rapidly over the past years as part of postcolonial discourse on theological "African" identity. The common denominator of these recent writings on African Initiated Christianity and the environment are normative claims that AICs should find a way back to their "African roots" in order to create a contextual eco-theology that speaks convincingly to the worldviews of their members (Anim 2019; Amanze 2016; Asamoah-Gyadu 2019; Oduro 2019; Sakupapa 2013). In these articles, environmental protection is depicted as a common aim for ATRs and AICs as nature plays a vital role in both traditions.[4] From a Ghanaian perspective, Asamoah-Gyadu writes that the spiritual beings in ATRs are believed to dwell in the biophysical environment (Asamoah-Gyadu 2019, p. xiii) thus always connecting the natural world to religious worldviews. Oduro confirms that, in ATRs rivers, forests, and animals are considered to be holy places or sacred beings in and of themselves (Oduro 2019, p. 61). Accordingly, "the environment" is understood in a broad sense in this chapter. It is not defined from the perspective of the natural sciences but analyzed as a spiritual category. Thus, perceptions of all features of nature including animals, water, land, or air are included in the analysis of the ecological concepts of AICs and ATRs.[5] Concepts of ecology in ATRs are of course as varied in their details as the people by whom they are practiced. Still, some broader shared concepts can be recognized in all the texts that were considered for this chapter, corresponding to exactly those raised in Bishop Lekganyane's speech.

The ZCC as a Safe Liminal Space – Theory and Hypothesis

The concept of a religious safe space supplies the theoretical background to the analysis in this paper. The concept relies on Christoffer Grundmann's article on Zionist movements in Southern Africa: "Heaven below here and now! The Zionist Churches in Southern Africa." In the article he writes that Zionist churches build a "safe haven" during worship for their members, observing that:

> It is also an expression of what "Zion" does mean to them, namely, to be sheltered from all that is intended to do harm. As long as the sons and daughters of Zion worship at their holy place they are in a safe haven, protected by the power present among them in their rituals and liturgical attributes. But once they take off their gowns and return the "holy staff" at the end of the service when leaving their "Zion" they emerge into a world full of danger and the constant threat of any kind of misfortune and disease.
>
> (Grundmann 2006)

Grundmann characterizes two opposing spaces or worlds, one as the safe space of "Zion" and the other as an opposing unsafe space for the church members where "any kind of misfortune or disease" could reach them. Although Grundmann does not elaborate further on this unsafe space,[6] it stands to reason that in an "African" worldview, these threats in the outside world are as much connected to spirituality as the safe haven inside the Zionist churches (Ellis and Ter Haar 2007). As such, Zionist churches are seen to provide a safe space against harmful spirits that could potentially attack anyone at any time outside the church. Indeed, these harmful or evil spirits are believed to stay active and threatening outside the church even though Zionist members have converted to a Christian church. The harmful spirits are not abandoned completely or uncovered as humbug, but merely disabled to an ineffective state inside the safe space of the church. This safety net, which can still hold power over the church members, is extended outside of church premises to all believers who obey the strict rules that regulate the everyday lives of believers. Observance of the rules, which forbid believers to drink or gamble, and that promote a healthy and holy lifestyle, is in Grundmann's opinion, a measure to keep evil spirits at bay (Grundmann 2006). This idea can already be found in Chidesters' description of AICs and account of the ZCC. Writing in 1992 he interprets the safe haven that the ZCC provided for their believers as a secure space against the disempowering structures of Apartheid (Chidester 1992, p. 135). Chidester also supports the idea that the ZCC innovates, instead of merely copies, beliefs of ATRs. Rather than focusing broadly on the relationship between the ZCC and ATRs, he considers the notions of purity that underpin the ZCC's dances, prohibitions of harmful practices, and star badge that members are asked to wear every day, even outside of the church. All of these were in his opinion provisions to keep members pure and secluded from the surrounding environment of poverty in the townships (Chidester 1992, p. 138).[7] Müller argues the same way in stating that the ZCC creates a world that provides a sense of belonging to its members in the living conditions of the townships. His particular view on the safe space that the ZCC creates is that it connects the borderlands of the rural and the urban spaces in which its members live (Müller 2015, p. 182). He equates these spaces with the extreme poverty and subsistence farming in the rural areas but upward mobility in the cities (Müller 2015, p. 181).

Our approach is to further develop these ideas of a new space that the ZCC creates and to connect it to the concept of liminal space as first described by van Gennep and later developed by Victor Turner (Turner 1969, p. 359). The liminal space is constituted by the two opposing, seemingly incompatible spaces that it connects. The essence of this concept, also referred to as "in-between space," (Cilliers 2013), is the field of tension that arises from the navigation between two antagonistic positions.

> [It] implies both integration of and resistance to whatever is either side of or outside of the in-between. In simple paradigmatic terms, one cannot occupy an in-between space or exist (in-) between two binary states without a resultant tension and/or mobility between both elements of the binary, which resist but also merge with the middle in-between.
>
> (Downey et al. 2016, p. 15)

This is the case for the AICs that claim for themselves a middle ground between the historic mission-introduced Christianity, and traditional African religion. Their current conflicted position between these two traditions might thus be *constitutive* for them. According to this theory, AICs would have to use concepts from ATRs and from historic mission churches to keep to their defining "in-between" position.

Ecology in the Zion Christian Church – A Speech Analysis

Participatory Observation, Context, and Content of the Speech

The speech under analysis was delivered by Bishop Lekganyane on 29/10/20 in the context of a conference on Churches in Southern Africa as Civil Society Actors for Ecological Sustainability. Bishop Lekganyane had been invited as a keynote speaker by the conference organizers from the German Protestant aid organization Bread for the World, the Research Programme on Religious Communities and Sustainable Development at Humboldt-University of Berlin, and the Theological Faculty at University of Pretoria. The speech was given in the Musaion auditorium of the University of Pretoria where the audience consisted of scientists and church leaders from different Christian backgrounds. They included representatives of AICs from several African countries, as well as university staff and students, some of whom were church members. This description constitutes important background information – it must be kept in mind that the speech was delivered in front of this mixed audience and that Bishop Lekganyane had been invited to the event to specifically speak on the environment. The speech was – probably for this reason – given entirely in English, unlike the mixture of Northern Sotho and English that Bishop Lekganyane usually uses in his speeches. Bishop Lekganyane had originally been invited to speak exclusively in front of the audience of the conference, but as soon as the administration of the University of Pretoria gained knowledge of the famous speaker, the university's VIP protocol, including heavy security measures, was installed and the lecture was relocated into a major lecture hall and opened to a wider audience. Only a strictly limited guest list avoided overcrowding of the venue. These circumstances underline the importance that Bishop Lekganyane – who usually only speaks on very limited occasions in public and rarely outside of church services – enjoys in South Africa. It also contextualizes his speech and points to the fact that the Bishop could have chosen any topic for his speech – he is such an important and recognizable figure in South Africa that any topic he chose to be fitting for the occasion would have been accepted by the audience and the conveners of the conference. The implication is that Bishop Lekganyane deliberately chose to embrace the topic of ecology as important for his church.

He had also publicly addressed the issue once before in a few sentences at the end of his New Year's sermon on 1 September 2019 which suggests that he was interested in ecology. At the end of that sermon he said:

> We need to take care of our local environment, as it is linked to our very survival. Solid waste pollution contributes to increasing infertility of land and polluted rivers has become so prevalent.
>
> (Lekganyane 2019b, p. 16)

He went on to say:

> Le tsebe gore go hlompha hlago le dihlolwa tša Modimo ke moeno wa kereke ya Sione. Le seke la nyatša methopo ya tlhago yeo re e filwego ke Modimo. Go lahla ditlakala ka mafastere a dinamelwa ga se go hlompha hlago.
>
> (Lekganyane 2019b, p. 16)

> Translated: You need to be aware that respecting nature and God's creations is a totem of the church of Zion. Please do not disrespect nature's creations that have been granted to us by God. Throwing rubbish through windows of cars is not a sign of respecting nature.

By "moeno," totem, he probably expressed that God's creation was very close to the core of the Church and did not refer to the literal meaning of a totem as often used in ATRs for animals or plants. Still, it is interesting that the Bishop used the word in connection with nature here and thus created the image of a connection between the traditional concept of a sacred totem (that is not to harm) and nature as God's creation. His audience can be assumed to have been familiar with the traditional concept of moeno. Bishop Lekganyane spoke of the local environment here as a source of survival for the members of the Church that he saw threatened by waste pollution. Theologically, he connected the appeal to his members not to dispose of their garbage on the roads with the image of God's creation that was granted to "us." These thoughts were elaborated much more in the speech that Bishop Lekganyane gave at the University of Pretoria.

In the beginning of the event, university staff had explained to the audience that the ZCC had asked for no pictures to be taken during the event. A delegation of the ZCC including elders of the church, the spokesman, and two of the sons of Bishop Lekganyane then came on stage together with one representative each of the organizing entities. The University of Pretoria was represented by various deans and the vice president. The audience rose as the Bishop entered the hall and only took their seats again when the group on stage sat down. Bishop Lekganyane's speech was preceded by words of introduction by the vice principal of the university, the dean of the faculty of Religion and Development, and by the spokesman at the time, Rev. Motolla[8], of the ZCC, who introduced Bishop Lekganyane to the audience. After Bishop Lekganyane's speech, no questions were allowed but a response was given by Coenie Burger, the former moderator of the Dutch Reformed Church. Words of thanks from all parties involved in the organization were added. The event was closed by a prayer from protestant reverend Thomas Stephan, who had been on stage as a representative of Bread for the World. The atmosphere on stage was very respectful and included many formal greetings and a strict observance of protocol. The audience was divided into the participants of the conference, who were seated in the front listening with interest, and a larger more excited contingent who had come in formal, festive clothing. Though not in full Church uniform, they were wearing their star badges and doeks (a hair-covering piece of clothing for women), while sitting on the sides and the back of the auditorium.

Despite the fact that Bishop Lekganyane spoke without much emotion or charismatic gestures, the speech was acclaimed by applause, ululation, and standing ovations of the audience. The ZCC members present at the auditorium greeted their leader after the speech with loud praising songs and prayers. Whenever he interrupted his speech and greeted the audience by saying: "Kgotso ebe le lena!", ("Peace be unto you!") the audience responded to his Sesotho greeting loudly and punctuated any pauses with much applause. Although the audience kept on singing even after he had left the stage, he did not come back out to greet church members again. The importance of his appearance to church members cannot be overstated. Case in point: The University of Pretoria's social media team expressed their surprise about the impact the event's official Facebook post had; according to them it got more likes and reposts[9] than anything else they had ever posted.

The role of every Bishop as hereditary heir to the church's founder and recipient of God's guidance is central in the ZCC (Müller 2015, pp. 60ff; 186). Interestingly, Müller compares the role of the Bishop to that of a traditional chief and rainmaker (2016, pp. 191–192). He describes how the Bishop accepts invitations from congregations to pray for rain on their land just like rainmakers in traditional rain ceremonies. According to Müller though, Bishop Lekganyane makes no claims to being a rainmaker himself (Müller 2016, p. 192). This description could serve as another example of how the tensions between the theologies of ATRs and AICs materialize when traditional concepts are reinterpreted by the ZCC: The Bishop *prays* for rain but doesn't perform a traditional ceremony and does not claim the title of a rainmaker. It could also give a hint on how (some) church members accept his role as engaging with and reflecting on nature.

The argument of the speech can be recounted and structured in the following way. After an introduction and words of thanks to the organizers of the event, Bishop Lekganyane spoke of nature as God's gift to people, while bestowing on them the responsibility to look after it (Lekganyane 2019a, p. 7). He went on to oppose the "Eurocentric foundation of identity" with an African identity which he said was built on Ubuntu (Lekganyane 2019a, p. 7). He explained that this meant that every person was connected to every other person and that this connection, stretching through time and space beyond the immediate surrounding of a person, determined the "ethical, social and legal judgement of human worth and conduct" (Lekganyane 2019a, p. 7). He said:

> Therefore, when you kill and destroy God's creation, you will run short of something essential to your being. While you may continue to exist,

> you will not earn the worth of being -ntu, and thus will lack the human form aspect of your existence.
>
> (Lekganyane 2019a, p. 7)

Lekganyane connected this individual responsibility to secure one's -ntu, or one's ability to be fully human, to community. While, according to him, every being came "out of relationships" to others, this community of relationships was "embedded in a natural environment" (Lekganyane 2019a, p. 7). He used this thread of thought to outline that harming the environment meant harming one's own ability to personhood, thus harming the whole community that was connected to oneself.

From this part about the concept of ubuntu Bishop Lekganyane went on to a theological interpretation and said that as we were God's legacy, violating the Earth was against this legacy and that 1.Cor 4,2 stated that the trusted must prove faithful by following God's legacy (Lekganyane 2019a, p. 8). Using almost the exact definition of sustainability as presented in the UN Brundtland report on sustainability (World Commission on Environment and Development, 1987), Bishop Lekganyane pointed to future generations of communities, whose needs must be considered.

The Bishop continued with a historical argument and related it back to the now biblically charged concept of ubuntu. He said that rural black communities had in the past been deprived of their land and that this had destroyed their connection to the natural environment and thus "a big portion of their Ubu" (Lekganyane 2019a, p. 8). This deprived people of their ability to "ubuntu," that is, to fully become and to be redeemed by keeping God's legacy safe. Bishop Lekganyane then described in detail how economies and politics exploit nature without limits and pointed out how drastically humans pollute and destroy the planet. Following this argument he called on governments to implement policies for ecological sustainability, on businesses to invest in green technologies, and on church leaders to preach "about the sacred nature of ecology" (Lekganyane 2019a, p. 10). The Bishop then connected the historical argument of land deprivation of black communities and their loss of connection to nature to the present by stating that land deprivation is happening again – through climate change (Lekganyane 2019a, p. 10). He closed by stating that redemption could be reached only through following God's commands, which includes taking care of God's creation. Complementary to his firm language and call for action to other society actors, in the middle of his speech, Lekganyane mentioned implicitly the actions that his own church takes for climate justice:

> When we speak about peace, we include what have been called environmental and social justice. When we pray for peace, we are very much aware that human conflicts are borne of contestations for natural

> resources. Therefore, within the message of peace we preach for peaceful and responsible sharing and usage of natural resources.
>
> (Lekganyane 2019a, p. 9)

Alongside preaching about ecological justice, the church also includes it in their prayer for peace. This prayer is at the heart of the church's theology, giving rise to the much used church greeting "Kgotso ebe le lena" (Peace be unto you). The prayer promises to charge ecological concerns with high theological importance in the ZCC.

Methodology

A qualitative content analysis[10] of the speech "Ecological Sustainability as a Topic of Redemption Within the Theology of the Zion Christian Church" by Bishop B.E. Lekganyane shows how ecological concepts that are prevalent in ATRs are reinterpreted in the theology of the ZCC. The analysis relies on the written speech that was published in the church magazine *The Messenger* in December 2019. The magazine usually contains all speeches by the Bishop given since the last issue and some articles of opinion or reports on church-related topics by other authors. As no Church website exists, the magazine is the most important media of official Church information, reaching millions of members.

According to the methodology of qualitative content analysis, words, expressions, and arguments that belonged in the semantic fields of the environment in ATRs and of ZCC theology were marked and coded (this includes any related words, e.g., water, land, creation, etc.). The coded parts were then ordered in higher-ranking categories, which were inducted as summarizing frames from an overview of the coded words. The inducted categories from Bishop Lekganyane's speech were then used in a second step as deductive categories to analyze texts about ATRs and ecology in order to find out in which parts of these categories they are the same, similar, or different. These outcomes were then assessed through the theoretical concept of a safe liminal space to analyze the relationship between AICs and ATRs as it presents itself regarding ecological concepts. For the interpretation of the data, information from the participatory observation that we could undertake during the event was added.

Results of the Qualitative Content Analysis

The following categories were inducted from the speech:

Origins of the Earth and Humans, Relationship Between Humans and Earth, Land Ownership/Deprivation, Responsibility for the Creation, Coping Strategies, Recipients of Ecological Care and Awareness, Sources for Ecological Knowledge and Behavior, Reasons for Environmental Destruction.

Some of these categories that can be retraced in the very short renarration of Bishop Lekganyane's speech fall in line with Asamoah-Gyadu's observations of ecological concepts that are prevalent in ATRs:

> In African ethical understanding, the earth is given to humanity as a gratuitous *gift* as *all human beings* possess an equal claim to it and the resources it offers. What this means is that any individual can only *hold land in trust for one's descendants* on behalf of the clan or ethnic group. Water sources, mineral resources, and forests are, in principle, *public property*. They are communally owned and *have to be cared for* and used as such" (Asamoah-Gyadu 2019, pp. xv–xvi).
>
> (The authors of this paper added the cursive typesetting.)

Table 10.1 illuminates how the inducted categories were employed in Bishop Lekganyane's speech in comparison to how they are used in descriptions of ATRs in literature:

The table shows that Bishop Lekganyane uses some concepts in his eco-theology that are similar to those prevalent in ATRs. Although the concept of the origin of the earth as a gift to humans by higher beings is very similar in both cases, none of the categories are used in exactly the *same* way. Instead, only three out of eight categories were used similarly, but were reinterpreted, while five concepts were used differently. Bishop Lekganyane extends, scientifies, contextualizes, or adapts the concepts into Christian patterns.

The concept of ubuntu that Bishop Lekganyane uses to describe the relationship between humans and the earth has become increasingly popular in South Africa after the end of apartheid and should not automatically be regarded as a traditional concept prevalent in ATRs.[11] That is not to diminish ubuntu as an important part of identity-building and belonging in today's South Africa – quite the opposite is the case. It is indeed a functioning common denominator that establishes a (South) African identity across language groups, and that connects people to each other and also to the environment. This context, however, should not hastily drive us to the conclusion that Lekganyane used a concept from ATRs as a core component of his speech.

The texts about ATRs used for this analysis present a different understanding of the relationship between humans and the Earth than the all-connectedness suggested by ubuntu. Specifically, the notion of a personified Earth cannot be found at all in Bishop Lekganayane's speech. Rather, his interpretation of ubuntu is shaped by a traditional understanding of land use and ownership. His description of land ownership/deprivation is interestingly similarly to the concept in ATRs. Bishop Lekganyane's claim that people who were deprived of their land and thus of their ability to fully become is connected to the idea that the land is where one's family and ancestors live(d). Land can, according to the understanding of ATRs that

Table 10.1 Categories in Bishop Lekganyane's speech

Inducted Categories	*Codes in Bishop Lekganyane's speech*	*Codes in ATRs*	*Same/similar/ different concept*	*Alteration between the concepts through:*
Origin of Earth and Humans	Creation is a gift by God to Humans.	Earth is a gift by God (Asamoah-Gyadu 2019, pp. xv–xvi).	very similar	–
Relationship Humans/ Earth	Ubuntu connection to all beings.	Concept of Earth as a mother (Oosthuizen 1988, p. 9); Personification of the earth and every component (Oduro 2019, p. 61).	different	Extension
Land Ownership/ Deprivation	Black communities in South Africa experienced dispossession of their land which lead to a breakdown of their relationship with nature; climate change is depriving people of their subsistence farmland today.	Holding land in trust for one's own descendants on behalf of the extended family (Asamoah-Gyadu 2019, pp. xv–xvi). Land cannot be owned, but only be leased (Asamoah-Gyadu 2019, p. xv).	similar	Contextualization to the life situation of the church members
Responsibility for the Creation	Responsibility by a legacy/commandment of God; only holding this commandment will redeem us.	Earth must be cared for as home to deities, a good relationship with whom is essential for an abundance of food and to prevent natural catastrophes (Asamoah-Gyadu 2019, pp. xiii–xiv).	different	Christian adaptation

Coping Strategies	Prayer for peace and ecological justice, taking care of what God has entrusted us with.	Fixed rules protecting the environment, e.g., clearing farmland only in distance to a river, fixed communal resting times for the land, etc.(Anim 2019, p. 114).	different	Christian adaptation
Recipients of Ecological Care and Awareness	All past and future generations of communities, "the human, animal and plant populations that together form the one whole community."	All human beings, descendants (Asamoah-Gyadu 2019, pp. xv–xvi).	similar	Extension to a wider circle of beings and generations to come
Sources for Ecological Knowledge and Behavior	The Bible Indirectly also the UN Brundtland Report	Oral tradition, myths, metaphors, hymns, rituals (Oduro 2019, p. 61)	different	Christian adaptation Scientification
Reasons for Environmental Destruction	Politics, businesses, and religious communities who exploit the Earth without limits, environmental tipping points and processes as e.g., melting of the glaciers or acidification of the oceans.	Harmful spirits (Asamoah-Gyadu 2019, p. xiv).	Very different	Scientification, Actualization

Asamoah-Gyadu describes, not be sold nor misappropriated. Exile of people from their land, whether in the unjust apartheid system or through climate change is, for Bishop Lekganyane, a problem, because the traditional connection of people with their roots is cut in the process.

The concept of responsibility for land is motivated differently. While responsibility in the ZCC's theology is derived from the Christian God's command to care for the creation, responsibility for the land in ATRs is due to the claims that one's descendants hold over the land. In both cases, neglect of the care for creation will result in two different and bad consequences: ungranted redemption, or bad harvests, respectively.

The strategies applied to coping with environmental threats are different. While Bishop Lekganyane mentions prayer, the texts about ATRs mention fixed rules for everyday life that protect the environment practically. This practical element of environmental protection was not mentioned in the speech (but only in a very limited way in the earlier sermon by the Bishop in which he mentioned the environment in a short part).

The recipients of ecological care are similarly in both traditions all humans, but Bishop Lekganyane widens the circle of beneficiaries even more from past generations to future ones as well, and from humans to all animals and plants.

The sources for ecological knowledge differ vastly between the traditions. While in the ZCC the Bible is clearly mentioned as a source of knowledge and Bishop Lekganyane implicitly cites the UN's Brundtland report on climate change, ATRs are informed by knowledge from descendants passed on orally in hymns or stories.

Bishop Lekganyane ascribes the responsibility for ecological destruction to politics and businesses exploiting the planet without limits, while texts about ATRs explain that harmful spirits are understood to be the reason for natural catastrophes.

Bishop Lekganyane altered most of the categories by *extending* them to a wider circle of people, by *actualizing them* to the current life situation of the ZCC church members, and by *adapting them* through a specifically *Christian* prism or by *scientification* of the understanding of environmental degradation. It is possible that the scientification of the arguments could be a consequence of the context in which the speech was held. Assuming that Bishop Lekganyane prepared the speech on the environment particularly for the event at the University of Pretoria, he might have chosen to mention scientific facts because he was speaking in front of a university-based audience that was used to scientific language. Still, he did approve publication of the speech in the *Messenger*, knowing that it would be circulated to all adherents of his church.

The only exception to the alterations of ATR concepts in Lekganyane's speech is the responsibility of politics and economies for the current destruction of the environment for which no equivalent can be found in the ATR literature.

In ATRs, practical approaches to protecting the environment, such as not fishing on certain days, or refraining from killing certain animals, are passed on (Asamoah-Gyadu 2019, pp. xiii–xiv) (Anim 2019, p. 114). Contrary to this approach, Bishop Lekganyane said that ecological injustice caused by climate change can be addressed by prayers, preaching, and calls for the ZCC to societal actors to action. According to this argument, a Christian practice – prayer – fills the void where the traditional coping strategies derived from ATRs are rejected by the church.

Discussion

The concept of the safe space at a liminal threshold elaborated previously and the results from Bishop Lekgayane's speech provide new insights in two different areas when brought together.

First, the hypothesis by Grundmann, Müller, and others that the ZCC offers a safe space for adherents, and our idea that the ZCC takes a liminal position between the historic mission churches and the ATRs, is affirmed by the speech under analysis here. The church can find new interpretations for ATR categories that remain active but are reduced to a nonthreatening form inside the ZCCs safe space. Indeed, ideas from ATRs remain an active background to the experiences and beliefs that ZCC members acquire and the Church must answer the question of whether they should be rejected or included in the lives of the believers. The speech on ecology by Bishop Lekganyane falls perfectly into the in-between space of "integration of and resistance to" the two opposing ideologies of the historic mission churches and ATRs that the ZCC navigates. Neither are assumed completely, but instead they are combined in a unique liminal theology in between both traditions. Bishop Lekganyane's speech can be understood to be informed by ATRs and Western Christianity alike. It constitutes an "African" Christian theology that provides a safe space for Church members against the perceived spiritual threats of climate change and the danger of a refused redemption. The spiritual threat that is described by Bishop Lekganyane here is that of falling short of one's possibility to ubuntu. Losing the vital connection to all beings and to one's own spiritual purpose is connected to the threat of losing one's land through climate change. The Church answers to this threat by offering a safe space against any spiritual dangers and by providing their members with the opportunity to fulfill their calling to ubuntu. Members do this by praying against conflicts that arise from climate change and by changing their behavior to keep God's legacy and to prevent the risk of losing their land and the spiritual connection to the life on it. The ZCC's eco-theology as displayed in Bishop Lekganyane's speech succeeds in walking the tightrope between the two traditions by altering the traditional environmental categories in the Christian approach. The ZCC keeps the conflicted position between the two traditions according to the theory that constitutes their identity.

Second, the theory also illuminates the speech itself from a new perspective. The spiritual threat of being deprived of one's ability to -ntu, is not abandoned by the Church. Instead, it is tackled inside the ZCC's safe space by prayers for peace, but also by a call for action against climate change and land loss directed to the government, other churches, and members of the ZCC (raised in Bishop Lekganyane's first speech with a short mention to the environment). Both speeches show different reasons behind the call for ecological action. In the earlier speech, waste disposal through the window of a car was argued to be an abuse of God's legacy, while the later speech argues that ecological destruction is a sign of not fulfilling one's purpose of being in community. Moreover, the analysis of the later speech taken together with the environmental remark in his earlier speech shows that there are four different reasons mentioned by Bishop Lekganyane that motivate ecological protection. The two theological reasons for ecological behavior are keeping God's legacy and commandment to environmental protection and avoiding the deprivation of one's ability to -ntu. The two practical reasons Bishop Lekganyane gave are that ecological destruction destroys our lives as humans and ecological crises foster human conflicts which need to be avoided to keep the Church's highest priority: peace. The reinterpretation of the spiritual threats of climate change in an "African" Christian perspective make the practical reasons for ecologically conscious behavior more accessible to believers by presenting a contextually embedded understanding.

Bishop Lekganyane's importance surely contributed a big part of the audience's excitement on environmental protection. Still, it can be assumed that the content of the message and the drastic words on climate change employed ("The time for action has come. No one can sit behind and watch." (Lekganyane 2019a, p. 10)) will find an echo in the lives of the ZCC members, who are used to giving their church's rules a high priority in their everyday practices. So far though, the ZCC did not elaborate any concrete programs on ecological conservation as a consequence of the theological interpretation of climate change by Bishop Lekganyane.

Conclusion

When speaking of climate and environmental destruction, AICs relate to concepts of nature that are prevalent in ATRs. They are even asked to do so normatively by scholars and development practitioners working on religion and the environment. In search of their own theological answers to environmental destruction and contextualized eco-theologies, AICs find themselves in a field of tension between the theological positions of ATRs and positions held by historic mission churches. AICs from the first and second founding wave in the late 19th and the early 20th centuries (Öhlmann et al. 2020) were created in opposition to historic mission churches and have kept it as part of their identity to interpret the gospel differently, in a way that is embedded in an African context.

On the other hand, AICs from the second founding wave like the ZCC, reject the dominating features of ATRs. Societal ecological discourses and the deteriorating state of the "nature" that congregants depend on for their living are pushing AICs to develop their eco-theologies further and more explicitly. The identity conflict of AICs between the two traditions that constitutes them can be observed in many different aspects of their religious practices. It becomes especially visible in the field of eco-theology as the nature-human relationship has been a strong feature of both the African Traditional and the historic mission churches' traditions. Whenever AICs aim for their own contextualized eco-theologies they will be confronted with the underlying identity conflict that Kwame Bediako (1997) described. It is important to view AICs as part of the ecumenical world Christian movement as they have their own, theologically unique and influential voices to add to the global ecological debate. At the same time, AICs should not too readily be asked to become change agents in the name of a trendy Western-dominated eco-theology. This instrumental view ignores the fact that when AICs open new debates about eco-theology, it raises fundamental questions of identity for them. At the same time, it is true that AICs are very influential societal actors and that they might act on ecological concerns without stating beforehand which eco-theology motivates them.

Functioning as liminal space between ATRs and historic mission churches, it is the task of the ZCC to create a safe space and a spiritual adaptation strategy against the (spiritual) threats of climate change for their believers, namely the disintegration of relations to past generations and the risk of being deprived of the ability to fully become. The ZCC offers a new theological perspective in which climate change induced land loss can be understood as a spiritual loss of the ability to fulfill what it means to be human. In this perspective, destroying the planet disconnects humans from their natural surroundings and climate change becomes embedded in an "African" Christian worldview, inside which it might be more intuitively understood and answered to.

This research responds to the aptly posed question of whether religions really are increasingly going green and if so, how this process takes place (Taylor et al. 2016, p. 296). The ZCC leadership seems to be actually embarking on a greening process. The case study accordingly supports those who argue in favor of the greening of religions hypothesis. Further research will be needed to investigate how the theological turn toward the greening of the ZCC will be put into practice in the future.

Notes

1 The last official census in South Africa that also regarded religious affiliation took place in 2001. It counted 4.9 Mio. members for the ZCC and displayed a growth in membership of almost 0.3% p.a. in comparison with 1996 (Statistics South Africa 2004, p. 25ff).

2 See (Öhlmann, Frost, and Gräb 2020, p. 4) for the wide definition of African Initiated Christianity used in this chapter.
3 Also note Jahnel 2016, who deconstructs the concept "African" theology as a construction of "other" theology in opposition to "western" theology in a very important and enlightening discourse analysis.
4 As Anim 2019, p. 118, Müller 2016, p. 17, Oosthuizen 1988, p. 9 and Oduro 2019, p. 61 have argued, among others.
5 See also Oduro, who uses "land," "earth," and "the environment" even interchangeably in his work (Oduro 2019, p. 60).
6 For a more in-depth description of spiritual threats and spiritually unsafe spaces, see Ashforth 1998.
7 Although the authors of this chapter do not agree with Chidester's seemingly one-sided negative image of townships, his observation of the ZCC as a liminal space is illuminating in this context and quoted for this reason.
8 Rev. Motolla has passed on in 2020 after this text was written. May he rest in peace.
9 377,706 views of the post were counted four days after the event, while the most popular posts before had received up to 20,000 views as according to the social media team of UP.
10 According to Mayring and Fenzl 2019.
11 See (Hobsbawm and Ranger 1983) and (Eze 2010) for the concept of invented traditions and further discussions on ubuntu as a tradition that became famous in South Africa after 1994.

References

Amanze, J.N., 2016. From "Dominion" to "In Communion": Ecotheology from an African Perspective. *Anglican EcoCare Journal of Ecotheology*, 2, 11–21.

Anim, E., 2019. Environmental Sustainability and Eco-Justice: Reflections from an African Pentecostal. *In:* L. Andrianos, M. Biehl, R. Gütter, J. Motte, A. Parlindungan, T. Sandner, J. Stork, and D. Werner, eds. *Kairos for Creation: Confessing Hope for the Earth*, 107–119. Solingen: foedus-verlag.

Asad, T., 1997. *Genealogies of Religion: Discipline and Reasons of Power in Christianity and Islam* [reprint]. Baltimore: Johns Hopkins University Press.

Asamoah-Gyadu, K., 2019. The Earth Is the Lord's: Mainstreaming Ecological Issues in African Theology. *In*: A. Ntreh, M.S. Aidoo, and D.N. Aryeh, eds. *Essays on The Land, Ecotheology and Traditions in Africa*. Eugene: Wipf & Stock, xi–xx.

Ashforth, A., 1998. Witchcraft, Violence, and Democracy in the New South Africa (Sorcellerie, Violence Et Démocratie Dans La Nouvelle Afrique Du Sud). *Cahiers d'Études Africaines*, 38 (150/152), 505–532.

Bahr, A.M.B., 2005. *Indigenous Religions*. New York: Chelsea House Publishers.

Bediako, K., 1997. *Christianity in Africa: The Renewal of a Non-Western Religion*. Edinburgh: Edinburgh University Press.

Chaplin, J., 2016. The Global Greening of Religion. *Palgrave Communications*, 2 (1), 1457.

Chidester, D., 1992. *Religions of South Africa. The Library of Religious Beliefs and Practices*. London: Routledge.

Cilliers, J., 2013. Between Fragments and Fullness: Worshipping in the in-Between Spaces of Africa. *HTS Teologiese Studies/Theological Studies*, 69 (2), 1–6.

Daneel, M.L., 2006. African Initiated Churches as Vehicles of Earth-Care in Africa. *In:* R.S. Gottlieb, ed. *The Oxford Handbook of Religion and Ecology*. Oxford: Oxford University Press, 535–567.

Downey, D., Kinane, I., and Parker, E., 2016. *Landscapes of Liminality: Between Space and Place*. London, New York: Rowman & Littlefield International Ltd.

Ellis, S., and Ter Haar, G., 2007. Religion and Politics: Taking African Epistomologies Seriously. *The Journal of Modern African Studies*, 45 (3), 385–401.

Eze, M.O., 2010. Ubuntu and the Making of South African Imaginary. *In:* M.O. Eze, ed. *Intellectual History in Contemporary South Africa*. New York: Palgrave Macmillan, 145–179.

Grundmann, C.H., 2006. Heaven Below Here and Now! The Zionist Churches in Southern Africa. *International Journal for the Study of the Christian Church*, 6 (3), 256–269.

Hobsbawm, E., and Ranger, T., eds., 1983. *The Invention of Tradition*. Cambridge: Cambridge University Press.

Jahnel, C., 2016. *Interkulturelle Theologie und Kulturwissenschaft: Untersucht am Beispiel afrikanischer Theologie*. Stuttgart: Verlag W. Kohlhammer.

Lekganyane, B.E., 2019a. Ecological Sustainability as a Topic of Redemption Within the Theology of the Zion Christian Church. Public Keynote Lecture. *The ZCC Messenger*, 105, 6–10.

Lekganyane, B.E., 2019b. Sermon Delivered by His Grace, the Right Reverend Bishop Dr. B.E. Lekganyane on the Consecration of the New Year in Zion City Moria, on Sunday 01 September 2019. *The ZCC Messenger*, 105, 13–16.

Lekganyane, B.E., 2019c. *The Theology of the Zion Christian Church and Its Role in the Development of Southern African Communities and Beyond*. Berlin, July 2.

Lugo, L. et al., 2010. Tolerance and Tension: Islam and Christianity in Sub-Saharan Africa. *Pew Research Center* [online]. Available from: www.pewforum.org/2010/04/15/executive-summary-islam-and-christianity-in-sub-saharan-africa/ [Accessed 21 March 2022].

Mayring, P., and Fenzl, T., 2019. Qualitative Inhaltsanalyse. *In*: N. Baur and J. Blasius, eds. *Handbuch Methoden der empirischen Sozialforschung*. Wiesbaden: Springer Fachmedien Wiesbaden, 633–648.

Mbiti, J.S., 1997. *African Religions and Philosophy*. Oxford: Heinemann.

Müller, R., 2015. The Zion Christian Church and Global Christianity: Negotiating a Tightrope Between Localisation and Globalisation. *Religion*, 45 (2), 174–190.

Müller, R., 2016. *African Pilgrimage: Ritual Travel in South Africa's Christianity of Zion*. London: Taylor and Francis.

Ngada, N.H., 1985. *Speaking for Ourselves: Members of African Independent Churches Report on Their Pilot Study of the History and Theology of Their Churches*. Braamfontein: Institute for Contextual Theology.

Ngada, N.H., 1999. Transformation of Theology in the African Indigenous Churches. *In*: C.W. Du Toit and N.H. Ngada, eds. *Hearing the AIC-Voice: Proceedings of a Conference Held by the Research Institute for Theology and Religion at Unisa on 2&3 September 1998*. Pretoria: Research Institute for Theology and Religion, 1–7.

Oduro, T., 2019. The Bible and Caring for the Land: A Theological Approach. *In*: A. Ntreh, M.S. Aidoo, and D.N. Aryeh, eds. *Essays on The Land, Ecotheology and Traditions in Africa*. Eugene: Wipf & Stock, 59–69.

Öhlmann, P., Frost, M.-L., and Gräb, W., 2020. Introduction: African Initiated Christianity and Sustainable Development. *In*: P. Öhlmann, M.-L. Frost, and W. Gräb, eds. *African Initiated Christianity and the Decolonization of Development: Sustainable Development in Pentecostal and Independent Churches*. Abingdon, Oxon, New York: Routledge, 1–29.

Olúpọ̀nà, J.K., 2014. *African Religions: A Very Short Introduction*. Very Short Introductions 377. Oxford: Oxford University Press.

Oosthuizen, G.C., 1988. Interpretation of Demonic Powers in Southern African Independent Churches. *Missiology: An International Review*, 16 (1), 3–22.

Sakupapa, T.C., 2013. Spirit and Ecology in the Context of African Theology. *Scriptura*, 111, 3, 422–430.

Statistics South Africa, 2004. *Census 2001: Primary Tables South Africa: Census'96 and 2001 Compared* [online]. Available from: http://arks.princeton.edu/ark:/88435/dsp0105741v10v [Accessed 22 March 2022].

Stork, J., and Öhlmann, P., 2019. *Ökologische Nachhaltigkeit in African Initiated Churches – Kurzstellungnahme 04*. Berlin: Humboldt-Universität zu Berlin.

Taylor, B., Van Wieren, G., and Zaleha, B., 2016. The Greening of Religion Hypothesis (Part One): From Lynn White, Jr. and Claims That Religions Can Promote Environmentally Destructive Attitudes and Behaviors to Assertions They Are Becoming Environmentally Friendly. *Journal for the Study of Religion, Nature and Culture*, 10 (3), 268–305.

Turner, V., 1969. *The Ritual Process: Structure and Anti-Structure*. Chicago: Aldine Publishing.

UNEP, 2016. *GEO-6 Regional Assessment for Africa* [online]. Available from: www.unep.org/resources/assessment/geo-6-regional-assessment-africa [Accessed 22 March 2022].

World Commission on Environment and Development, 1987. *Our Common Future*. Oxford: Oxford University Press.

Part IV

Religious-Societal Tensions

11 Kosher Electricity and Sustainability

Building Block or Stumbling Stone?

Lior Herman

Introduction

In recent years, energy transformations and the socio-technical dimensions of energy have come to the forefront of academic research and public awareness (e.g., Cherp et al. 2018). In these contexts, religion seems to play a marginal and ambiguous role, offering a unique cultural-technical perspective on the issue of energy transformations (Koehrsen 2015; Leonard and Pepper 2015; Parker 2015). Yet, as a system of faith, worship, and morality, religion can assume a critical and powerful role, guiding and directing the actions of individuals, organizations, and states. Put differently, sustainable energy transformations require considerable changes in individual and collective preferences and behavior, in addition to political support and extensive resources. If the interests of religious players intersect with energy-related issues, these players can wield a considerable influence, of either a positive or negative nature, over such transformation processes.

This chapter examines the case study of Kosher Electricity (KE), exploring how the social acceptance of various means of electricity production affect individuals and communities, and the ensuing ramifications for sustainability. KE is electricity produced in accordance with the Sabbath observance. Indeed, according to Jewish law, human participation in the production and usage of electricity during the Sabbath is prohibited, as part of the wider restriction on performing work during that day. Since the Israeli electricity system is not fully automatized, many ultra-orthodox Jews regard electricity produced during the Sabbath as non-kosher. The article asks whether religious actors contribute to or hinder sustainability transitions. Contrary to existing studies, which have determined that religious actors' participation has negligible effects (e.g., Koehrsen 2015), this research discerns that religious agency has extensive influence and impact on sustainability, in a variety of ways. While issues of sustainability are abundant and consistent with Judaism, ultra-orthodox actors are driven by concerns of Sabbath observance, which can have both positive and negative impacts on sustainability. Thus, the struggle over KE exposes not only several religious-societal

DOI: 10.4324/9781003017967-15

tensions but in addition inner-religious tensions. Indeed, these different categories of tension are not independent but rather interlinked.

This chapter is structured as follows: the next section provides a brief literature review. The ensuing two sections present the methodology underlying this research, followed by a discussion of KE and how it relates to the Sabbath. Thereafter, the chapter explores the case study of KE, its implications for sustainable transition, and the role of religious actors. The final section draws conclusions, and discusses three religious-societal tensions and one inner-religious tension emerging from the nexus between the production and use of electricity during the Sabbath and sustainable transition.

State of the Art: Religion and Sustainable Transition

Sustainability transitions involve, among other things, implementing sustainable ways of energy production and consumption (Ramos-Mejía et al. 2018). The success of such a transition requires structural and fundamental changes in individual and collective practices, together with alterations in infrastructure. Since fossil energy consumption bears a substantial negative footprint on sustainability, transition toward renewable energy is an important (though insufficient in itself) element in fostering greater sustainability. Renewable energy is energy replenished by nature and obtained directly or indirectly from renewable sources such as solar radiation, wind, geothermal heat, rain, and waves (Ellabban et al. 2014). In this context, religion is regarded as a potentially powerful resource in garnering the necessary behavioral change. As a system of faith, worship, and moral foundation, religion, and together with it, religious actors, is well-positioned to mediate between values and human behavior, directing and influencing people's actions to facilitate sustainable energy transition (Gottlieb 2006; Laurendeau 2011).

Eco-theology

Nevertheless, the literature considering the role of religion in sustainable transition remains limited and lacks empirical grounding. This literature can be broadly divided into two strands. One strand is concerned with the nexus between religion and ecology (Gottlieb 2006; Johnston 2010). It examines religions' attitudes toward environment, focusing on how various religions can contribute to sustainable change. Given the mounting interest in climate change and the environment, there is a concomitant effort, particularly by religious actors, to identify ways and means whereby religious teachings and values fundamentally support environmental objectives and how religious values, institutions, and resources can be practically harnessed for the benefit of the environment and renewable energy transitions (Harper 2011; Laurendeau 2011). For example, the growth of religious environmental awareness in the United States is accompanied by a rising tendency to invoke the terms "stewardship" and "justice" in reference to man's relationship with the natural world and to encourage environment-friendly practices (Harper 2011). The eclectic role of religion vis-à-vis the environment includes religion's moral imperative to challenge prevailing consumerist values (Gottlieb 2008; Simkins 2008); its utilization of religious facilities and resources to promote green practices (Koehrsen 2015); its contribution to a more comprehensive definition of sustainability based on religious traditions (Tucker 2008); and its provision of the language, imagery, and metaphor that integrates indigenous values into the sustainability narrative, hence facilitating the intersection between the sustainability discourses of marginalized peoples and the global North (Johnston 2010). It should be noted that environmental beliefs and support of renewable energy can be mutually exclusive. Indeed, we should not assume that environmental beliefs necessarily engender support of renewable energy, given that research has yielded conflicting results in this regard (Olson-Hazboun et al. 2016).

The theological emphasis of the ecotheological approach enriches our understanding of religious attitudes toward the sustainability transformation debate. It is useful in understanding the underlying arguments of various faiths for and against sustainability, as well as theological contradictions and pragmatic avenues for change. However, this literature is limited in several ways. First, its arguments remain largely hypothetical and functional. Thus, it supposes that religion can facilitate change yet falls short in empirically supporting this claim. Second, as this literature is theologically driven, it regards sustainability largely as a fixed construct and consequently does not relate to advances in sustainable transitions research or prevailing debates in these fields (Koehrsen 2015). Finally, the literature ascribes limited agency to religion or religious actors in facilitating sustainable transition. Religious actors can be messengers and mobilizers of religious constituencies via various means, ranging from speeches, writing, symbols, and example setting (Rasmussen 2011). They can also act as facilitators and arbitrators of moral challenges relating to transitions (Laurendeau 2011). However, these roles appear to be rather limited (El Jurdi et al. 2017) compared to the broad array of roles and functions ascribed to other actors who are engaged in facilitating socio-technical transformations (e.g., interest groups, coalitions, political activism, etc.). In the absence of empirical foundations, it remains unclear how religious actors in practice affect change or what happens when religious prescriptions clash with other religious cosmologies or political and economic practices.

In recent years, there has also been growing interest in the specific relationship between Judaism and sustainability. This literature, which is mostly grounded in the eco-theology literature, emphasizes universal and Jewish duties toward nature. It attempts to reconcile conflicts between, on the one hand, conceptions of man as above nature and the reconfiguration of nature for man's utility and, on the other, humanity's duties to protect, respect, and preserve nature as God's creation (Gerstenfeld 1998; Benstein 2006). Yet, scholarly work on Judaism and sustainability suffers from similar problems to those that beset the more general eco-theology literature. As such, it fails to provide empirical analysis and lacks focus on agency and leaves RET underexplored.

Socio-Technical Transition

The second literature strand regards sustainability transitions as socio-technical transitions. This literature discusses sustainable transitions as a function of both social and technical spheres, identifying transitions as the co-production and multifaceted interplay between technology, industry, markets, policy, civil society, and culture (Geels 2010; Geels et al. 2017). This literature accords religion little attention; it is often explored as part of the cultural component, which is merely identified as an element of the socio-technical regime and not examined in significant depth. A study of the Rathu Chu hydropower project in Sikkim, India, for example, demonstrates how hydroelectric project developers and proponents ran afoul of a coalition of environmentalists and Buddhist priests opposing the threat to Sikkim's most sacred river (Huber and Joshi 2015). Another example – a study examining the practice of solar cooking in different cultural contexts via six cases in two countries, namely, Burkina Faso and India – demonstrates how both culture and religion can either impede or facilitate the adoption of solar cooking practices (Otte 2014). In contrast, Koehrsen (2015) found the role of religion in an instance of energy transition to be tenuous at best. He examined the impact of religion on energy transition in the city of Emden, Germany, and the role of religion as a local subsystem of regional innovation systems. Koehrsen determined that religion was in fact subsumed by other, more important, such as the political and economic subsystems. This finding points to the significance of local variation with regard to religion's role in energy transformations. Thus, with a few exceptions, according to this strand of literature, religion remains largely passive and in the background of sustainability transitions (Koehrsen 2018).

Methodology

KE has yet to receive academic attention. Nevertheless, a large body of theological literature, mainly in Hebrew, addresses the religious aspects of electricity usage and production during the Sabbath. I draw on this literature to explain the concept and reasoning behind KE. The empirical analysis was conducted during 2018–2019 using complementary methods. First, data was gathered from written and broadcasted media, including national, local, and religious newspapers as well as television and online videos. This data was used to trace the chain and sequence of events relating to KE, in addition to determining the views and positions of various stakeholders. Media data was also examined as background material for the interviews and in formulating questions. Second, primary documents were collected from several actors, among them government offices, parliament (the Knesset), and stakeholders. These primary documents include protocols of meetings, position papers, correspondence, plans, and reports. Finally, semi-structured interviews were conducted with nine stakeholders, among them government officials, ultra-orthodox activists, ultra-orthodox consumers, a generator operator, and KE business actors. These interviews were transcribed and selected on the basis of relevancy (e.g., key decision makers or main market actors in the field), and availability (e.g., illegal generator operator and consumers). Some interviewees requested to remain anonymous and therefore interviewees are referred to herein by numbers. Different data sources were cross-checked and, in cases of inconsistency, omitted from the analysis.

Kosher Electricity

The observance of the Sabbath, the seventh day, is a central pillar of the Jewish faith. On this day, God ceased the work of creation and rested; thus, the Torah prescribes its observance as an inter-generational covenant between God and the Israelites. Indeed, the Torah repeatedly prohibits the performance of any kind of work on this day of rest (e.g., Exodus 31:12–17). Every year there are 60 days of Sabbath and other holidays (during which the same rules proscribing work apply), amounting to some 1,560 hours. The prohibition of work during these days includes a long list of activities, as expounded by the Halakha, the collective body of Jewish religious laws derived from the Written and Oral Torah, as well as rabbinical rulings (Responsa) from later generations. The Halakha lists 39 categories of activities (*lamed tet avot melakha*) from which Jews must refrain during the Sabbath.

There is a strong consensus within different branches of Jewish orthodoxy, both ultra-orthodox and national religious, that the use of electricity during the Sabbath is prohibited if it is directly operated by a Jew. For example, it is prohibited to turn on the light by pressing a switch. At the same time, the use of electricity would be permitted if it is carried out without active involvement. For example, it is permitted to use pre-set timers (known as Sabbath Clocks), which automatically turn lights on and off during the Sabbath. It is also permitted to leave the air-conditioning system turned on before the Sabbath begins and to enjoy it throughout the day, so long as it is not turned off during the Sabbath and its settings remain unchanged.

Despite the consensus among orthodox Jews that the operation of electricity is prohibited, disagreement prevails with regard to which of the 39 categories of activities mandates this prohibition. For example, many rabbis have taken the view that turning on an incandescent lightbulb violates the prohibition against igniting fire. Though the tungsten wire is not burning inside the bulb, it creates force which generates heat and light and these are restricted in similar cases by the Halakha and Torah (Broyde and Jachter

1991). Rabbi Avraham Yeshaya Karelitz, known as the Chazon Ish (1878–1953) and one of the most influential leaders of ultra-orthodox Judaism in the early days of the state of Israel (Kaplan 1992), argued that closing an electrical circuit is equivalent to completing a product, which is referred to as "striking the final hammer blow" (*makkeh bepatish*) and is also prohibited. He similarly likened the closing and opening of an electric circuit to the prohibited acts of building and destroying (Kletenik 2010). A more controversial interpretation is that which compares an electric current to the restricted category of creating anything new (*molid*) (Broyde and Jachter 1991; Ben Yaakov 2010).

The problem of using electricity on the Sabbath is not limited to consumption. Even if indirect use of electricity is permitted under certain conditions, electricity must still be produced and delivered in ways consistent with the Halakha. However, modern, large-scale production, transmission, and distribution of electricity involves human labor at various stages. If a Jew works during the Sabbath to produce electricity, consequently violating the laws regarding the day of rest, other Jews are not permitted to benefit from that electricity. Since certain places, such as hospitals, require a continuous supply of electricity in order to operate life-saving machinery, the majority of orthodox authorities have permitted the use of electricity generated during the Sabbath if it is produced for the purpose of saving lives (*pikuach nefesh*).[1] Thus, according to the common view and practice, even though Jews are involved in the production of electricity during the Sabbath, the consumption of this electricity is permitted because it is required for the operation of hospitals, outpatients at home (e.g., storage of medicine in refrigerators and other life-saving equipment) and other necessary activities (*mevashel lachole*).

Kosher Electricity: A Building Block or Stumbling Stone?

While most orthodox and ultra-orthodox Jews consume electricity produced during the Sabbath in accordance with the principle outlined previously, increasing numbers of ultra-orthodox communities reject this behavior.[2] Such communities follow the rulings of Chazon Ish, who made this a cardinal matter, ruling already in 1948 (the year of the establishment of the State of Israel) that using non-KE is a religious offence of the highest order: It disgraces the very name of God because it reveals people's hypocritical faith (*chilul hashem*). Indeed, a Jew cannot enjoy the fruit of forbidden work performed by another Jew. Even if electricity is needed to save lives, others should not be using it, because this leads to more work than is strictly necessary (*shema yirbe*). Furthermore, these communities argue that those Jews who work on the Sabbath to produce electricity intended for saving lives are being misled: They are unaware that in practice they produce more electricity than necessary (Darzi 1998; Morgenstern 2012). As noted by a KE consumer, "It is like burning a Torah scroll and then lighting a cigarette from the ashes" (Interview 3). This "holier than thou" or overly strict approach suggests that KE invokes inner-religious tensions, as has also been noted with regard to the Majlis Ulama in Indonesia (Jamil 2022).

The rejection of electricity produced by the national utility company, the Israel Electric Company (IEC), has acute negative consequences for sustainability. These KE communities, initially consisting of followers of the Chazon Ish and increasingly also spreading to other parts of the ultra-orthodox population, Lithuanian and other ultra-orthodox streams (including non-Ashkenazi Jews), includes some 50,000 households, amounting to some 250,000 people. Rather than avoiding electricity completely during the Sabbath, and thus inadvertently contributing to the reduction of energy consumption, pollution, and emissions, these KE communities implement a highly unsustainable solution. Every weekend, and before holidays, community members disconnect from the national electricity grid for the duration of the Sabbath or holiday, relying instead on diesel generators and batteries to supply electricity to their homes and institutions (for example synagogues, yeshivas).

Diesel generator sets are the most common solution employed by KE followers living together as a community. Their usage reflects a religious-societal tension between religious KE communities and the state. Until now, there has been no regulation permitting these generators, and therefore their operation and use is illegal (Ynet 2007). Nevertheless, and despite the fact that only the IEC is allowed to disconnect and connect customers to the national grid, no actions have been taken by the state against the operators of these generators. The generators are often installed in densely populated areas, where there is limited space and no advance planning for their location or connection to households and institutions. Diesel generator sets are highly polluting and generate toxic pollutants, such as particulate matters and nitrogen oxide (Aderibigbe et al. 2017). The generators also create constant noise pollution throughout the Sabbath. Consuming KE is likewise highly expensive – this is particularly relevant considering that the ultra-orthodox community at large is poorer than the general population in relative terms – and purchasing KE can double (or more) the overall monthly expenditure on electricity for households and institutions. Nevertheless, KE consumers who follow the rulings of the Chazon Ish and the rabbinical authorities that succeeded him accept the financial burden and environmental costs, viewing them as a necessity and "welcoming expenditure for the dignity of the Sabbath" (Interview 2). "People believe that what they spend on the Sabbath, returns as a blessing" (Interview 3).

Many KE consumers also use batteries. Batteries are widely used when there is no access to diesel generators, as in the cases of smaller KE communities or individual KE consumers living outside such communities. Some also use the batteries as a backup for generators in the case of failure (Interviews 3; 8). The batteries differ from generators in terms of costs: They involve an initial costly expenditure yet entail lower operational costs. Until recent years, car batteries were the most common solution: they were connected to various electric appliances with wires. These batteries were not only dangerous but also generated pollution inside homes from chemicals such as mercury and lead. Today, lithium batteries connected directly to the main electricity switch at home are often used (Interviews 2; 7).

Why Has the Promotion of KE Been Successful?

There are several reasons why ultra-orthodox actors have been so successful in promoting KE among their communities, with the ensuing negative repercussions for sustainability. First, ultra-orthodox communities possess several relevant distinguishing characteristics.[3] They are strictly observant and tend to follow the Halakha stringently (particularly Ashkenazi, ultra-orthodox Jews). The Rabbi is considered not only a spiritual and religious authority but also an authority on public and personal matters. As such, rabbis wield great influence and act as the supreme arbiters in their communities. Such communities are also deliberately segregated from the "outer world," physically and spiritually. In so doing, they attempt to shield themselves from exposure to undesired external values: This is achieved via separate housing, education systems, and ultra-orthodox mass media (newspapers, websites, radio). These characteristics allow religious actors, such as rabbis, to exert significant influence over the behavior of their constituencies, as was particularly true in the case of the Chazon Ish (Kaplan 1992). This isolation results in little awareness of environmentalism and sustainability issues. Consequently, there is little opposition to solutions that are highly unsustainable. Lack of awareness regarding sustainability is also the result of the centuries-old ultra-orthodox tradition that focuses on studying Torah and Halakha and pays little attention to the physical vanities of the world. This tradition was reinforced by the lack of a Jewish state and the largely urban character of pre-Holocaust European Jewry, which was largely excluded from engaging in civil governance and did not nurture an attachment to nature and related values (Shelhav and Kaplan 2003; Wolpe 2012). Hence, although the sustainability discourse and its relationship to Judaism has penetrated modern orthodox and non-orthodox Judaism, it has failed to affect ultra-orthodoxy (Interviews 1, 6). As one ultra-orthodox environmental activist noted, "Our public is not connected to global environmental issues. We prefer to leave the global concern for the future of the world to the leader of the world [God], may He be blessed, and we have no one else to rely on but our Lord in heaven" (Interview 6).

Second, the concentration of ultra-orthodox communities makes them important actors at the municipal level in many cities throughout Israel. Indeed, constituting relatively large constituencies within specific cities and towns enables KE religious actors to influence local politics in several ways. In some cities, where the ultra-orthodox account for the majority of residents, such as Bnei Brak, Kiryat Sefer, Beit Shemesh, and Beitar Ilit, they are involved in municipal politics and administration. In these municipalities, such players are able to use the local government to promote the installation of diesel generators, for example, by appropriating land for the generators and installing the necessary infrastructure to connect customers, even though such activities are illegal (Ynet 2013b). In one instance, in Bnei Brak, the municipality seized an area designated as a public park and allowed the illegal installation of a large compound of KE diesel generators. The resulting massive pollution and noise led several citizens to sue the municipality, which initially refused to move the farms, even after the court ruled against it (Ynet 2014). Ultra-orthodox actors are also able influence municipal politics in mixed cities, such Jerusalem and Zichron Ya'akov. In these cities, the municipality may not promote KE diesel generators but turns a blind eye, refusing to remove generators or take action against their operators (Israel Broadcast Authority 2014).

Finally, as segregated communities, ultra-orthodox communities tend to seek independent solutions. Indeed, some community entrepreneurs establish their own diesel generator sets and are thus not dependent on other actors, like the Israel Electric Company (IEC) (Interview 7).[4] These ventures are usually developed as a not-for-profit business (*gemach*), which also contributes to lesser enforcement by the tax authorities. Since these are illegal ventures, the operators have little incentive to meet environmental standards and regulations (Interview 6). Yet, as mentioned, electricity produced by the generators is very costly and this may deter consumers from purchasing. Rabbinical rulings that expenditure on KE electricity is an expenditure for the Sabbath helps to reframe and positively change people's attitudes toward purchasing KE in several ways, which furthers the societal-religious tension. Framing is instrumental in legitimizing the actions of production and purchasing of KE, by placing the Sabbath above the state's secular law. Making the purchase of KE an expenditure for the Sabbath facilitates a change in people's preferences: It becomes a meritorious deed (*mitzvah*) (Interviews 3; 7; 8). Framing is also instrumental in raising donations from the community to purchase KE for community institutions, such as synagogues and yeshivas, which consume significant amounts of electricity for air conditioning and lighting (Interview 3). These institutions traditionally name and honor those who donate the KE for a particular Sabbath. Thus, KE communities create an informal market, allowing both supply and demand to function properly.

Tensions and Drivers for Change

Since 2010, changes have begun to take place within KE communities and other actors in Israel that may lead to a transformation which is less polluting and produces less emissions. Nevertheless, future development will have different consequences for sustainability transitions. These changes result from several factors, which have altered some preferences among religious, state, and market actors.

The success of the generators in providing a solution created greater demand for KE, beyond the circles of the Chazon Ish's followers and the Lithuanian stream (Hutta 2018; Ynet 2018). While the majority of ultra-orthodox and orthodox Jews previously viewed non-KE as a necessary vice for the sake of saving lives, the generators demonstrated that a relatively convenient alternative exists, one which is consistent with Sabbath observance. Thus, demand for KE began to increase among parts of a population that accounts for a fifth of Israel's residents (ultra-orthodox Jews and orthodox Jews comprise 9% and 10% of the total population respectively, and 14% and 16% of the total Jewish population) (Central Bureau of Statistics 2018).

At the same time, existing KE communities started to voice growing dissatisfaction with the negative aspects of the diesel generators. Health and safety cases, such as electrocution of children playing next to dangerously installed electricity cables in streets and between houses or wires at home, became widely known among the ultra-orthodox community (Ynet 2007). In addition, certain changes began to affect ultra-orthodox attitudes toward the unsustainability of diesel generators. On the one hand, an increase in the number of generator sets led to growing resentment at the considerable noise and pollution they create in the densely populated ultra-orthodox neighborhoods. Generator operators took some measures to address these problems – installing silencers and tall chimneys – but these did not eliminate the problems (Interview 3). On the other hand, some degree of awareness regarding environmental and sustainability issues has begun to infiltrate the ultra-orthodox community. Greater demand for cleaner streets and problems of waste, sewage, and pollution treatment have all raised awareness regarding the need to increase sustainability as a way to improve living conditions. One result of this has been the establishment of Haredim Le'Sviva, an ultra-orthodox NGO dedicated to promoting sustainability (Israel 21c 2010). However, it should be stressed that awareness of sustainability among the ultra-orthodox community and willingness to act in this regard remains limited; most community members rank religious concerns and material needs far above sustainability matters (Interviews 2; 3; 7; 8). Even if an organization such as Haredim Le'Sviva views KE as a critical environmental problem that is unique to the ultra-orthodox community (Israel 21c 2010), the organization will not advocate a solution which may compromise Sabbath observance or undermine rabbinical authority (Interview 6). Furthermore, rising expenditure on KE, due to both the increase in oil prices and cases of monopoly power abuse by generator operators, further exerted pressures on rabbis to act to improve the situation (Ynet 2011a).

Societal-religious tension and drivers for a more sustainable change also include protests by secular activists living in proximity to KE communities, complaining about the health and safety hazards created by the generators. Citizens have appealed to the Ministries of Energy and the Environment as well as local and national politicians (Israel Broadcast Authority 2014). The IEC has also recognized the importance of this issue for two reasons. First, several IEC employees were electrified when performing maintenance works in ultra-orthodox neighborhoods due to KE generators connected to the grid (Ynet 2011a). Second, the IEC recognized the loss of revenues from KE customers who disconnect during Sabbaths and holidays. It therefore sought for solutions which will enable the company to provide customers with KE electricity instead of the generators (Ynet 2013a). The economic interest of the IEC recently intensified as a reform in the national electricity market introduced private actors and prohibited the IEC from increasing its production power. Thus, all actors have become aware of the need for a solution that will provide an appropriate legal framework, which meets both health and safety and environmental standards, is consistent with electricity market regulatory rules, and takes into consideration the interests of the IEC.

Driven mostly by rabbis from the Lithuanian stream, community leaders and activists recently began pushing for a KE solution at the national level. The Electricity Administration at the Ministry of Energy refused to negotiate a solution based on religious parameters (Interviews 4; 7), as a senior government official commented: "Had we been willing to negotiate this issue [i.e., KE] as a religious issue, rather than a consumer issue, we would have lost the fight from the beginning" (Interview 4). Nevertheless, these religious actors were successful in putting KE electricity on the national agenda for a number of reasons. First, as mentioned previously, other actors were also interested in finding an appropriate solution. Second, the ultra-orthodox community is represented in the Knesset by several parties. In contrast to other parties, parliamentarians representing the ultra-orthodox are preselected directly by rabbis; these rabbis make the key decisions and set the policies later advocated by the parties in the Knesset. Ultra-orthodox parties have gained considerable electoral strength throughout the decades and play a key role in the formation of almost all government coalitions (Sandler and Kampinsky 2018). Hence, religious actors directly influence national as well as local politics. Third, ultra-orthodox activists have also proved to be well-organized and technologically knowledgeable. A case in point is the establishment of an NGO called the Committee for Matters Related to Energy-Use on the Sabbath. This committee was established by rabbis from the Lithuanian stream in order to provide logistical and planning support for the establishment of private power stations for KE; to monitor electricity production which violates Sabbath rules; to advocate and negotiate KE solutions with the government and other actors; and to create public awareness of KE (Schneider 2016). Another important aspect relating to the committee's work is the establishment of an interest-free loan fund for users of KE and the establishment of private KE power stations.

Over the last decade, even before the establishment of the committee, ultra-orthodox rabbis and activists demonstrated a remarkable degree of knowledge concerning the technical aspects of electricity production and the various available solutions. This know-how accorded ultra-orthodox actors credibility in their negotiations with governmental and market actors (Hutta 2018; Kol Hazman 2018; Interview 4).

Possible Solutions

Several solutions to the KE issue and the accompanying tensions have been proposed over the past decade. In 2011, ultra-orthodox rabbis and parliamentarians, together with the IEC (which was concerned about possible consumerist sanctions imposed by the well-organized ultra-orthodox community), convinced then-Minister of Energy Uzi Landau to propose a law requiring electricity produced during the Sabbath to be kosher throughout the whole of Israel (News 1 2011). At the time, Israel was beginning its transition to mass electricity production from natural gas, phasing out coal-based production. Compared with the situation under coal-based production, such a solution was more sustainable and would have displaced the polluting diesel generators. However, the law's provision for supervision of electricity production during the Sabbath by the national Chief Rabbinate sparked criticism from two different groups. Secular groups campaigned against the proposed law and what they viewed as religious coercion. Together with a public campaign, they collected 25,000 signatures on a petition against the law (Ynet 2011b). Surprisingly, the proposed law exposed interreligious tension, and was also opposed by the ultra-orthodox, because they do not accept the religious authority of the Chief Rabbinate, which is a state-run orthodox institution and is considered insufficiently stringent on religious matters (Interviews 4; 7). In the face of this opposition, the minister withdrew the proposed law. Other efforts included failed attempts for the IEC to supply safer generators or replace Jewish workers with non-Jewish workers during the Sabbath. The latter solution was largely accepted by both ultra-orthodox rabbis and IEC management but foundered due to strong opposition by the company's trade union. Employees working during the Sabbath receive almost double payment for their work, and they were not willing to lose this benefit (Ynet 2013a).

A major turning point occurred in July 2018, as part of the reform of the IEC. Following decades of attempts to reform the national electricity monopoly, the government was able to pass a reform of the company, cancelling its monopoly status and opening the electricity market to competition in production and delivery. As part of the reform, the IEC would no longer manage the national electricity system. It also entailed the retirement of almost 2,000 employees (Bank of Israel 2018). The reform was a golden opportunity for ultra-orthodox parties, which threatened to sabotage the bill if a solution for KE was not included within it. The ensuing political crisis brought together politicians and regulators to renegotiate a solution for KE, and eventually a compromise was reached. Under the compromise, the government committed to finding a number of KE solutions, which will then be discussed and accepted by all stakeholders (Eilat 2018). Each of these proposed solutions has various sustainability implications.

The first proposed solution is to legalize KE generators by creating appropriate regulation and standards for the operation of generators fueled by either diesel or natural gas (condensed, liquefied). This solution has been discussed for several years and an agreement has been reached that the IEC will conduct a pilot project in the city of Ramat Bet Shemesh or neighborhoods in Jerusalem or Bnei Brak (Felter-Eitan 2018). Ultra-orthodox actors, such as the Committee for Matters Related to Energy-Use on the Sabbath as well as local rabbis and activists throughout Israel, favor this solution in the short term because it can address the need for KE relatively quickly. Compared to costly solutions such as the full automation of IEC production or the establishment of KE power stations, the installation of generators is cheaper, and it is easier to find appropriate land for their installation, either within cities or on their outskirts. Ultra-orthodox rabbis regard the generators as a partial and complementary response, alongside other solutions (Felter-Eitan 2018). While the government agreed to hold a generator pilot test, it viewed this as a highly unsustainable solution, contrary to its efforts to decrease emissions and pollution. Research evaluating KE production options, concerning the average scenario of tradeoff between installing one power station or eight generators (considering also different fuels), showed that a power station will decrease pollution from generators from 100% to less than 10%. Pollution from a power station will decrease even further when calculated for various other likely operational scenarios during day and night times. Thus, the generator option is likely to augment the current, unsustainable situation, emitting high levels of pollution within urban areas (Ministry of Environment 2018; Skivin 2018). An examination of existing KE generators placed in the Har Habanim neighborhood in Bnei Brak revealed significant pollution, electromagnetic radiation, and noise even when generators are installed in open areas outside neighborhoods (Maruani 2014).

The second proposed solution is the establishment of peaking power plants. These are power plants that usually operate only when the demand for electricity is high. While these peaking power plants can operate on a variety of fuels, they are usually gas turbines, which automatically begin to work when supply by a base load power plant does not meet demand (Masters 2013). This proposed solution would establish peaking power plants in cities with a sizeable ultra-orthodox community, either homogenous ultra-orthodox or mixed cities. The plants will be completely automatic and entail no human involvement. They will be connected to the main grid and contribute to the general power supply during the week. Over weekends, they will be disconnected from the grid and operate in those cities in a mini-grid format, ensuring that the electricity supplied is kosher and not mixed with electricity produced elsewhere. Peaking power plants improve sustainability in the short run, because they will run on natural gas and thus replace existing generators. Nevertheless, over the long term they will produce more energy and reduce incentives to install renewable energy solutions. Although ultra-orthodox actors would prefer a nation-wide solution that will fully automate the electricity system, they recognize that such a solution is less feasible from technical and financial perspectives and may also clash with security considerations. Thus, ultra-orthodox actors prefer the peaking power solution over other alternatives (Gafni 2018, Interviews 4; 7), and few actors, such as parliamentarians from the Sephardi Shas ultra-orthodox party, attempt to push for full automation.

The third proposed solution seeks to replace fossil-based electricity production with renewable energy. While governmental actors, such as the Ministry of Environmental Protection and The Ministry of Energy, as well as environmental NGOs prefer this solution (Interviews 4; 7), it is not promoted by key actors in the ultra-orthodox community. Since appropriate wind and water conditions are of lesser relevance in most of Israel, solar energy is the only pertinent renewable solution. Indeed, solar energy is attractive, given the excellent solar radiation conditions in Israel throughout most of the year. Solar solutions would promote the transition toward a more sustainable energy system and are consistent with Israel's commitment to increasing renewable energy's share of its total energy mix. It is also a solution consistent with Sabbath observance, without any need for human involvement in production and supply. Ultra-orthodox actors, such as the Committee for Matters Related to Energy-Use on the Sabbath and parliamentarians, do not oppose solar solutions. However, they do not believe that solar panels can provide the energy availability and reliability necessary for the ultra-orthodox community. Their reasoning is three-fold. First, they argue that Israel is too small and lacks open areas near cities that can be utilized to construct sufficiently large solar energy fields. Second, they maintain that the ultra-orthodox community resides in highly dense urban settings and consequently often in high-rising buildings. In these settings, there is not enough roof space to generate sufficient electricity, and parts of these roofs are already taken up by solar water heaters. Third, they do not trust that existing storage solutions will be sufficient to support them throughout the duration of Sabbath (25 hours) or in the event of a two-day holiday or a holiday and Shabbat together (48 hours). They also deem existing storage solutions too expensive, and while recognizing that storage technology is rapidly advancing, they insist on finding an immediate solution (Gafni 2018; Ministry of Environment 2018).

While ultra-orthodox actors are not advancing solar energy as a national or local KE solution, changes are occurring as part of a bottom-up process. In light of the considerable reduction in the price of solar panels in recent years, coupled with governmental incentives, market actors have recognized the potential economic opportunities of selling solar energy systems to the ultra-orthodox community (Interviews 1; 2; 9). Hundreds of such systems are now sold every year to ultra-orthodox and orthodox households, and to a lesser extent to synagogues and yeshivas. Solar energy entrepreneurs report that demand is mostly motivated by economic considerations: reducing electricity costs. However, consumers realize that the system also provides a KE solution, and this in turn stimulates a growing demand for KE among the orthodox community (Interviews 2; 9). Some entrepreneurs likewise provide unique KE components to the solar energy systems that add additional smart-house features, allowing for better kosher consumption of the electricity produced, as it automatically controls various appliances and uses. One interviewee reported that the smart-house KE solution has a major positive, sustainable spillover effect, and it is now sold by the company in various African countries (Interview 1). In certain locations, KE consumers using solar energy systems sell any excess solar energy produced during the week to KE generator operators.

A range of acceptable solutions to the KE problem are currently being advanced. These have different impacts on the transition toward a more sustainable energy system. Solutions such as generators and peaking power plants have negative consequences for sustainability in terms of emissions, pollution, radiation, and noise. If powered by natural gas, new generators will certainly improve the existing situation in the short term, yet they will still have considerable negative effects. Adopting these solutions also entail tradeoff consequences for finding large- scale renewable energy and storage solutions. At the same time, bottom-up solutions, driven by both economic and KE interests, push for greater deployment of solar energy, thus offering a positive contribution to a more sustainable energy system.

Conclusions

Recent surveys have estimated that 84% of the world's population identify with a religious group (Hackett and Grim 2012; Hackett and Stonawski 2017). Given the ongoing importance of religion, religious actors are well positioned to influence the beliefs and actions of individuals with regard to sustainability transition. While previous research has found that religious actors wield little to negligible influence in this respect (Koehrsen 2015), the current contribution argues that although ultra-orthodox agency is not averse to sustainability, it nevertheless places the Sabbath, and consequently KE, above sustainability, even when reconciliation between the two is possible. Ultra-orthodox Jewish agency in the context of KE is not *dark green* motivated, in the sense of considering nature as sacred and worthy of reverent care (Taylor 2010). Indeed, the sacredness of the Sabbath overrides all other considerations, even if sustainability is considered important. Thus, ultra-orthodox preferences tend to have a negative impact on sustainability, although the recent and growing adoption of solar energy systems has had a positive effect. In religious agency terms, ultra-orthodox agency in the context of KE derives from both individual (e.g., rabbis) and collective sources (e.g., NGOs and religious parties) and is consistent with previous findings in sustainable transition and religious agency literature (Pesch 2015; Koehrsen 2018). At the same time, the case of KE also reflects on some of the limitations in the eco-theology literature. The empirical evidence runs against the hypothetical and functional arguments largely assumed in parts of this literature (e.g., Gottlieb 2008; Harper 2011; Laurendeau 2011) that religious institutions and agents will be harnessed for the benefit of sustainability. Furthermore, while socio-technical transition literature findings identify religious agents as largely passive regarding sustainable transition, religious actors in the case of KE have been active in influencing and slowing transition.

The case of KE exposes several tensions relating to sustainability. Religious-societal tensions are most evident and are informed by at least three distinct yet related types of tension. First among these are the religious-societal tensions between KE communities, on the one hand, and secular and non-ultra-orthodox Jews, on the other. These tensions concern the permissibility of producing and using electricity during the Sabbath, and the adoption, placement, and financing of alternative solutions consistent with KE. Ultra-orthodox rabbis, activists, and politicians play an important role in intensifying these tensions, with rabbis assuming a particular agency role in terms of framing and projecting community identity. Nevertheless, religious actors are not immune to internal community pressure. When resistance to the adverse impact of diesel generators increased, rabbis and other ultra-orthodox actors played a positive role in reducing tensions, albeit reframing the problem in terms of community well-being rather than sustainability and environmentalism. Second, tension exists between ultra-orthodox KE communities and the state. This tension is most evident in the existence of an illegal KE market, which is religiously legitimized by rabbis. Need, rather than ideological or religious compromise, pushes KE communities to attempt to legalize this unlawful behavior by leveraging ultra-orthodox power in the state's institutions, most notably the Knesset and city councils. The cases of Halal recycled wastewater in Indonesia and KE in Israel are both cases in which religious actors are powerful and have an influence over politics. These cases differ from each other since in the Halal recycled wastewater case, politicians seek legitimacy from religious leaders (Jamil 2022), while in the case of KE, religious leaders actively direct politicians' behavior. This situation differs from Western European politics, where religious actors are less powerful and consequently have less impact on sustainability transitions.

A third religious-societal tension exists with regard to the energy sector. While differences in religious interpretations concerning the use of electricity date back to the discovery of electricity, these differences have recently intensified in three dimensions. The first dimension concerns greater awareness regarding the *kosherness* of electricity along the various segments of the electricity production chain. Ultra-orthodox demands extend from theological debates on the notion of electricity and its consumption to the acts of production, transmission, and distribution. These demands transform KE into a national problem rather than an issue that can be resolved on an individual or community basis. The second dimension relates to the choice of the energy resources used to produce electricity and the means of production. This dimension has significant negative implications on sustainability, because KE electricity proponents have immediate preferences for more polluting solutions. The third dimension concerns authority. In contrast to other cases in which religious actors' preferences were influenced by knowledge deriving from external scientific and technical authorities (Gojowczyk 2022), ultra-orthodox rabbis and activists have demonstrated a remarkable degree of learning and have researched the technical and scientific aspects of electricity production. This independent capacity often clashes with traditional actors possessing such authority, such as government agencies. Renewable energy solutions to the problem of KE are cases in point: ultra-orthodox actors refuse to accept the scientific and technical authority of the Ministry of Energy concerning renewable energy production and storage. As noted in an official position paper of the Committee for Matters Related to Energy-Use on Shabbat, outlining the scientific and technical consistency of various solutions with KE:

> As concerns solar energy, there is difficulty in maintaining continuous supply, not to mention the wide area necessary, which is not found in populated areas. The area on the roofs is also negligible in relation to the amount of consumption required. As for wind energy, there is difficulty in maintaining continuous supply.
>
> (Committee for Matters Related to Energy-Use on Shabbat 2017)

This issue was also apparent in the 2018 government compromise following the IEC reform crisis, which sought simultaneous multiple solutions due to disagreement concerning the feasibility of renewable energy to meet KE needs (Eilat 2018).

Against these three religious-societal tensions, a fourth inner-religious tension unfolds, revealing that these types of tensions are not necessarily dichotomous and that tensions are often linked. Inner-religious tensions are evident in the case of KE in several ways. One example is ultra-orthodox communities' refusal to accept KE solutions offered by orthodox institutions (e.g., the Chief Rabbinate). Another concerns the interpretation of sustainability and the environment, and the preferences accorded to them over other issues. Jewish theological interpretations attribute great importance to sustainability. However, these founder when juxtaposed with Sabbath observance, posing a problem also for ultra-orthodox environmentalists. Thus, there is a need for integration and communication between religious environmentalism and sustainability literature. Environmental and sustainability activists must improve their understanding of religious nuances and interpretations, and the ways they inform the preferences of religious actors. Such understanding can help future framing and fine-tuning of both religious and sustainability messages and solutions, and may reconcile these tensions.

Notes

1 According to this principle, the preservation of human life overrides almost all other religious considerations, including the rules of Sabbath observance.
2 Production of electricity during the Sabbath is less of a problem outside Israel because electricity is produced by non-Jews and predominantly for the use of non-Jews. Thus, Jews can enjoy this electricity inadvertently and unintentionally.
3 These characteristics are general to ultra-orthodox communities. However, ultra-orthodox Jewry is not unsusceptible to external influences, particularly the clash between modernity and religious conservatism, as well as to conflicts emanating from living within a democratic secular state.
4 Kiryat Sefer, an ultra-orthodox settlement in the West Bank, is an exception due to the fact that Israeli law does not apply in the West Bank. In 1999, the IEC installed its own diesel generator farm to provide KE for this relatively new town (established in 1990).

References

Aderibigbe, M. et al., 2017. Retrofitting Diesel Engine Generators for a Sustainable Environment. *Journal of Engineering and Applied Science*, 12 (12), 3242–3246.

Bank of Israel, 2018. *The Reform in the Electricity Company: Summary and Evaluation*. http://www.boi.org.il/he/NewsAndPublications/PressReleases/Documents/%D7%94%D7%A8%D7%A4%D7%95%D7%A8%D7%9E%D7%94%20%D7%91%D7%97%D7%91%D7%A8%D7%AA%20%D7%94%D7%97%D7%A9%D7%9E%D7%9C.pdf [Accessed 31 August 2022].

Ben Yaakov, I., 2010. *Electricity and Halakha: Halakhic Responses to Electricity, 1875–1940*. Beer Sheva: Ben Gurion University.

Benstein, J., 2006. *The Way into Judaism and the Environment*. Woodstock: Jewish Lights Publishing.

Broyde, M., and Jachter, H., 1991. The Use of Electricity on Shabbat and Yom Tov. *Journal of Halacha & Contemporary Society*, 21 (26), 4–47.

Central Bureau of Statistics, 2018. *Religion and Self-Definition of Extent of Religiosity: Selected Data from the Society in Israel, Report No. 10* [online]. Available from: www.cbs.gov.il/he/mediarelease/DocLib/2018/195/32_18_195b.pdf [Accessed 4 April 2022].

Cherp, A. et al., 2018. Integrating Techno-Economic, Socio-Technical and Political Perspectives on National Energy Transitions: A Meta-Theoretical Framework. *Energy Research & Social Science*, 37 (Supplement C), 175–190.

Committee for Matters Related to Energy-Use on Shabbat, 2017. *Electricity Provision During Shabbat to the Ultra-Orthodox Population: Regulatory and Technological Aspects*. Bnei Brak.

Darzi, Y., 1998. *Shevut Yitzhak*. Jerusalem: Yeshivat Haran.

Eilat, A., 2018. *Letter to MK Eitan Cabel: Priniciples for Consideration of Kosher Electricity and Possible Solutions*. Jerusalem.

El Jurdi, H.A. et al., 2017. Harnessing the Power of Religion: Broadening Sustainability Research and Practice in the Advancement of Ecology. *Journal of Macromarketing*, 37 (1), 7–24.

Ellabban, O.H. et al., 2014. Renewable Energy Resources: Current Status, Future Prospects and Their Enabling Technology. *Renewable and Sustainable Energy Reviews*, 39, 748–764.

Felter-Eitan, N., 2018. *Meeting Summary: Kosher Electricity Policy – Deputy Finance Minister, IEC, Electricity Authority, 24 June*. Jerusalem.

Gafni, M., 2018. *Meeting Summary, Knesset Finance Committee, Chairman Moshe Gafni, Concerning Kosher Electricity Solutions, 2 May*. Jerusalem. Document no. 82596.

Geels, F.W., 2010. Ontologies, Socio-Technical Transitions (to Sustainability), and the Multi-Level Perspective. *Research Policy*, 39 (4), 495–510.

Geels, F.W. et al., 2017. The Socio-Technical Dynamics of Low-Carbon Transitions. *Joule*, 1 (3), 463–479.

Gerstenfeld, M., 1998. *Judaism, Environmentalism and the Environment: Mapping and Analysis*. Jerusalem: Jerusalem Institute for Israel Studies.

Gojowczyk, J., 2022. From Global Goal to Local Practice. Potential Lines of Tension in Religious Environmentalism in Catholic Religious Orders. *In:* J. Koehrsen, J. Blanc, and F. Huber, eds. *Religious Environmental Activism: Emerging Conflicts and Tensions in Earth Stewardship*. London: Routledge.

Gottlieb, R.S., 2006. *The Oxford Handbook of Religion and Ecology*. Oxford: Oxford University Press.

Gottlieb, R.S., 2008. You Gonna Be Here Long? Religion and Sustainability. *Worldviews: Global Religions, Culture, and Ecology*, 12 (2–3), 163–178.

Hackett, C., and Grim, J.B., 2012. *The Global Religious Landscape*. Washington, DC: Pew Research Center.

Hackett, C., and Stonawski, M., 2017. *The Changing Global Religious Landscape*. Washington, DC: Pew Research Center.

Harper, F., 2011. Greening Faith: Turning Belief into Action for the Earth. *Zygon*, 46 (4), 957–971.

Huber, A., and Joshi, D., 2015. Hydropower, Anti-Politics, and the Opening of New Political Spaces in the Eastern Himalayas. *World Development*, 76, 13–25.

Hutta, B., 2018. *The Million Petition for the Holy Sabbath* [online]. Available from: www.youtube.com/watch?v=7UnS59JbDt8 [Accessed 1 May 2019].

Israel 21c, 2010. *Bringing the Environment to Israel's Ultra-Orthodox* [online]. Available from: www.israel21c.org/bringing-the-environment-to-israels-ultra-orthodox/ [Accessed 1 May 2019].

Israel Broadcast Authority, 2014. "Kosher" Electricity: Life-Threatening. *Channel 1* [online]. Available from: www.youtube.com/watch?v=2mbZVoUndWY [Accessed 1 May 2019].

Jamil, S., 2022. Halal Wastewater Recycling. Environmental Solution or Religious Complication? *In:* J. Koehrsen, J. Blanc, and F. Huber, eds. *Religious Environmental Activism: Emerging Conflicts and Tensions in Earth Stewardship*. London: Routledge.

Johnston, L., 2010. The Religious Dimensions of Sustainability: Institutional Religions, Civil Society, and International Politics Since the Turn of the Twentieth Century. *Religion Compass*, 4 (3), 176–189.

Kaplan, L.J., 1992. The Hazon Ish: Haredi Critic of Traditional Orthodoxy. *In:* J. Wertheimer, ed. *The Uses of Tradition: Jewish Continuity in the Modern Era*. New York, Cambridge: Jewish Theological Seminary and Harvard University, 145–173.

Kletenik, M., 2010. Alarm Systems in Halakha. *Tradition: A Journal of Orthodox Jewish Thought*, 43 (4), 57–68.

Koehrsen, J., 2015. Does Religion Promote Environmental Sustainability? – Exploring the Role of Religion in Local Energy Transitions. *Social Compass*, 62 (3), 296–310.

Koehrsen, J., 2018. Religious Agency in Sustainability Transitions: Between Experimentation, Upscaling, and Regime Support. *Environmental Innovation and Societal Transitions*, 27, 4–15.

Kol, H., 2018. *The End of the Generator Era* [online]. Available from: www.kolhazman.co.il/324131 [Accessed 1 May 2019].

Laurendeau, N.M., 2011. An Energy Primer: From Thermodynamics to Theology. *Zygon*, 46 (4), 890–914.

Leonard, R.J., and Pepper, M.D., 2015. Les attitudes face aux changements climatiques et les actions pour la décroissance énergétique des chrétiens pratiquants: Les effets des persuasions religieuses et du capital social. *Social Compass*, 62 (3), 326–343.

Maruani, T., 2014. *Plan BB/830: Regulating Shabbat generators in Har Habanim, 18 May*. Environmental Document. Bnei Brak: Bnei Brak Municipality.

Masters, G.M., 2013. *Renewable and Efficient Electric Power Systems*. Hoboken, NJ: John Wiley & Sons.

Ministry of Environment, 2018. *Meeting Summary Concerning Kosher Electricity at the Planning Department, Ministry of Environment, May 13*. Tel Aviv. Document no. DI_290518081630830.

Morgenstern, Y.M., 2012. *Electricity on Shabbat in Out Times, Water on Shabbat in Out Times: A Description of the Production of Electricity and Water Delivery in Israel – A Comprehensive Halakhic Inquiry of Questions of Electricity and Water on Shabbat*. Jerusalem: Otzar Hachochma [Hebrew].

News 1, 2011. *A Foothold for the Rabbinate in the Electricity Sector* [online]. Available from: www.news1.co.il/Archive/001-D-284613-00.html [Accessed 1 May 2019].

Olson-Hazboun, S.K. et al., 2016. Public Views on Renewable Energy in the Rocky Mountain Region of the United States: Distinct Attitudes, Exposure, and Other Key Predictors of Wind Energy. *Energy Research & Social Science*, 21, 167–179.

Otte, P.P., 2014. A (New) Cultural Turn Toward Solar Cooking – Evidence from Six Case Studies Across India and Burkina Faso. *Energy Research & Social Science*, 2, 49–58.

Parker, C.G., 2015. Religion et transition énergétique: une étude en Amérique du Sud. *Social Compass*, 62 (3), 344–361.

Pesch, U., 2015. Tracing Discursive Space: Agency and Change in Sustainability Transitions. *Technological Forecasting and Social Change*, 90, 379–388.

Ramos-Mejía, M. et al., 2018. Sustainability Transitions in the Developing World: Challenges of Socio-Technical Transformations Unfolding in Contexts of Poverty. *Environmental Science & Policy*, 84, 217–223.

Rasmussen, L.L., 2011. Energy: The Challenges to and from Religion. *Zygon*, 46 (4), 985–1002.

Sandler, S., and Kampinsky, A., 2018. Israel's Religious Parties. *In:* R.O Freedman, ed. *Contemporary Israel: Domestic Politics, Foreign Policy, and Security Challenges*. New York: Routledge, 77–96.

Schneider, A., 2016. *Certificate of NGO Registration*. Jerusalem: Ministry of Justice.

Shelhav, Y., and Kaplan, M., 2003. *The Ultra-Orthodox Community and the Environment*. Jerusalem: Jerusalem Institute for the Study of Israel.

Simkins, R.A., 2008. Religion and the Environment. *Journal of Religion and Society*, 10, 5–26.

Skivin, D., 2018. *Comparative Analysis of Air Pollution from a City Power Plant Compared to Neighborhood Generators*. Beer Sheva: Ben-Gurion University.

Taylor, B., 2010. *Dark Green Religion: Nature Spirituality and the Planetary Future*. Berkeley: University of California Press.

Tucker, M.E., 2008. World Religions, the Earth Charter, and Sustainability. *Worldviews: Global Religions, Culture, and Ecology*, 12 (2–3), 115–128.

Wolpe, R., 2012. The True Way to Loving God: Nature in the Haskala. *The University of Toronto Journal of Jewish Thought*, 3, 1–28.

Ynet, 2007. *Not Everything Is "Kosher" in Ultra-Orthodox Generators* [online]. Available from: www.ynet.co.il/articles/0,7340,L-3457852,00.html [Accessed 1 May 2019].

Ynet, 2011a. *Protests Against Kosher Electricity Bill Expand* [online]. Available from: www.ynet.co.il/articles/0,7340,L-4163001,00.html [Accessed 1 May 2019].

Ynet, 2011b. *The Pressure Worked: Landau Withdraws the "Kosher Electricity Bill."* [online]. Available from: www.ynet.co.il/articles/0,7340,L-4163351,00.html [Accessed 1 May 2019].

Ynet, 2013a. *IEC to Observe Sabbath* [online]. Available from: www.ynet.co.il/articles/0,7340,L-4397553,00.html [Accessed 1 May 2019].

Ynet, 2013b. *Kosher Electricity in Bnei Brak: 120 Shekels for a Weekend* [online]. Available from: www.ynet.co.il/articles/0,7340,L-4354462,00.html [Accessed 1 May 2019].

Ynet, 2014. *The "Kosher Electricity" Farm Will Be Evacuated from the Center of Bnei Brak* [online]. Available from: www.ynet.co.il/articles/0,7340,L-4503710,00.html [Accessed 1 May 2019].

Ynet, 2018. *The Half a Million Petition: Demanding "Kosher Electricity" on the Sabbath* [online]. Available from: www.ynet.co.il/articles/0,7340,L-5311031,00.html [Accessed 1 May 2019].

12 The Negotiation of Self-Identity in Swiss Biodynamic Wine-Crafting

Facets of a Sentient and Practitioner-Based Sustainable Agronomy

Alexandre Grandjean

Introduction: A *Sentient* and *Intimate* Version of Sustainable Agriculture

Frank is a Swiss peasant raising livestock and cropping medicinal herbs and grape vines. He is involved in local organizations promoting bio-organic agriculture. In recent years, he has turned toward a holistic and cosmological approach to agriculture: biodynamics. This esoterically driven agronomy is inspired by Rudolf Steiner's conferences on agriculture. In the transcription of these conferences, *The Agriculture Course* (Steiner 2006 [1924]), the vegetal is defined as an entity in motion which is acted upon by earthly and cosmic "forces" (Choné 2013, p. 22). The peasant is then invited to consider intimately the visible and invisible dimension of plant growth. In these conferences, Rudolf Steiner, the charismatic founder of the anthroposophical movement, promotes a *virtuosi* cosmology grounded in various esoteric traditions such as Goetheanism, Paracelcism, Alchemy as well as other influences from Theosophy (Besson 2011, pp. 44–47; Choné 2013).

Yet, for Frank as for other practitioners I encountered, this view on plants is targeted pragmatically toward an attitude of *care* and protection from so-called "pests" as well as climatic imperilments. During our interview, Frank explained how he conceived his plants as reacting to alternative healing practices, just like human bodies do. Using herbal teas and plant decoctions as well as following the overall guidelines of naturopathy, he assumed these would have direct effects on his plants:

> You listen to yourself, and you will always act upon your plants as you would with yourself. If there is some cold, well what do you do? You do a decoction with willow's barks. It is a vegetal aspirin. So you do a decoction, you add it to your mixture and you sprinkle it on your vine plant. And then it warms her.[1] She will make it. If you cut her members, it is like if I injure you. You'll need to heal. So I am tiring you and you don't have time to care of your grapes, and stuff like that. . . . So

DOI: 10.4324/9781003017967-16

> each time I prune you, I am weakening you. So you need to think over it. You'll have to give [the plant] a sedative to relax her, because you stressed her. And then you will do a little [tisane of] valerian.
>
> (Interview 05.09.2017)[2]

His quote presents a sentient and intimate version of sustainable agriculture. One that is strongly entangled with issues such as localness, emphasis on subjectivity, sensorial experimentations, holistic values and cosmology as well as relationality with plants and soils. Indeed, in this version, the "success" of economic activity is not necessarily envisioned as solely monetary, but also values notions of "naturalness," as well as a quest for autonomy, well-being, and to a return to more "authentic" tastes (Fourcade 2011; Teil et al. 2012).

Starting in the year 2000, biodynamic farming has received increased attention in the Swiss wine-crafting milieu.[3] It is either depicted as a radical engagement in agroecology by its promoters, or attacked by its deniers as "cultic," "superstitious," "magic," and not as "sustainable" as it claims to be. As a result, biodynamic wine-crafting tends to be seen as a professional segment (Strauss 1992, p. 82), distinct from other sustainable agronomies such as the state-funded program for "integrated production" or "standard" bio-organic farming. The goal of this chapter is to highlight that the process of making and unmaking identity boundaries is multileveled (Wimmer 2008). Swiss biodynamic wine-crafters constitute an informal community connected by shared agricultural practices and overall criticism of intensive and agrochemical-based agronomies. Yet, as an informal community its members also negotiate their self-identity in a two-folded manner: On the one hand, they claim that biodynamic guidelines are both more "sustainable" and more suitable for wine-crafting's quality over quantity production. They thus tend to present their agronomy as featuring *do-it-yourself*, *low-tech*, *sentient*, and *practitioner-based* distinctive dimensions. On the other hand, biodynamic wine-crafters negotiate their identity as separate to anthroposophy by claiming autonomy over biodynamic guidelines in regard to more "orthodox" interpretations of Rudolf Steiner's legacy.

This setting catalyzes multiple lines of demarcation (which can turn to potential tensions) that I will analyze: (1) demarcation over "sustainability" implying a position in regard to wider professional concerns and (2) demarcation over "spirituality" which implies presenting oneself as distant from institutional and established stakeholders of biodynamics' guidelines and cosmology. I will however also point out through facets of this sentient and practitioner-based agronomy, that these demarcation lines are situationally and individually negotiated by wine-crafters. Indeed, as a non-homogeneous population, biodynamic wine-crafters advocate various new social practices, cultural artefacts, and professional *ethos* that social scientists should be reflexive about. Indeed, through a contemporary rhetoric of "spirituality," self-experimentation, and self-expression, biodynamic wine-crafters tend to bridge secular and religious domains.

In the first section, I present outlines of biodynamic farming in regard to "standard" secular bio-organic agriculture and integrated production. In the second section, I present my methodology to study the wine-crafting milieu as a field of multi-layered lines of demarcation (tensions). In the third section, I detail a theoretical framework which enables me to link together cosmological, sentient, and material dimensions of biodynamics. This theoretical framework is then related to field observations on how a "sentient ecology" (Ingold 2000, p. 5) is enacted in contrast to the dominant *high-tech* and expert-based approach to sustainable wine-crafting. I then present ethnographic observations on the formulation of so-called "biodynamic preparations." These empirical insights demonstrate how novices in biodynamics are invited to give more credit to their "selves," bodily intuitions, and senses and to a specific eco-spiritual worldview ("cosmic holism"). In the conclusion, I discuss the case study of biodynamic wine-crafting in Switzerland as an illustration of moving social boundaries between religious and secular dimensions of ecological issues.

Context: Biodynamic Wine-Crafting and the Swiss "Organic" Farming Movement

As a social phenomenon, biodynamic wine-crafting is becoming more visible in public spaces such as the media, in wine rankings, or in the stalls of wine retailers. For 20 years, the number of wine-crafters certified in biodynamics has risen from three in 1997 to 60 by 2018.[4] Though it still remains a marginal phenomenon in comparison with the 5,000 wine-crafting domains in Switzerland (OFAG 2016, p. 7), this growth is significant and reflects changes in how the activity of grape-cropping is being conceived under sustainable criteria. These – on another analytical level – incorporate holistic and spiritual elements. Interestingly, starting in the 2010s, biodynamic wine-crafting has acquired social legitimacy in Swiss public spaces, illustrating for many journalists and wine *connoisseurs* a contemporary approach that can be "respectful of the environment" or even linked to a revival of Western "traditional ecological knowledges" (TEK) A notable example is that biodynamic wine-crafters have been compared or compare themselves to "druids."[5]

In Swiss vineyards, biodynamic wine-crafters as well as members of the generic "organic" farming movement contrast pragmatically – and to some extent counterculturally – with industrial, agrochemical, and intensive models of modern farming (Barton 2018, 2011). More precisely, they present a radical alternative to the dominant sustainable models in Switzerland known as "integrated production" and "targeted resource management." In the media, biodynamics is often framed as a continuity of "standard" bio-organic farming which accounts for the majority of farms and wine-crafting domains certified in "organic" farming (230 in 2017). This relationship between "standard" bio-organic and biodynamic farming is the one of companionship: In order to be certified as biodynamic (Demeter), one first has to undertake the same 3-year "reconversion" cursus as for bio-organic. Being certified in bio-organic farming facilitates access to a biodynamic certification. This fosters the view that biodynamic is a more radical version of "organic" farming.

In the field biodynamic practitioners share with the "organic" farming movement, there is an overall representation of "nature" as a "harmonious," "balanced," and mostly "self-resilient" entity (LeVasseur 2017, pp. 27–28). They also have numerous agroecological repertoires of action in common: "organic" treatment products such as herbal teas, decoctions, solutions of copper and sulfur to replace conventional agrochemistry and biodiversity management strategies, or "auxiliary species" used to prevent pests and parasites. When following the specifications of the Demeter certification, wine-crafters are allowed to use fewer treatment products, at a lower dose, as well as adding homeopathically diluted preparations to their repertoire of agroecological actions: what is known as the "500" and the "501," which I detail further on. These dilutions are then set in motion ("dynamized") and sprinkled on plants under certain – atmospheric and ritual – conditions, following an astrological calendar.

Another distinctive feature between wine-crafters in "standard" bio-organic and in biodynamics is their relation to founding figures and their disposition toward religious/spiritual dimensions. Indeed, only rarely are bio-organic wine-crafters asked to position themselves in regard to texts written by the English agronomical engineer Albert Howard or notably by the couple Hans and Maria Müller who implemented bio-organic guidelines in Switzerland in addition to idealized rural Christian conservative values (Besson 2011, pp. 111–117). For biodynamic wine-crafters, Rudolf Steiner is mentioned as an important figure. Yet, my interviewees all tended to present themselves as distant from his esoteric backdrop, and, as we will see further on, considered him as an inspiring figure among other prominent ones. This view of Rudolf Steiner's legacy was notably explained by one of my informants who stressed his subjective "well-being" apprehension of practicing biodynamic guidelines:

> I am not saying: "Yeah, Steiner is the only guy who's right." We don't care, we all have our truths. It's not about that. I think what is interesting is the point of view that he is giving back to humanity, because to me, he has not really invented stuff. There's stuff that comes from other cultures and so on. But he gave it back, and overall, I am benefiting from it and today I am practicing without asking myself too many questions, while being fit in my sneakers.
>
> (Interview 06.09.2017)

Methodology: Studying Biodynamic Wine-Crafting as a "Field of Tension"

The empirical material of this chapter results from a case study I conducted in the biodynamic community in French-speaking Switzerland (Romandie) between 2017 and 2020. I sampled wine-crafters who are engaged in biodynamic and other holistic guidelines. Participant observation was carried out on important occasions, such as practical and theoretical workshops, pruning, harvesting, vinifying, and "open cellars" days and so on. Semi-structured interviews were conducted with 40 wine-crafters, the majority of whom had been certified by the Swiss Demeter label. In order to situate and contrast specificities of this group, eight other wine-crafters working under the specifications of "integrated production" and "standard" bio-organic agriculture were also interviewed. All participants were based in four French-speaking cantons (Vaud, Valais, Neuchâtel, and Jura)[6] and were found though their certification labels, by "snowball sampling" overall recommendations and word of mouth. Contact with prospective participants was initiated via an e-mail asking to discuss environmental and sustainable actions undertaken in the last 10 to 20 years. For those who agreed, I came to visit them in their wine-crafting domains.

In this chapter, I study the ways in which the Swiss biodynamic wine-crafting community constitutes a self-defined professional segment (Strauss 1992, p. 82). On the one hand, I find it to be marked by the features of "integrated production" such as a *high-tech* and expert-based approach to grape-cropping. On the other, I find that it prioritizes access to pragmatic and cosmological resources from anthroposophical stakeholders, while keeping its distance from overtly *virtuosi* forms of "spirituality." This makes biodynamic wine-crafting a "field of tension" (Koehrsen et al., this volume) in which potential conflicts can arise. In this paper, I rely on different kinds of data and observations to apprehend the boundary-making strategy of this professional segment: life-path narratives, textual and statistical data collection, observed situation of interactions and organizational networks. Through ethnographic and embedded perspectives, the researcher experiments with and learns about "non-verbal" and sentient forms of practical knowledge (Krzywoszynska 2015) – sometimes classed as "spiritual" by social actors. These examples tend to directly bridge distinctions between what is understood as secular and religious dimensions, as well as subjective and objective forms of knowledge-building.

Theoretical Framework: The Material and Sensory Facets of Biodynamics

The study of agriculture, and especially biodynamics, has so far received little attention from the religion and ecology/sustainability debate. Two noteworthy exceptions are the historical study by Dan McKanan (2018) on Steiner's legacy in contemporary environmental movements, and Todd LeVasseur's study on "religious agrarianism" (2017). In the present book, the chapter by Stéphanie Majerus brings about another welcome perspective on how biodynamic farming is implemented in a milieu more connected to anthroposophical stakeholders (this volume). I argue that the approach of material culture and Tim Ingold's insights on "sentient ecology" (Ingold 2000, p. 5) are of great support to apprehend biodynamic wine-crafting as a two-fold phenomenon. On the one side, biodynamics provides an eco-spiritual worldview to practitioners, and on the other one it also purports and articulates specific *low-tech* and *do-it-yourself* sustainable practices to agriculture. It should be noted that one aspect cannot be separated from the other, even if they are negotiated separately on the ground by social actors. I will explore this point further by gathering some empirical material on how biodynamic wine-crafters negotiate their self-identity within "integrated production" on the basis of sustainable practices, while also dealing with anthroposophical stakeholders on spiritual and cosmological concerns.

Overcoming the "Worldview/Practice" in the Religion and Sustainability/Ecology Nexus

In the religion and ecology/sustainability nexus, religious actors and institutions are often relegated to the role of disseminators of ecological values and worldviews (cf., Jenkins and Chapple 2011, p. 477; Koehrsen 2018). In the realm of alternative spirituality (Bloch 1998), social actors are at best conceived as being part of *Dark Green Religion* when promoting *biocentric* or *ecocentric* conceptions in which "nature is sacred, has intrinsic values, and is therefore due reverent care" (Taylor 2010, p. 10). Yet, no strong relationship between spiritual actors and further material engagement in sustainability programs is usually established. In a way, the religion and ecology/sustainability debate tends to foster a "worldview/practice" divide. In the literature, while individuals and institutions all too easily adopt an ecospiritual worldview and/or promote green theology aiming at personal change, they rarely address an individual's ability to *materialize* this in concrete sustainable practices. Setting aside Marxist insights on access to means of production or financial resources (which provide more agency to religious/spiritual actors on sustainable concerns), I will consider phenomenological and sensory-led approaches to explore this epistemological tension.

The case study of biodynamic farming enables scholars to consider "sustainability" and "spirituality" as two contextually grounded social categories. Neither of them rely solely on abstract or textual normative (re)sources, but are conceived under a "lived and embodied perspective" (McGuire 2008): wine-crafters, as land owners and entrepreneurs, are socialized to new cosmological concepts, agricultural practices, and bodily apprehensions to care for microbiological life in their soils and the "vitality" of their plants and local biotopes.[7] As shown by Jean Foyer, adhering to biodynamic guidelines is an entrance path for many wine-crafters to give more credit to scientifically "unconventional" forms of knowledge and ecological ontologies (Foyer 2018). As such, it is important to consider biodynamic wine-crafters, first as part of a professional segment and then as potential spiritual actors. Or more precisely, as professional actors by which religious and spiritual references are "popularized" (Knoblauch 2008, 2014) into socially acceptable responses to their daily and professional concerns. In this context, "Spirituality" can be seen as a polysemous notion used by social actors to describe their specific ontological "selves" (worldview and ethos) and to detail their sustainably orientated engagements such as "organic" plant treatments (Grandjean 2021).

Biodynamic Wine-Crafting and the Embodiment of a "Sentient Ecology"

Scholars focusing on "material culture" and "lived religion" (Csordas 1994; McGuire 2008; Meyer 2008; Morgan 2010) have provided new interpretations, raising awareness about the fundamental role corporality and material artefacts play in religious and spiritual phenomena. In these approaches, religion/spirituality or even notions such as "belief" or "faith" are not normatively defined, but rather conceived as specific iterated material and sentient settings, such as explained by David Morgan:

> Forms of materiality – sensations, things, spaces, and performance – are a matrix in which belief happens as touching and seeing, hearing and tasting, feeling and emotion, as will and action, as imagination and intuition. Moreover, religion happens not in spaces and performances as indifferent containers, but as them, carved out of, overlaid, or running against prevailing modes of place and time. Materiality refers to more than a concrete object or to this or that feeling. Sensation is an integrated process, interweaving the different senses and incorporating memory, and emotion into the relationships human beings have with the physical world.
>
> (Morgan 2010, p. 8)

Biodynamic farming – as a specific cosmology *and* as a set of agroecological practices – relies not only on abstract spaces and "forms of materiality." Following on David Morgan's insights, biodynamic farming is grounded and experienced within specific material and experiential settings. Indeed, biodynamic farming is an applied eco-spirituality invested emotionally, symbolically, relationally, and practically in material settings: soils, plants, animals, landscapes, mooning, seasonality, and so on. In other words, biodynamic farming can be understood as a *dwelling* practice (Ingold 2000, 2008). More precisely:

> What it means is that the forms people build, whether in the imagination or on the ground, arise within the current of their involved activity, in the specific relational contexts of their practical engagement with their surroundings.
>
> (Ingold 2000, p. 186)

According to Tim Ingold, the "worldview/practice" divide, just like the modernist "body/mind" distinction, is anthropologically irrelevant. Grounding his argument on the phenomenological philosophy of Maurice Merleau-Ponty, he states that there is no radical cognitive disjunction between individuals, their sensory and bodily experiences, and their direct environment. An illustration of this is Frank's quote (see introduction). In his analogical and empathic approach to the vegetal world, Frank portrays his vine plants as reacting like human bodies, or more precisely, like his own body. In this case, imagination and bodily experiences are intertwined in specific agricultural interventions such as using herbal teas and decoctions.

For Tim Ingold, what differentiates individuals is the relationality "constituted by the presence of the organism-person, indissolubly body and mind, in a richly structured environment" (Ingold 2000, p. 353). Wine-crafters are engaged in an everyday environment, full of meaningful information and components which individuals are interacting with through direct practices and representations. Tim Ingold states that individuals, by interacting with their environment, develop specific "sensory skills" which raise *awareness* on practical information: Distinguishing the texture of a vine stalk, the smell of soils, or the taste of wine fermented by "natural" yeast are but a few examples of the "sensory skills" practitioners develop to pursue their professional activity. Interestingly, the acquisition of such "skills," as well as the new cognitive approach entrained by biodynamics, was often qualified as a "spiritual" process by my interviewees. Notably, one of them remarked that, according to him: "It is time to open oneself, it is time to let go of everything we could learn to keep some space for everything we do not know yet." For him, biodynamics was defined as "biology in motion" and therefore, one had to constantly adapt his own knowledges depending on the year, situation, and sensory output.

This attitude of *dwelling* and developing "sensory skills" contributes to what Tim Ingold terms "sentient ecology."[8] Interestingly, this notion finds specific resonance in biodynamic wine-crafting. Indeed, during workshops or in biodynamic handbooks, emphasis is given to raising awareness about places, local biotopes, and distant cosmological "influences." To illustrate these claims, the handbook delivered by the Demeter certification to every wine-crafter, specifies that:

> Three points are essential "for biodynamic farming":
>
> - It demands an engagement and a **personal implication** of the domain's stakeholder in work.
> - It is essential to acquire an appropriate knowledge of life, a **sense of life** and of forces acting within it.
> - It is necessary to remind oneself that agriculture is a **science of adaption, of individualization of processes** and methods. *(My translation, bold in the original text Masson 2015, p. 15)*

The acquisition of "appropriate knowledge of life" or a "sense of life and forces acting within it" is a clear reference to Steiner's cosmological legacy. Yet, interestingly, emphasis given to "personal implication" and agriculture as "a science of adaption" and "individualization of processes" also stress upon what Tim Ingold termed as a "sentient ecology."

Biodynamics as *Low-Tech* and Sensory Practices of Grape-Cropping

As previously stated, biodynamic farming contrasts with the industrial, agrochemical-based intensive models of modern farming. Following a theoretical framework that seeks to abolish a common divide between "worldviews" and "practices," I now want to stress the pivotal roles of "sensory skills" and material artefacts in embodying and enacting eco-spiritual worldviews and agroecological practices. This first empirical section explores how "sentient ecology" involves wine-crafters adopting and adapting new agronomical practices situating them apart from *high-tech* and *expert-based* approaches to grape-cropping ("integrated production").

Connecting the Theoretical Framework With Field Observations

One of my informants provides a poetical description to introduce "sentient ecology" and wine-crafting on his website. Not necessarily claiming to follow biodynamic guidelines, he nonetheless describes a life-course that pushed him into a logic of "individualization" which assigns sensory *awareness* to every plot and plant. Interestingly, during our interview he described this *ethos* as the "state of mind of organic" – a notion that resonates with the text on his website:

> Since I did not study oenology, nor have I followed viticulture classes, I can only speak by experience. That is the experience of the one that [works with] his vines, who listens, who observes, who takes notes and

who memorizes. The one who feels. And also the one who does not understand, but who shall one day understand. . . . Herbicides disappeared twenty years ago, fertilizers too, and the moon and her[9] cycles replaced them. The pruning adapts to each slot. To each grape variety. Because what is necessary for one piece of land is not for another – wind, sun, depth and quality of the soil, humidity of the place, exposure – influence and predestinate. And now I know, and I sense it too, that every strain is different. Just like every human being is.[10]

This quote highlights that the biodynamic wine-crafters I encountered are engaged in a practice of "sensory ethnography" (Pink 2009) with their plants and land, much like anthropologists in the field. As the previous quote illustrates, they "take notes" and "memorize" over years, engaged in a radical participant observation of their plants and soils. For instance, one of my interviewees in the canton of Vaud stated that he had stopped using electrical pruning shears to prune his vines. He explained that, even though it took more time and would be more painstaking, he preferred that option as it allowed him to better feel how much force he should use to cut vine stalks. This would allow him to grasp bodily – or intuitively – their state of "vitality." Having decided to work within the specifications of biodynamics, he was thus seeking *low-tech* solutions as a means to pay more attention to information that would help him to prevent pests and diseases from damaging his harvest (such as mildew or odium attacks).

This anecdote reinforces what anthropologist Anna Krzywosynka noticed during her fieldwork in bio-organic vineyards in northern Italy: "In vine pruning, like in other farming practices, the acquisition of skill – that is the understanding of the 'meaning'- the ability to go on – was not dependent on the assimilation of abstract rules, expressed in languages." Rather, these skills "resided in the capacity to make situated and knowledgeable decisions about intervention into the plants, informed by previous experiences with vines in this particular place" (Krzywoszynska 2015, p. 315). This description follows the skill insights of Tim Ingold, who argued that, "skilled practice cannot be reduced to a formula" (Ingold 2000, p. 353). Now, how can claiming to ground an "individualized" agronomy over these sensory skills constitute part of the self-identity of the biodynamic wine-crafting community?

Situation of Tensions and Their Negotiations: Demarcation Lines Over "Sustainability"

On the ground, I could observe little interaction featuring explicit tensions between biodynamic wine-crafters and "conventional" ones. Not that tension was absent – pioneers in biodynamic wine-crafting often described how stigmatized they were in the early 2000s. This situation, however, has eased over the last 5 years as biodynamics has become more widely recognized.

During exploratory interviews with wine-crafters in integrated production, they explained why they personally would not engage in "standard" bio-organic or biodynamic farming. For them, these approaches were a topic of interest, but only something they might consider in due course. For instance, during a tasting session one wine-crafter was asked by another participant whether he was following biodynamic guidelines or not. He answered that although he was thinking about it, going bio-organic would be the first step. He concluded, somewhat humorously that he had to "believe first in stars . . . and in disasters."[11] This ambivalent relation toward biodynamic wine-crafting was also expressed by other wine-crafters who instead favored high-tech sustainable practices such as installing solar panels, heat pumps, or investing in new electric trucks and "resistant" hybrid grape varieties developed by the agronomical research center. They would also mention the negative aspects of biodynamic wine-crafting: not being able to manage cropping "risks" in time, the increased need for manual labor, and the constraints of organizing labor planning in accordance with the lunar calendar.

Only once in a local festival dedicated to products from the Vaudois "terroir" did I encounter a conflictual interaction between two wine-crafters. One was a respected figure of "integrated production" in a region in which bio-organic and biodynamic approaches were at that time almost inexistent. The other was hired by a public and communal domain to implant biodynamic practices. When I introduced the two to each other, we started tasting wines and discussing vinification techniques. Rapidly, the two wine-crafters started a debate about whether "conventional" or biodynamic wine-crafting was more "sustainable." The first wine-crafter argued that using agrochemical inputs with limited doses avoided tamping the soil with trucks and reduced both the treatment required and the resulting CO_2 emissions. Also, according to him, using so-called "organic" treatments such as sulfur and copper would pollute soils and waterways with "heavy metals." The second wine-crafter counter-argued that copper, sulfur, and CO_2 were not major pollutants. According to him, enhancing biodiversity and the bio-chemical life of soils would better trap CO_2 as well as effectively dissolve "heavy metals." This could be directly and sensorially grasped by detailing the soil's texture and smell, counting the number of worms, or even by tasting the wine.

This sequence illustrates demarcation lines that are grounded not only on a religious/secular axis, but also on technical and scientific framings on what should be classified as "pollution" and an effective "sustainable" agenda. Taken together with a "sentient ecology" relying on claims of "sensory skills," these tensions and divergences appear as two opposing versions of "sustainability" over grape-cropping: on the one hand an expert-based and high-tech approach, and on the other a practitioner-based agronomy which demands frequent interventions in vineyards relying on low-tech and DIY techniques that highlight sensory observations as a means of action. In both cases, uncertainty over delineating "efficient" and "sustainable" practices led wine-crafters to be reflexive in their own taxonomies over what constitutes more or less serious pollutants. Other means of validating one's agronomy were then at stake. Biodynamic wine-crafters frequently spoke about how important it was to be an "exemplar" in the wine-crafting milieu: namely, showing that biodynamic guidelines are "efficient," "sustainable," and result in "wine excellence." Amusingly, one of my interviewees, though he was letting spontaneous plants (adventitia) spread between the vine rows, explained to me that he would mow the three first visible rows next to the road to fit the so-called "Swiss expectations" of tidiness and pride over hard working (cf., Lambek 2015).

Biodynamics as Do-It-Yourself: Becoming the Producer of One's Treatment Products

In the following ethnographic description, I focus on one aspect of biodynamic guidelines: producing of one's own treatment products in a do-it-yourself logic. Doing so enables me to accentuate what has been exposed previously on the delineation of the biodynamic wine-crafting community as distinct from "conventional" approaches to grape-cropping. It also allows me to analyze another set of lines of demarcation (tensions) regarding anthroposophical stakeholders in biodynamics. In this ethnographic description, I present situations in which novices – the author included – are trained by advanced practitioners in situated and sensory forms of knowledge about "biodynamic preparations." I start by giving contextual information. Then, through notes taken during the workshop, I present situations in which sensory skills are transmitted by nonverbal, implicit forms of both practical and cosmological knowledge. Finally, I analyze several understated elements such as the relation of biodynamic guidelines to scientific measurement tools and I further expose Steiner's cosmological legacy. Doing so helps to understand how the biodynamic wine-crafter community negotiates its self-identity, as well as selects, adapts, and translates biodynamics in a new guise.

Contextual Information on the Ethnographic Description

The sequence takes place in an anthroposophical community for people with different disabilities ("handicaps"). Part of the Camphill movement,[12] it is situated near Lake Geneva (canton of Vaud). A workshop was given to transmit background knowledge about "biodynamic preparations," the main element distinguishing biodynamic farming from "standard" bio-organic farming. A total of six novices and four "advanced practitioners" (two Francophone and two Germanophone) took part in this workshop. The head of the community's land service supervised the session which was the second half of a two-part workshop. The first part had been given 6 months earlier in which participants learned how to formulate the preparations used in biodynamics (those with codenames "500" to "507," see infra). The prepared elements were buried in specific spots and left 6 months to be "transformed" by cosmic and earthly forces.

Interestingly, none of the participants had been wine-crafters. I subsequently learned that this was because they had had their own workshop which also included collecting and elaborating collectively their preparations. This separation between wine-crafters and "usual" biodynamic practitioners is a sign of a divergence that exposes boundary-making strategies between the wine-crafting community and practitioners inspired – with more or less intensity – by anthroposophy. Indeed, there is a perception that wine-crafters do not fully apply biodynamics: First, they do not use all the "biodynamic preparations" – they often limit themselves to "500" and "501," while avoiding complex composting techniques. Second, they are grounded in a monoculture which contradicts the ideal of the farm as a self-sufficient "organism" initially promoted by Rudolf Steiner. Moreover, in anthroposophy the consumption of alcohol is not recommended, though religious scholars know that there are always gaps between norms and practices (cf., Blanc, this volume). These three elements distinguish the wine-crafting population from other common nonprofessional biodynamic practitioners (the so-called "Sunday gardeners") as well as advanced practitioners. Despite this, these actors still enjoy a mutually beneficial relationship: Wine-crafters have access to pragmatic resources, while anthroposophical stakeholders take advantage of the overall visibility and legitimacy given to biodynamic wine-crafting in the public space. The ethnographic description that follows highlights the central role played by "advanced practitioners" ("experts") by assessing their authority to provide an "orthodox" anthroposophical framing of the practice, both in a sensory and representational guise.

Ethnographic Description: Unearthing and Sorting Out the "500"

The public view of biodynamics is a picture of cow horns being filled with cow manure or crushed quartz (silica). This iconography presents biodynamics at a threshold between folk peasant wisdoms (almanac's, proverbs, or "grandmothers' decoctions"), experimental agronomies, and periscientific knowledge[13] (Foyer 2018; Pineau 2019). These two preparations are commonly referred to as the "500" and the "501." The "502" up to the "507" are preparations used in compost which involve inserting specific plants into animal containers (stag's bladder, cow's viscera, skulls, and many more). In biodynamics, these preparations are buried for several months in specific places which are said to bear "ambiances" and have a "memory." During these months, earthly and cosmological "forces" are said to act upon the preparations and transform them. The "500" and "501" are homeopathically inserted into rainwater. The transformations of the manure or the silica are often described in the jargon as becoming "colloidal substances" (Masson 2015, p. 18), namely solvents that easily dissolve in water. During preparation, specific requirements have to be met. Some may appear more as ritual prescriptions, while others are more agronomically grounded. For instance, in some versions of biodynamics described to me, the manure must come from a cow that had already calved once and who had been well-fed with sufficient good quality (and of course, organically grown) straw and food.

I received this explanation during the workshop by a well-dressed man in his fifties, wearing tight glasses, with his shirt fashionably tucked into his pants. We were in an anthroposophy-related institution for people "in situations of handicap" in the canton of Vaud. As one of the four "advanced practitioners" he was helping to lead a workshop about the formulation of biodynamic preparations. His outfit automatically presented him as an "expert," since other participants were wearing casual working pants and t-shirts. He spoke with a metropolitan French accent and his well-formed syntax gave him the same *habitus* as those who have undergone higher education in France. In contrast to other advanced practitioners, he was a trained agronomical engineer, yet an "alternative" one who would also share his interpretations of Steiner's writings on preparations. When I explained that I came from the University of Lausanne, he greeted me with sarcasm, asking ironically when academics had become interested in biodynamics.[14]

During that workshop, we unearthed the "biodynamic preparations" going from site to site (ponds, forest clearings, meadows, etc.) to collect each of them. At one point, we all gathered around our host while unearthing cow horns (the "500"). Also, while showing us one spot in the field which had already been dug up, the host explained that one week ago they had unearthed cow horns with the group of wine-crafters:

> The result was overtly positive. The majority of horns had been completely transformed. We are going to compare this cuvee with the one we are unearthing today. And we will observe what happened to the odors, colors and textures. A few weeks ago, we [unearthed] three to four horns, but we saw that the process did not materialize correctly. So now we are starting to get our bearings on these differences and nuances.

He showed us another spot, marked with wooden tokens, where we would find the other cow horns. Two of the male participants plus the head of the land service started to shovel. One female participant was collecting the horns with her bare hands, having to distinguish them in the soil as they appeared. The aforementioned expert took a specific posture at one angle of the digging hole. Withdrawing to help the others at first, he then started collecting some soil samples in his hands and giving details to one participant about how important it was for the earth not to be too dry and to find a good day to unearth the horns. He mentioned that "it is important that there was still an ambiance" [*meanwhile he flaked the earth in his hand*].

Other participants were not listening to him, instead talking about the rainfall of previous weeks. While digging, one participant mentioned that he found an ants' nest. No one reacted to that information, even though it suggested that some horns would not have transformed properly.

A few minutes later, the expert joined his partner and the head of the land service who were putting the horn in a green plastic basket. "It presents well" said the head of the land service while getting rid of the remaining earth on one horn. The expert then took another horn in his hand. "We see some of them in which we . . ." [*he turns the horn through variable angles, mimicking disappointment*] "so . . . it is not something that is well distinguished compared to the others" [*he smiles to its interlocutor, the head of the land service*]. His partner adds: "But we take what we have." To which he nods: "Yes, indeed. And then we look at the results." In an understated interaction, they mention the possibility that not all the horns transformed.

* * *

In the afternoon, we all gathered in a "hangar" (a storage place for machines and straw balls) to sort all the collected preparations. Each participant took one preparation and, with the help of the advanced biodynamic practitioners, started to take them out of their protecting coating made of animal parts. We were asked to pay attention to the smell and texture of the preparation. My role was to extract some oak bark out of piglet skulls. While doing so, I was asked to note that the remaining flesh on the skull did not smell rotten.

We then all gathered around one table to sort out the cow manure we had previously unearthed. We were told to hit the horn on the table and then use a metal stick to help us hollow it out. It made rhythmic sounds that broke the silence amidst the concentrating participants. We had to sort the preparations that did transform from those that did not. Emphasis had to be given to the texture and smell – which were supposed to be like humus. The effluvium of the manure had to be eradicated. Sometimes the smell was so strong that I thought I was going to vomit. When the preparation was either too dry or exhibited a greenish liquid texture, it had to be thrown away in a separate bucket. During the process of sorting out the preparations, participants were constantly smelling them, touching them, or observing them through different angles. The novices – myself among them – were frequently asking others to help them with dubious and unclear cases. I have to confess that I was probably a bad sorter, as the experts had to do my selection again.

Biodynamics in a "Lived Perspective": The Demarcation Lines Over "Cosmic Holism"

The previous ethnographic description briefly introduces the reader to a "lived and embedded religion" perspective of biodynamics. It only focuses on one element of the practice, that is, to become the producer of one's own treatment product. Comments on Steiner's cosmological legacy were to be discussed later during a specific evening event given by two of the "advanced practitioners." This ethnographic description highlights the claims that often prevail about biodynamics standing as a "do-it-yourself" approach to agriculture. Interestingly, biodynamics is often construed as a quest for autonomy from agrochemistry firms selling treatment products. The description, however, has brought insights on the sensory *awareness* that is passed on to novices on material artefacts such as biodynamic preparations. The workshop featured two different kinds of participant, novices being initiated to the practice, and "experts" being able to state with assurance to others if the preparation was "presenting well" or not. Part of their authority was given by their embodied knowledge and skills, but also by their capacity to provide agronomical interpretation of Steiner's writings. Indeed, *The Agricultural Course* is often considered cryptic by wine-crafters, who, as a result, seek "vulgarizers" or "counselors" to train them in the pragmatic know-how of formulating and "dynamizing" biodynamic preparations.

Regarding my theoretical framework, I have so far detailed how biodynamics articulate specific low-tech and do-it-yourself sustainable practices to agriculture. Overall, this approach differs from the dominant and "conventional" approach to agriculture and wine-crafting. I still need to explain how biodynamics also provide a *virtuosi* eco-spiritual worldview to practitioners. As I have mentioned, the cosmology of biodynamics cannot be separated from its material dimension and the emphasis it gives to "sensory skills." Wine-crafters, nonetheless, tend to separate these two aspects and focus on agricultural practices rather than on the esoteric cosmology of biodynamics.

Beyond Sensory-skills: "The Ever-Widening Webs of Interaction"

The implicit part of the previous ethnographic description was *what* caused the transformation of the preparations. An eco-spiritual aspect, also standing as a demarcation line (tension), can be grasped within this very understated dimension. While participants were all familiar with the broad *corpus* of biodynamics, the "experts" were constantly framing it within an anthroposophical cosmology. The use of plants and animals needed for the preparation was framed as a "sacrifice." This act, said to be "violent" was in turn described as a human action aiming to help "nature" find its "balance" again and therefore "renew" itself. On the other hand, one could just as easily link these transformations only to biochemical processes. In the case of such a secular interpretation, the choice of the date or the ritual observances would play no explicit role in the transformation of the preparations. Yet, during the workshop, by mentioning "forces" and "influences," or even speaking of a "reconnection with the cosmos," the experts suggested the existence of such supersensitive entities.[15]

In order to get a better understanding of this dimension and its point of divergence, it is important to consider Steiner's original text. In *The Agricultural Course* (1924), Rudolf Steiner depicts an organicist and vitalist conception of the cosmos inspired by alchemical traditions from the Renaissance (Faivre 1996; von Stuckrad 2008). According to Steiner, modernity and industrialization lead to a threefold degeneration – that of soils and seeds, nutrition, and human morality (Steiner 2006 [1924], pp. 16–17) – which should be counteracted by biodynamics. In his view, plants are conceived as entities at the threshold between two astrological "forces": the earthly ones, personified by so-called "Ahrimanian" principles, and the cosmological ones, enacted by "Luciferian" influences (cf., Brendbekken 2002). Both principles (or forces) are part of a polarity between life and death, as well as between the material and spiritual realm. In a system of typical correspondences of western esotericism (Faivre 1996) and of *analogical ontologies* (Breda 2016, 2019; Descola 2005, p. 280), these principles are associated with the planets of the solar system (macrocosmos) and with alchemical substances such as carbon, nitrogen, or silica (microcosmos). Therefore, by combining the influence of the planets on specific days with preparations fostering these alchemical substances, the practitioner would support plants' growth.[16]

Each preparation (the "500" and "501") corresponds to one of the two poles between earthly and cosmic forces. They have to be homeopathically diluted in water, activated ("dynamized"), and then sprinkled on plants once a year to guarantee a certain "balance" between these "forces." Bearing this process in mind, the advanced practitioners were enacting what historian Dan McKanan termed "cosmic holism" through their very situated practice (McKanan 2018, p. 226):

> By "cosmic holism" I refer to the dimensions of anthroposophy and of biodynamics that cannot be found in mainstream organics: the homeopathic preparations, the astrological planting calendar, the alchemical vocabulary, the notion that Christ's blood still lives in the soil, the ideal of the farm as a living organism, and the conviction that the farmer's spiritual striving ensures the health of the farm and the nutritional values of its produce.
>
> (McKanan 2018, p. 226)

These anthroposophically inspired practices and narratives support "imaginations to include ever-widening webs of interconnection" (McKanan 2018, p. 226). In formulating their treatment product, biodynamic novices were invited to consider that their spiritually *sentient* "practices were not only meaningful but also efficacious, connecting them to a real source of power in their lives" (McGuire 2008, p. 137). Interestingly, this process is validated by and mediated to participants through a strong material and sensory culture as illustrated by the previous ethnographic description. In a sense, this kind of workshop provides pragmatic resources for biodynamic practitioners – wine-crafters included. It however remains unclear in my ethnographic description to what extent the cosmological legacy of Rudolf Steiner is being readapted and translated by the biodynamic wine-crafting community into socially more acceptable "popularized" references.

The Negotiation Work Over Biodynamic Wine-Crafting's Autonomy

On the ground, this "cosmic holism" is one of the demarcation lines between the biodynamic wine-crafting community and anthroposophical stakeholders. It has much to do with the "worldview/practice" divide I unpacked in the theoretical part: Anthroposophical stakeholders consider biodynamics as a zone of mutual percolation between the material enactment of "preparations" and the broader cosmological insights of anthroposophy. This percolation is particularly prominent in biodynamic handbooks in which specific agricultural recommendations are combined with cosmological insights. One example is the book edited by Jean-Michel Florin – the current headmaster of the agricultural section of the Goethanum in Dornach – on biodynamic wine-crafting (2017). In this book, pragmatic advice is combined with a mythological exegesis on the divine figure of Dionysus. In a similar vein, the book features pedagogical insights on Johannes Wolfgang Goethe's phenomenological approach to the vegetal world and to the "archetypal plants" which inspired Rudolf Steiner. The book demonstrates how eclecticism and holism appear as strong characteristics of biodynamic farming in contrast to the specialization logic of contemporary agronomical sciences which is split into numerous fields of study such as botany, pedology, and entomology, and so on.

Faced with "cosmic holism," wine-crafters tend to secularize the guidelines of biodynamics by avoiding anthroposophical jargon and categorizing their practices into secular agronomical specializations such as pedology (the science of soils), and presenting biodynamics as "science in becoming" (Champion 1993). Indeed, among the biodynamic wine-crafting community, interpretations differed markedly from the anthroposophical and alchemical cosmology of biodynamics. Wine-crafters predominantly considered biodynamic preparations advantageous because they give "information" or "impulsions" about their plants and soil. Some characterized these preparations as "bio-reactors," in which specific bacteria could dissolve in heated rainwater and "dynamize" it to favor spawning. Wine-crafters would then spray the water solution on plants to share the "information" and "impulsions" with them. Many would declare not "knowing why it works" but would nonetheless explain that it would. In these cases, "sensory skills" would be drawn upon to validate the "efficiency" of these preparations. These included changes to soil texture and smell, light reflection from foliage, "terroirs" as an expression of wine, and practitioner's inner experiments detailing how "dynamizing" biodynamic preparations could "boost" or "re-enchant" their profession. Emphasis on the this-worldly sensorial dimension of biodynamic preparations constitutes a clever way to avoid engaging in the complex cosmology of Rudolf Steiner. It can be described as "secularization by omission."

On the "worldview/practice" divide, biodynamic wine-crafters tend to evade the worldview aspect by secularizing their explanations of the efficiency of biodynamic preparations. On the other hand, they fully engage in the low-tech, do-it-yourself, and sensory approach promoted by anthroposophical stakeholders. One way to further secularize biodynamics is by framing it as distant to anthroposophy, thus seeking referential autonomy for this esoterically driven agronomy. For instance, one wine-crafter I interviewed – who used to work as a protestant chaplain before taking up his father's domain – was eager to explain that biodynamics was not only Steiner's legacy, but a collective one that would not necessarily be bound to anthroposophy. He told me:

> If we read *The Agricultural Course*, Steiner only gives vague indications. "You take a bucket, as big as a milking bucket. And here, you had half a corn, one horn and then you have to brew." If there weren't entire generations that refined it afterward, no one would speak about that.
>
> (Interview, Neuchâtel, 07.07.2017)

Another wine-crafter involved in neo-shamanistic practices was promoting an alternative approach to biodynamic "preparations." Inspired by spiritual agricultural experimentation in ecovillages such as Findhorn, Scotland, or Perelandra, Virginia, he developed what he called "elixirs" – water he would "charge with intentions." To him, "Steiner would totally accept that we change this kind of method given the evolution of today's society." Indeed, the guidelines of biodynamics were only one "stage" in his life-path – and certainly did not constitute a stabilized method. In this vein, he considered anthroposophical stakeholders as being too conservative in the defense of orthopraxis. The same observation was made by another interviewee involved in freemasonry. He detailed how he would meditate every Wednesday afternoon in his vineyards and use energy channeling crystals during the vinification process. He criticized the Demeter certification label for not recognizing his use of essential oils in preventing "pests" and "parasites."

Overall, when asked about their relationship with notorious anthroposophical stakeholders, wine-crafters would be elusive about it. One mentioned how difficult it was to delineate who counts as an "anthroposophist" and how to distinguish them from a "biodynamist" who disavows anthroposophy. "It is not written on their face" he explained to me while directly adding that "I have very good friends who have attended a Steiner-Waldorf school for instance. So I know a bunch [of them]. I mean that they are anthroposophists, but they have been a little bit inhibited by[it]." Interestingly, for him, being an anthroposophist stands more as a "sensibility" or a "state of mind" rather than a formal belonging to an anthroposophical institution. In a sense it implies more of an attribute of "believing without belonging" as famously studied by Grace Davie (1990).

All these examples are illustrations of the numerous ways biodynamic wine-crafters self-define as distant to anthroposophy. On the one hand, they tend to position biodynamics as autonomous from the anthroposophical worldview, yet on the other, they rely on anthroposophical stakeholders such as "counselors" or "vulgarizers" to provide pragmatic resources. Notably, these stakeholders are non-homogenous and tend to promote different interpretations of Steiner's legacy and the pragmatic formulation and "dynamization" of biodynamic preparations. In this case, it is interesting to note that the self-identity and logic of boundary-making are not only multi-levelled (Wimmer 2008), but also situational and relational. If biodynamic wine-crafters tend to present themselves as more "spiritual" or sensorially *aware* than their colleagues in integrated production, they nonetheless describe themselves as not being *excessively* "spiritual." Recurrent expressions on the ground such as mentioning "ayatollahs of biodynamics" or "radical steinerians" are also notions that pertain to this marking of boundaries between a socially acceptable – according to them – approach to biodynamics in regard to *virtuoso* ones.

Conclusion: The Moving Boundaries of the Religious and Secular

This contribution has highlighted a two-fold tendency in biodynamics farming: On the one hand, it involves a *low-tech*, *do-it-yourself sentient* approach to agriculture. On the other, it relies on an esoteric backdrop deriving from Rudolf Steiner's anthroposophical legacy. It is important, however, to consider these two aspects not as separated but actually to a certain extent entangled: I have shown that Swiss biodynamic wine-crafters negotiate their self-identity as being "in-between" or in "balance," demarcating themselves from "conventional" colleagues but also from anthroposophical stakeholders. Indeed, biodynamic wine-crafters do differ from these two profiles in several ways: first on what counts as a "sustainable" agricultural practice, and second on what constitutes biodynamics as part or apart of Rudolf Steiner's cosmological legacy ("cosmic holism"). Both the wine-crafting milieu, as well as the biodynamic milieu exhibit these potential demarcation lines (tensions).

Overall, I have highlighted the ability of social actors to negotiate these demarcation lines given that I observed little conflict on the ground. If there were conflicts and tensions in the early 2000s, they now seem to have dissipated. That being the case, we can now question whether some values of self-experimentation and self-expression bridge secular and religious dimensions. Indeed, they are sometimes labeled as "spiritual" by social actors involved in the do-it-yourself and low-tech approaches in biodynamic wine-crafting. As suggested by Boaz Huss, the social category of "spirituality" "does indeed create novel taxonomies and shape new lifestyles, social practices, and cultural artifacts that blur and undermine the modernist distinction between the religious and the secular" (2014, p. 51). The case study of biodynamic wine-crafting and wine-crafters precisely documents how the religious, secular, and subjective and objective dimensions intertwine in new sociocultural forms. Furthermore, in contexts of ecological struggle, these sociocultural forms also tend to favor new "holistic" models to apprehend new ecosystemic complexities. Steiner's cosmology then provides one holistic model that wine-crafters can adapt and individualize into personal agronomies.

Notes

1 In French, "la vigne" (vine plant) is a feminine noun.
2 Every quote stated in this paper has been translated from French to English.
3 I chose to translate the French word of "vigneron" with "wine-crafter" in English to stress the crafting skills they promote and enact daily (cf., Sennet 2009).
4 Cf. RTS, "Le Canton de Neuchâtel parmi les champions de la viticulture biologique", 26.07.2018: www.rts.ch/info/regions/neuchatel/9724436-le-canton-de-neuchatel-parmi-les-champions-de-la-viticulture-biologique.html [Accessed 12 April 2019].
5 Cf. RTS, "Le grand défi de la biodynamie dans le secteur du vin", 16.10.2017: www.rts.ch/play/radio/ici-la-suisse/audio/ici-la-suisse-le-grand-defi-de-la-biodynamie-dans-le-secteur-du-vin?id=8983382&station=a83f29dee7a5d0d3f9fccdb9c92161b1afb512db [Accessed 12 April 2019].
6 It is hard to give an exact number of wine-crafters certified or not by Demeter. Different modalities of certification exist, and the milieu is rapidly changing. It also needs to be noted that alternative certifications specifically for biodynamic wine-crafting have arisen in France and Germany, which some Swiss wine-crafters are also engaged with. For instance, *Renaissance des appellations* is one of these, being led by one charismatic promoter of biodynamic wine-crafting, Nicolas Jolly (cf., Garcia-Parpet 2014).
7 Given that engaging in biodynamics bears many features of "conversion" phenomena, it is important to consider what Tanya Luhrmann said, that: "Conversion is a complex process and above all else a learning process. Converts do not make the transition from nonbeliever to believer simply by speaking – by acquiring new concepts and words . . . they must come to believe emotionally that those new concepts and words are true" (Luhrmann 2008, p. 519).
8 "[Sentient ecology] is knowledge not of a formal, authorized kind, transmissible in contexts outside those of its practical application. On the contrary, it is based in feeling, consisting in the skills, sensitivities and orientations that have developed through long experience of conducting one's life in a particular environment" (Ingold 2000, p. 5).
9 "Moon" is a feminine word in French.
10 http://christopheschenk.ch/a-propos-de-moi/ [Accessed 17 April 2019]
11 "*Il faudrait d'abord que je croie dans les astres . . . et les désastres,*" in French this is a play on words using "astres" (stars) and "désastres" (disasters).

12 The Camphill movement is an international organization of anthroposophically inspired therapeutic communities for people in situation of handicap (cf., McKanan 2018, pp. 159–174).

13 By "periscientific," I follow Jean Foyer's statement that: "we refer to a regime of peripherical knowledge, which borrows to science a certain formalism and theoretical basis, thus it is not validated by official sciences of the academia and of the principal scientific institutions." (My translation, Foyer 2018, p. 310.)

14 I interpret this sarcasm as a result of a situation in which those who are often depicted as doing "magic," "superstitious," "parascientific," or being "cultic" (cf., Onfray 2015, p. 119, for instance) could reverse the social stigmas and depict the "urban (secular) intellectuals" as being too *materialistic* and not spiritual enough. It probably also was a means to mark to other participants witnessing the scene his specific status, probably motivated by the recent formation of his counseling business in coaching farmers in their "reconversion" process to biodynamics.

15 The French tradition of pragmatic sociology, and especially the work of Elisabeth Claverie (2003), has well featured how supersentive entities such as the "Virgin Mary" are rendered *present* to pilgrims through numerous linguistic, material, visual, and performative apparatus.

16 One of my interviewees framed it as "doing cosmetic," hence giving a shape to plants.

References

Barton, G.A., 2018. *The Global History of Organic Farming.* Oxford: Oxford University Press.

Besson, Y., 2011. *Les fondateurs de l'argiculture biologique: Albert Howard, Rudolf Steiner, Maria and Hans Müller, Hans Peter Rusch, Masanobu Fukuoka.* Paris: Sang de la Terre.

Bloch, J.P., 1998. Alternative Spirituality and Environmentalism. *Review of Religious Research*, 40 (1), 55–73.

Breda, N., 2016. The Plant in Between: Analogism and Entanglement in an Italian Community of Anthroposophists. *ANUAC*, 5 (2), 131–157.

Breda, N., 2019. Are Anthroposophists Environmentalists? *Public Anthropologist*, 1, 208–223.

Brendbekken, M., 2002. Beyond Vodou and Anthroposophy in the Dominican-Haitian Borderlands. *Social Analysis: The International Journal of Social and Cultural Practices*, 46 (3), 31–74.

Champion, F., 1993. La Croyance en l'Alliance de la science et de la religion dans les nouveaux courants mystiques et ésoétiruqes. *Archives de sciences sociales des religions*, 82, 205–222.

Choné, A., 2013. Les fondements de l'écologie spirituelle chez Rudolf Steiner. *Politica Hermetica* (27), 15–35.

Claverie, E., 2003. *Les guerres de la vierge: Une anthropologie des apparitions.* Paris: Gallimard.

Csordas, T.J., 1994. *The Sacred Self: A Cultural Phenomenology of Charismatic Healing.* Berkeley: University of California Press.

Davie, G., 1990. Believing Without Belonging: Is This the Future of Religion in Britain? *Social Compass*, 37 (4), 455–469.

Descola, P., 2005. *Par-delà nature et culture.* Paris: Gallimard.

Faivre, A., 1996. *Accès de l'ésotérisme occidental (T.1)*. Paris: Gallimard.

Florin, J.-M., 2017. *Viticulture biodynamique: Nouvelles voies pour la culture de la vigne*. Dornach: Editions du Goetheanum.

Fourcade, M., 2011. Cents and Sensibility: Economic Valuation and the Nature of "Nature". *American Journal of Sociology*, 116 (6), 1721–1777.

Foyer, J., 2018. Syncrétisme des savoirs dans la viticulutre biodynamique: Incorporation dans l'expérience et le sensible et trajectoire initiatique. *Revue d'anthropologie des connaissances*, 12 (2), 289–321.

Garcia-Parpet, M.-F., 2014. Vers une nouvelle redéfinition des vins d'appellation d'origine? *In:* A. Cardona et. al., eds., *Dynamiques des agricultures biologiques*. Paris: Editions Quae, 57–74.

Grandjean, A., 2021. Biodynamic Wine-Crafting in Switzerland: The Translation and Adaptation of Rudolf Steiner's Cosmology into Dark Green Agronomies. *Journal for the Study of Religion, Nature and Culture*, 15 (3), 317–343.

Huss, B., 2014. Spirituality: The Emergence of a New Cultural Category and Its Challenge to the Religious and the Secular. *Journal of Contemporary Religion*, 29 (1), 47–60.

Ingold, T., 2000. *The Perception of the Environment: Essays on Livelihood, Dwelling and Skill*. London, New York: Routledge.

Ingold, T., 2008. Bindings Against Boundaries: Entanglements of Life in an Open World. *Environment and Planning A*, 40, 1796–1810.

Jenkins, W., and Chapple, C.K., 2011. Religion and Environment. *The Annual Review of Environment and Ressources*, 36, 441–463.

Knoblauch, H., 2008. Spirituality and Popular Religion in Europe. *Social Compass*, 55 (2), 140–153.

Knoblauch, H., 2014. Popular Spirituality. *In:* E. Hense, F. Jespers, and P. Nissen, eds. *Present-Day Spiritualities: Contrasts and Overlaps*. Leiden/Boston: Brill, 81–102.

Koehrsen, J., 2018. Religious Agency in Sustainability Transitions: Between Experimentation, Upscaling, and Regime Support. *Environmental Innovation and Societal Transitions*, 27, 4–15.

Krzywoszynska, A., 2015. On Being a Foreign Body in the Field, or How Reflexivity around Translation Can Take Us Beyond Language. *AREA*, 47 (3), 311–318.

Lambek, M., 2015. Le Bonheur Suisse, Again. *Journal of Ethnographic Theory*, 5 (3), 111–134.

LeVasseur, T., 2017. *Religious Agrarianism and the Return of Place: From Values to Practice in Sustainable Agriculture*. Albany: Sunny Press.

Luhrmann, T.M., 2008. Metakinesis: How God Becomes Intimate in Contemporary U.S. Christianity. *American Anthropologist*, 106 (3), 518–528.

Masson, P., 2015. *Guide pratique pour l'agriculture biodynamique*. Château: Editions BioDynamie Services.

McGuire, M.B., 2008. *Lived Religion: Faith and Practice in Everyday Life*. Oxford: Oxford University Press.

McKanan, D., 2018. *Eco-Alchemy: Anthroposophy and the History and Future of Environmentalism*. Oakland: University of California Press.

Meyer, B., 2008. Media and the Senses in the Making of Religious Experience: An Introduction. *Material Religion*, 4 (2), 124–134.

Morgan, D., 2010. Introduction: The Matter of Belief. *In*: D. Morgan, ed. *Religion and Material Culture: The Matter of Belief*. London: Routledge, 1–17.

OFAG, 2016. *Rapport sur le système de contrôle des vins: Contrôle de la vendange et du commerce des vins*. Berne: Département fédéral de l'économie et de la recherché.

Onfray, M., 2015. *Cosmos: Une ontologie matérialiste*. Paris: Flammarion.

Pineau, C., 2019. *La corne de vache et le microscope: Le vin " nature ", entre sciences, croyances et radicalités*. Paris: La Découverte.

Pink, S., 2009. *Doing Sensory Ethnography*. London: Sage.

Sennet, R., 2009. *The Craftman*. Newhaven: Yale University Press.

Steiner, R., 2006 [1924]. *Agriculture: Fondements spirituels de la méthode biodynamique*. Genève: Editions anthroposophiques romandes.

Strauss, A., 1992. *La trame de la négociation: Sociologie qualitative et interactionnisme*. Paris: l'Harmattan.

Taylor, B., 2010. *Dark Green Religion: Nature, Spirituality and the Planetary Future*. Berkeley: University of California Press.

Teil, G., Grant, H.J., and Grant, L.L., 2012. No Such Thing as Terroir?: Objectivities and the Regimes of Existence of Objects. *Science, Technology, and Human Values*, 37 (5), 478–505.

von Stuckrad, K., 2008. Esoteric Discourse and the European History of Religion: In Search of an Interpretational Framework. *Scripta Instituti Donneriani Aboensis*, 20, 217–236.

Wimmer, A., 2008. The Making and Unmaking of Ethnic Boundaries: A Multilevel Process Theory. *American Journal of Sociology*, 113 (4), 970–1022.

13 The Green, the Secular, and the Religious

The Legitimacy of Religious Environmentalism in Global Climate Politics

Katharina Glaab

Introduction

Climate change is one of the biggest challenges of our times. Extreme weather events as well as rising sea levels and temperatures will have major impacts on societies, and particularly the most vulnerable will suffer (IPCC 2014). In global climate politics, states, business, and civil society aim to find political solutions to curb emissions and create a more sustainable future. Among a diverse group of civil society groups comprising indigenous, farmers, business, women's groups and environmental NGOs, religious groups have become important actors in global climate politics. Not only since the prominent publication of the papal encyclical on the environment in 2015 have faith-based actors' (FBAs) contributions to environmental politics debates received increasing attention in national and international contexts. However, the United Nations (UN) and the United Nations Framework Convention of Climate Change (UNFCCC) – the international arena where political agreements on climate change are negotiated – are often described to be particularly "secular" spaces, where FBAs struggle "to be taken seriously" (Haynes 2014, p. 23), thus opening questions about the role of religion within this international space. But while the role of FBAs as actors in global climate politics seems to be getting more attention (Glaab 2017; Glaab et al. 2018; see Krantz 2022), the question in which way religious environmentalism plays out and may or may not create tensions in a presumably secular space has been widely overlooked.

This chapter aims at critically discussing the practices, legitimacy, and tensions of religious environmentalism in global climate politics. Taking a Bourdieusian approach, this chapter understands the UNFCCC as a "field" in which different actors compete to challenge and change the "doxa" – the fundamental, common-sensical beliefs and values – of what climate change means. Starting with the Bourdieusian observation that people act on different cultural capital, it investigates how "religious" and "secular" classifications matter and inform practices at the UNFCCC. It discusses how FBAs act within the presumably secular field of global climate politics,

DOI: 10.4324/9781003017967-17

in which way they differ from other civil society actors, and how they make and perceive differentiations in this context. The chapter focuses on the deliberations of FBAs in the context of the UNFCCC. Building on 3 years of research as an observer at the global climate change negotiations, it analyzes field notes and interview material of FBAs active at the scene. Thus, this chapter aims to contribute to the discussion on the increasing role of FBAs in climate change debates and sheds light on the political struggles and tensions between what is presumed to be "secular" and "religious" in the field of global climate politics ultimately challenging the presumed boundaries.

In the following, the chapter will first discuss whether global climate politics can be described as a secular field. Introducing Bourdieu's thinking about the field and doxa, it outlines how these notions can help to understand how varied cultural capital defines the role of the religious and secular at the UNFCCC. Second, it will describe the methodological approach of the study and introduce the empirical material and the methods applied for the following analysis. By drawing on Bourdieusian thinking, the chapter will analyze in a third step how FBAs position themselves within the field of global climate negotiations, how they make and perceive differentiations between the secular and religious and potentially challenge the secular doxa at the UNFCCC. Finally, the chapter will discuss how the analysis exposes tensions of religious environmentalism and evaluate the results of the study in light of its theoretical framework.

State of the Art: The Place of Religion in the International Arena

Global climate politics takes place within a global governance structure in which the UN represents a major institutional arena for international collaboration. It is the place where governments, civil society, and other actors have the opportunity to meet, discuss, and find solutions and agreements on collective problems. The UN hosts a diverse set of organizations that represent as varied issues such as gender, religion, environment, or security, but it is nevertheless often described as "a demonstrably secular organization" as it is "founded on nonreligious values" (Haynes 2014, p. 24). Having its roots within a Judeo-Christian tradition (Carette and Miall 2017), it "underpin[s] and reflect[s] the characteristics and global spread of a post-Westphalian, West-directed and focused international order" (Haynes 2014, p. 24). This traditional setting leads to a seemingly clear distinction between secular and religious organizations by the UN (Knox 2002). Scholars have pointed out that secularization processes are far more complex than simple dichotomies between the "secular" and "religious" can picture (Lehmann 2016). Yet, this categorization still offers a potentially useful analytical heuristic to grasp phenomena empirically while recognizing their overlapping character.

The presumed distinction between the secular and the religious becomes particularly evident when it comes to issues of the environment. Science and scientific evidence play a key role in discussions on the environment. As the secular is often associated with reason and rationalism, while the religious is associated with irrationalism and emotions, the concept of science is seemingly clearly situated within the realm of the secular (Glaab 2018, p. 175f). This may have the consequence that natural scientists and environmentalists alike find the contributions of religious actors unhelpful in debates on environmental issues (Wilson 2012). Although ethical and normative dimensions of climate change are increasingly taken into account in scientific or public debates (Hulme 2009; Klinsky et al. 2017), the debates at the international level are still dominated by scientific expert discussions that barely include alternative forms of knowledge (Lidskog and Sundqvist 2015). In the case of global climate politics this means that there has been a strong focus on scientific assessments, quantitative modelling, technical, or economic solutions at the UNFCCC negotiations at the cost of discussions on justice, equity, or representation (Klinsky et al. 2017). Considerations of the ethical dimensions of climate change have been introduced by various actors – encompassing indigenous, women, or youth groups, but also environmental or humanitarian organizations. In this context, there is also growing acknowledgment that religion has a part to play in our response to climate change (Hulme 2017) and can contribute to existing discussions on ethical and normative dimensions of climate change.

Despite the seemingly clear embeddedness of global governance institutions within a global secular order, the relationship between the secular and the religious is a lot more ambiguous at the UN. Bettiza and Dionigi (2015) therefore describe the UN as a non-aggressive secular organization as "the UN has generally been neutral and open toward the public participation of religious groups and communities in its debates and activities" (p. 629). While religion does not appear to be present at first sight, the presence of faith-based organizations (FBOs) speaks another language. Over the last 25 years the number of religious non-governmental organizations registered at the UN has risen considerably from a few dozen to over 300, making up around 10% of all NGOs registered at the UN (Beinlich and Braungart 2018; Haynes 2014; Carette and Miall 2017). This is seen in context of a broader discussion of religious resurgence, which – according to Haynes (2017, p. 1067) – is "characterised by growing prominence of ethical and moral (often overlapping with faith-based) concerns in debates about values, norms, and behaviours." One of the results of this religious resurgence within the context of global governance is that religious perspectives are more often articulated and heard when it comes to questions of morals and ethics in global politics (Haynes 2017, p. 1067), tying in and adding to broader normative debates on that matter. This leads to the UN having a "paradoxical relationship with religion" (Carette and Miall 2017). While it is formally a secular organization, religious perspectives are represented by

FBAs that take part in policy processes and often contribute to the functioning of the UN and the implementation of policies. Therefore, one cannot assume a strict division at the UN between what is presumed to be "secular" and "religious." Instead, the interaction between FBAs and other actors at the UN is much more complex and may play out along multiple lines (Haynes 2014; see also Glaab 2018).

Theoretical Framework: Bourdieu and the Global Climate Negotiations as a "Secular Field"

In order to better understand the ambiguous relationship between the religious and the secular in these international settings, Bourdieu's thinking about fields of cultural production as fields of power offers a fruitful approach to the study of FBAs at the UNFCCC (see also Huber 2022 for a Bourdieusian approach to religious environmentalism). According to Bourdieu (1994), a field is a specific social context in which people act based on certain dispositions. While Bourdieu described this in relation to the housing market (2005), artistic work (2006), or even religion (1991), a field can encompass various arenas and can be informed by diverse cultural, economic, or religious relationships. In essence, what is produced within these – may it be real estate, art, or religion – is situated within powerful social relations of a diverse set of actors that are all involved in its production. Depending on in which field one is situated, power can be experienced very differently. While a religious leader may have a legitimate authority to act within her local faith community, she may not have that legitimacy within a scientific community.

With this broad understanding, the notion of the field can also be translated into other areas. Then, the UNFCCC depicts a specific social context in which diverse actors act according to their particular understanding and interpretation of how to perform within the setting, ultimately producing a certain form of climate politics. It encompasses not only the work of the Secretariat of the organization as its bureaucratic machine room, but also the meetings, diplomatic exchanges, and negotiations of states and observers to find a political solution to govern climate change. It is characterized by a social setting, in which the same people representing diverse governments, businesses, and civil society organizations meet regularly over years or sometimes even decades.

These people act on the knowledge that is (re)produced within the field. Bourdieu suggests that there is a specific cultural capital that defines those actors' decisions and actions. Bourdieu (1986) argues that people mark their hierarchical standing by means of education, language, appearance, and other forms of social classification in the everyday. Cultural capital functions as a mechanism of power and necessarily creates "insiders" and "outsiders," "included" and "excluded." It sustains societal power relations as cultural capital creates unequal relations by way of cultural and societal practices of distinction. These hierarchies of power are sustained through the exercise of symbolic power, as people tacitly accept the cultural and societal domination within the field. This depicts the "doxa" or what is taken for granted within a society. It differs from what Bourdieu terms "heterodoxy" – when there are multiple understandings that are contested – and "orthodoxy" – when there is a collective and consensual understanding. Doxa stands out as it is tacitly self-evident to people and therewith particularly powerful in guiding actions. Yet, doxic understandings can become heterodox when unquestioned and self-evident issues are discussed, challenged, and negotiated again.

What Bourdieu observed in his study of French society can also be analyzed in other settings such as international conferences. Bourdieu emphasizes how people make classifications all the time and the same can be said for the specific field of the UNFCCC. First of all, the Secretariat distinguishes between insiders and outsiders through their system of registration. While "parties" have access to all areas and meetings, "observers" face restrictions in access to meetings and ability to make their voices heard within the conference setting. Next to this formal distinction, informal distinctions between parties and observers alike play out along the lines of experience, legitimacy, state or institutional representation, and which transcend the formal boundaries that the Secretariat sets through its policies. Symbolic power and exclusion are then exercised not only in and through formal differences in power, but also in everyday practices such as diplomatic rituals, conference proceedings, or informal gatherings that we can also observe at international conference settings. The meetings of the UNFCCC proceed in a routinized way and require a lot of informal knowledge of the institutional setting and its working procedures. For instance, as a newcomer it is nearly impossible to understand the role and function of the different work streams, plenary meetings, working groups, or side events without prior knowledge or a knowledgeable network that helps to navigate through the endless possibilities of engagement at those conferences. Indeed, Berger (1999) described how a cosmopolitan secular elite that is familiar with that environment dominates international settings such as the UN.

When it comes to the role of religion within the field of climate politics, the literature suggests that the doxa at the UNFCCC is a secular one (Berger 1999; Knox 2002; Haynes 2014), creating secular insiders and religious outsiders. But do power relations manifest only along that simple hierarchy or are the classifications that are based on cultural capital more fluid and ambiguous ultimately challenging the presumed doxa? It has been observed that there is an "increasing entanglement of religious, scientific and political issues" (Glaab 2018, p. 176) at the UNFCCC that reflect the increasing tensions and struggles for legitimacy between those areas. Therefore, in the following, the chapter will interrogate how the presumably secular field of global climate politics structures representation and actions of faith-based actors, how the clear classifications become more ambiguous, and how they sometimes can create tensions between "religious" and "secular" actors.

Methods and Data: Studying Faith-Based Actors in the "UNFCCC Field"

To study the role of faith-based actors within the secular field of climate change politics, the focus of this chapter is on the case of the UNFCCC as a field of practice. The UNFCCC is the main global governance arena to negotiate a political agreement to curb climate change. With the choice of this specific field, the aim is not to infer something representative about the role of religion at the UN. As discussed previously, power and the role of religion can play out very differently in other political fields (see also Baumgart-Ochse and Wolf 2018 for a comparative study). Neither does this case aim to generalize about the role of religion in global environmental politics as such. In line with the theoretical approach, this case provides in-depth knowledge of the role of FBAs in the specific field of the UNFCCC which may also resonate with findings on religious environmentalism in other fields of practice.

The study focused on practices of FBAs at the UNFCCC which in contrast to the term faith-based organizations encompasses organizations and individuals. In order to qualify as "faith-based" actors for the purpose of this study, organizations or individuals needed to have a clear religious or spiritual background or identify as faith-based in their mission statement (for a similar approach, see Berger 2003). For pragmatic reasons, it did not look at all actors inhabiting the field of global climate change politics. Therewith, the study is limited to understanding the field from the perspective of FBAs without taking into account the perceptions of government, business, or other civil society actors. The focus is then not on understanding the power relations of all actors within the field and its consequences for climate politics, but the specific classifications that FBAs make and act upon in order to understand the tensions of religious environmentalism within a presumably secular field.

The chapter is based on an in-depth study of FBAs at the UNFCCC over a period of 3 years. Between 2013 and 2016, the author conducted participant observation at major conferences and interim meetings of the UNFCCC. In that context, the author conducted 21 semi-structured interviews with FBAs at the meetings or in follow-up skype talks. Among those interviews, 15 interview partners were of Christian faith. This does not come as a surprise as it has been observed before that Christian Western-based organizations comprise the majority of NGOs at the UN (Berger 2003), because these organizations tend to have the financial and personal capacity to fill out their representative role at the UN. According to Haynes (2014), between 58% and 75% of the more than 300 registered religious NGOs at the UN are Christian and Northern organizations. The dominance of Christian faith organizations in the set of interview partners is therefore representative of this bias. In addition to interviews, field notes were written and web material gathered. As I attended the interfaith focal group meetings at the negotiation site and followed the discussions and organization of FBAs' email group, I was well informed about activities, meetings, and debates. This diverse and encompassing material provided the basis for a content analysis, in which codes were developed "in vivo." These were organized in "secular" and "religious" language and practices to understand the composition of the field, the classifications that are based on cultural capital, the doxa present, and ultimately the ambiguities and tensions of religious environmentalism at the UNFCCC.

Results: The "Religious" in the Secular Field of Global Climate Politics

When it comes to the role of religion at the UNFCCC, the observation of an earlier *Religion Counts* report – that people often do not think about religion as having a place within the UN – still applies (Knox 2002). Yet when one zooms into this field of practice, it becomes evident that there is not only a growing presence of FBAs at the presumably secular UN but also at the UNFCCC. Here, FBAs are made up of a diverse set of actors. The position of FBAs cannot easily be categorized according to their formal distinction into states and observers. As religious groups do not have their own constituency at the UNFCCC,[1] FBAs mostly register as part of broader civil society who have access as observers to the conferences. While it is difficult to pin down the exact number of FBAs at the climate change conferences due to unclear self-ascription and changing attendance, an earlier study estimates that around 30 organizations, both small organizations and large alliances, actively take part in the UNFCCC processes (Glaab et al. 2018, p. 51). This number may appear small in light of growing civil society engagement, but many organizations participate with a number of delegates or represent large conglomerates such as the World Council of Churches. Yet, most of these organizations do not belong to the big civil society actors, but represent small- and middle-sized organizations "without much . . . individual financial, diplomatic, or ideological leverage" (Haynes 2014, p. 3).

However, the focus on FBAs as civil society actors with observer status neither accounts for states with a state religion such as Israel or Saudi-Arabia nor for individual actors of faith that can be part of state delegations, as well as exceptions in the international system such as the Holy See, which has a special observer state status as the diplomatic representation of the Vatican State. This means that faith-based participation takes place across and beyond formal markers that distinguish between faith-based and secular actors along states as "insiders" and observers as "outsiders." The division between what is presumed to be "secular" and what to be "religious" is therefore more complex and ambiguous. In fact, within the specific social field of the UNFCCC, distinction materializes in various forms and not only in relation to governments or other civil society actors.

Particularly the question of legitimacy of the religious in a secular field becomes important as it indicates whether religious actors are accepted and how potential tensions between religious and secular actors unfold. Legitimacy enables and naturalizes doxic understandings and it does not only matter to make legitimate claims, but these claims are related to being perceived as a legitimate actor in the social field. In the case of the UNFCCC – conceptualized as a social field of power, which gives space for public deliberation and negotiation – what course of action is decided on to stop climate change is related to the legitimacy of the cultural capital of FBAs in that context. Broadly speaking, there are two different ways that FBAs appear legitimate within the UNFCCC: one, by hiding the religious identity and adapting to the secular environment, and two, by emphasizing their religious background and distinguishing themselves from the secular environment.

When adapting to the secular field, FBAs' religious identity takes a back seat. In order to make their voices heard, actors tacitly and strategically foreground other aspects that matter within the secular environment. This corresponds with Haynes' (2014, p. 23) observation that they

> must learn to adapt to UN norms and conventions, in order to be heard and accepted. This means that to be significant players in global public policy debates, they must necessarily adopt and adapt to the terms and rationale of liberal – that is, nonreligious – discourse, even when they do not agree with it.

Those FBAs act like most other civil society actors and take part in nongovernmental organizations (NGO) network meetings, consult with state delegations, or observe plenary or working group meetings. They engage in and contribute to the work of civil society groups such as the Climate Action Network (CAN) by emphasizing their professionalism, experience, and technical knowledge of the process. A member of ACT Alliance highlights that "in the advocacy work we are also technical and I can enter into discussion about the paragraphs in the Kyoto Protocol. And I think I could match the lobbyists as well" (Interview ACT Alliance). This professionalism matters in certain contexts and a SAFCEI representative recognizes the need to adapt to different environments and people: "If I am talking to our negotiators or people in the government, I really need to speak their language" (Interview SAFCEI). Similarly, a representative of the Holy See emphasizes the need to "operationalize" religious concepts in a side event just before the papal encyclical on the environment was published (own observation, Bonn 2015). And the Catholic development organization CIDSE suggested ways how religious concepts of the papal encyclical such as human dignity or solidarity can be translated into a more appealing human rights language (CIDSE Poverty and Climate Justice Group 2015).

These instances illustrate that some FBAs recognize that their cultural religious capital does not count in a presumed secular context. In order to be intelligible and helpful to the work at the UNFCCC, those FBAs strategically but also sometimes tacitly adapt to the doxic understanding of this environment as a secular one by focusing on technical and scientific contributions. In order to be included, they need a specific cultural capital that is based on their long-standing professional experience that allows them to know how the process works and how to navigate it. In addition, it requires the cultural capital to translate and adapt to this context – what Koehrsen and Heuser (2020) term "boundary bridging competence." This applies particularly for FBAs that work as NGOs on development issues more broadly such as Christian Aid, Bread for the World, or the Association of World Council of Churches related Development Organizations in Europe (APRODEV). Some of these FBA representatives do not even self-identify as religious or faith-based at all, and likewise, they are also not perceived in religious terms by other civil society actors (Interview material). The cultural practices of distinguishing between the religious and secular therewith dissolve by adapting to the secular doxa at the UNFCCC. This form of adaptation is something that some FBAs are particularly concerned about. While they see the need and the worth of engaging with secular actors and to act in a more political way, they fear that the distinctions between secular and religious groups completely dissolve and FBAs could even become co-opted. A representative argued that "there's also a risk that the faith-based organizations, if they're not careful, they get drawn too much into NGO language" (Interview Network of Engaged Buddhists), warning that a language that is almost identical to NGO language goes over people's heads and the congregation do not recognize themselves in it anymore. This reflects a fear that their religious capital might be weakened. Yet, if their position of power within the UNFCCC field derives mainly from their religious capital, a weakening of this capital might also lead to a loss of legitimacy at the UNFCCC.

When distinguishing oneself from the secular environment, other FBAs do not sideline their religious identity and explicitly foreground their faith background. This can be done both through language and performance. Some FBAs do not adapt to the dominant way of communication by avoiding the overtly technical and scientific language and by using religious concepts or theological arguments in debates instead. Campaigns by faith actors would for instance explicitly invoke religious symbols and language such as the "Fast for the climate" or the "Climate pilgrimage" that took place before the important climate conference in Paris in 2015. Using emotive language is, of course, not unique to FBAs – after all, a lot of the campaigns of environment and development organizations explicitly aim to engage people on an emotional level. But while many of these FBA or civil society-led campaigns take place outside official meetings of international institutions, FBAs also use this form of language within the institutional context. In official meetings and the plenary of the conferences, FBAs would speak about the normative and moral challenges of climate change and highlight the personal experiences of those people who suffer from its effects through the use of emotive language (Glaab 2017). So it largely depends on the individual actor whether faith language or technical language is used; it depends on "who you are" (Interview Act Alliance).

Often FBAs adjust their language depending on the context and who is addressed: "When we talk with Sweden, there's no use to have any faith language at all. When we talk with Peru, it's very good to have a faith language" (Interview ACT Alliance). This tactic of lobbying states can similarly be observed in interaction with other civil society actors. Some FBAs would play down their faith background in meetings with secular environmental groups such as CAN but stick to their faith background in meetings with other FBAs. Prayers are a normal practice in inter-faith meetings but not practiced in secular network meetings. This indicates that within the presumably secular field of the UNFCCC, there are subfields in which the dominant secular doxa can be circumvented. However, lately the "faith factor" seems to have gained momentum in the secular context, too. There are, for instance, representatives from faith organizations – such as CIDSE (until 2018) and Christian Aid – on the board of directors of CAN (CAN 2019). And there is an increasing recognition among secular civil society actors that faith can play an important role in the fight against climate change as former UNFCCC secretary Christina Figueres (2014) exemplifies in her call to religious actors to act on climate change. While this may point to an increased legitimacy of religious capital in the UNFCCC field, it is often perceived to have an instrumental view of the religious as another variable that can be used to create climate action. That perception also speaks to fears of cooptation of FBAs through secular actors as discussed previously.

Some FBAs also show the distinction from the secular through performance and their appearance at the conferences. While the traditional business outfit seems to be the standard for the "secular cosmopolitan elite" (Berger 1999) in international conference settings, some actors chose a different dress to distinguish themselves from these. Some FBAs wear religious cloth or symbols at the meetings in order to show their religious identity. While one could argue that the religious appearance within a secular surrounding is a disadvantage, it can also be an advantage when it comes to questions of legitimacy. As one interviewee observed, sometimes religious actors have more legitimacy than other civil society or state actors:

> Many governments are critical to NGOs, but churches, a bishop is always welcome. You do not say "No" to a bishop, but you can say "No" to Oxfam. . . . many countries think that it's more neutral to talk with a church.
>
> (Interview ACT Alliance)

Another representative argued that

> It's crazy. They [the religious authorities] can literally phone the President and ask for a meeting and they'll get it. If they say they want to . . . be one of their negotiators and wear a pink badge? Immediately, [they] said yes. I mean we find that people trust religious leaders.
>
> (Interview Interfaith Liaison representative)

In this context, the distinction as religious plays out as an advantage as religious actors are perceived to be more legitimate due to their presumed neutral status in politics. The secular assumption that religion acts outside the public political sphere comes as a benefit in this context, as they are not seen to have a political message and can be trusted. This advantage of religious capital within a dominantly secular field seems surprising at first, but it suggests that religious capital can be converted into social capital that is central within that field. This would indicate that capital is not really religious here, but derives its standing from its ability to be used to build social relations.

Discussion: FBAs at the UNFCCC and Tensions Between Secular and Religious Environmentalism

The UNFCCC field provides an interesting setting to understand the tensions of religious environmentalism. As part of the UN architecture it is embedded in a presumably secular international order, which assumes that technical and scientific knowledge dominates and religious perspectives are sidelined. In this context, where religion can become a marker of distinction, the question of the legitimacy of religious capital and its potential contribution to climate change politics becomes pertinent. One could argue that we can observe a challenge to the secular order by the mere presence of FBAs at the climate change negotiations. The Religion and Ecology literature would for instance assume that religion can make a difference here. Yet the analysis showed that FBAs engage in various ways with the dominant secular doxa at the UNFCCC. Some FBAs accept that their religious capital does not count in certain contexts and adapt to dominant secular reasoning and practices. So when they engage with other civil society organizations or presumably secular states, they "speak their language" (Interview SAFCEI) – as an interviewee put it – and foreground their technical expertise of the process based on their experience and knowledge. The same FBAs would emphasize their religious capital in contexts where it seems to be legitimate to speak with a religious voice. When meeting with other faith-based actors of their network or with presumably religious states or state representatives, they speak from a religious perspective instead of (only) evidence-based science. Other FBAs would not adapt to the secular context but use their religious capital regardless of which context. The knowledge of how to act in certain contexts is a specific form of cultural capital and to

FBAs there seem to be different understandings of where and when religious capital is legitimate, leading to various subfields within the UNFCCC field where the dominant doxa is followed or circumvented.

From a Bourdieusian perspective, religion is not the only marker of distinction in this field, but different markers create diverse and fuzzy forms of legitimacy of religious capital. While some FBAs see this as their specific cultural capital, others see the distinction and their capital rather along professional experience or technical knowledge. Yet, while religious capital can be argued to not having strong value in secular environmentalist contexts, it can sometimes be even more legitimate than the social capital of civil society actors due to the perception of FBAs as neutral and apolitical actors. While there is an adaptation of FBAs to the dominant secular environment, it also shows that the distinction around secular and religious markers is not that simple. The question is whether the different classification-making practices of FBAs lead to the dissolution of the dominant doxa. The presence of FBAs is not just a challenge to the secular order, but from a Bourdieusian perspective it can even be interpreted as a dissolution of the secular doxa that privileges rationalism and scientific and technical knowledge in international conference settings. When this thinking is not self-evident anymore, doxa dissolves into heterodoxy and orthodoxy. We can observe more heterodox understandings, where these diverse beliefs and opinions are debated and negotiated. Indeed, there are more and more efforts to include FBAs in political processes by the UNFCCC Secretariat (Interview UNFCCC Secretariat) and other civil society actors. In addition, normative issues are increasingly addressed and there is more room to speak of justice and fairness issues from a religious perspective. However, secular thinking is still orthodox as the belief that the religious is not supposed to take part in political negotiations is still strongly presented at the UNFCCC.

Tensions do not play out through the distinction of the religious and the secular as such but arise in the interrelation between their heterodox and orthodox understandings. Although the dominant doxa of secularism and rationalism at the UNFCCC seems to be challenged by the presence of religious actors and more heterodox thinking, many participants still do not see a role for religion in international climate politics. This may create tensions when secular actors are hesitant to work with religious concepts and actors, or when secular and religious actors simply do not understand each other. The adaption to secular language may limit potential tensions with civil society actors as the distinctions become unrecognizable. However, it can create tensions among FBAs themselves as they are seen to lose legitimacy among other FBAs when giving up their specific cultural religious capital. As the fields are interrelated, this might ultimately also diminish their legitimacy within the field of the UNFCCC. With the growing role of FBAs in global climate politics and the intensified collaboration with other civil society actors, these tensions will potentially become even more important. It will be interesting to observe in which way FBAs will use their specific cultural capital to influence decisions and how they might further challenge the secular doxa at the UNFCCC.

Note

1 There are nine official constituencies at the UN: business and industry NGOs, environmental NGOs, indigenous peoples organizations, local government and municipal authorities, research and independent NGOs, trade union NGOs, women and gender constituency, youth NGOs, and farmers and agricultural NGOs (UNFCCC 2017). Being part of a constituency enables the representation of certain views and allows, for instance, to speak in the plenary.

References

Baumgart-Ochse, C., and Wolf, K.-D., eds., 2018. *Religious NGOs at the United Nations. Polarizers or Mediators?* London: Routledge.

Beinlich, A.-K., and Braungart, C., 2018. Religious NGOs at the UN. A Quantitative Overview. *In*: C. Baumgart-Ochse and K.-D. Wolf, eds. *Religious NGOs at the United Nations: Polarizers or Mediators?* London: Routledge, 26–46.

Berger, J., 2003. Religious Nongovernmental Organizations: An Exploratory Analysis. *Voluntas: International Journal of Voluntary and Nonprofit Organizations*, 14 (1), 15–39.

Berger, P.L., 1999. *The Desecularization of the World: Resurgent Religion and World Politics*. Washington, Grand Rapids: W.B. Eerdmans Publishing.

Bettiza, G., and Dionigi, F., 2015. How do Religious Norms Diffuse? Institutional Translation and International Change in a Post-Secular World Society. *European Journal of International Relations*, 21 (3), 621–646.

Bourdieu, P., 1986 [1979]. *Distinction – A Social Critique of the Judgement of Taste*. London: Routledge.

Bourdieu, P., 1991. Genesis and Structure of the Religious Field. *Comparative Social Research*, 13, 1–44.

Bourdieu, P., 1994. *The Field of Cultural Production*. Columbia University Press.

Bourdieu, P., 2005. *The Social Structures of the Economy*. Cambridge: Polity Press.

Bourdieu, P., 2006. *The Rules of Art. Genesis and Structure of the Literary Field*. Stanford, CA: Stanford University Press.

Carette, J., and Miall, H., 2017. *Religion, NGOs and the United Nations: Visible and Invisible Actors in Power*. London: Bloomsbury.

CIDSE Poverty and Climate Justice Group, 2015. Paris, for the People and the Planet. The Encyclical Laudato Si' – What It Means for the COP21 Negotiations and Beyond. Policy Paper.

Climate Action Network (CAN), 2019. *Board of Directors, 2018*–2019 [online]. Available from: www.climatenetwork.org/can-international-board [Accessed 12 August 2019].

Figueres, C., 2014. Faith Leader Need to Find Their Voice on Climate Change. *The Guardian*. Available from: www.theguardian.com/environment/2014/may/07/faith-leaders-voice-climate-change [Accessed 14 August 2019].

Glaab, K., 2017. A Climate for Justice? Faith-Based Advocacy on Climate Change at the United Nations. *Globalizations*, 14 (7), 1110–1124.

Glaab, K., 2018. Faithful Translation? Shifting the Boundaries of the Religious and the Secular in the Global Climate Change Debate. *In*: T. Berger and A. Esguerra, eds. *World Politics in Translation – Power, Relationality, and Difference in Global Cooperation*. London: Routledge, 175–190.

Glaab, K., Fuchs, D., and Friederich, J., 2018. Religious NGOs at the UNFCCC: A Specific Contribution to Global Climate Politics? *In*: C. Baumgart-Ochse and K.-D. Wolf, eds. *Religious NGOs at the United Nations: Polarizers or Mediators?* London: Routledge, 47–63.

Haynes, J., 2014. *Faith-Based Organizations at the United Nations*. New York: Palgrave Macmillan.

Haynes, J., 2017. Global Governance and the United Nations: The 'Clash of Civilisations' and the 'Rise' of RNGOs. *Globalizations*, 14 (6), 1060–1068.

Huber, F., 2022. Environmentalism in the Religious Field. Established vs. Newcomers in Switzerland. *In:* J. Koehrsen, J. Blanc, and F. Huber, eds. *Religious Environmental Activism: Emerging Conflicts and Tensions in Earth Stewardship*. London: Routledge.

Hulme, M., 2009. *Why We Disagree About Climate Change: Understanding Controversy, Inaction and Opportunity*. Cambridge: Cambridge University Press.

Hulme, M., 2017. Climate Change and the Significance of Religion. *Economic and Political Weekly*, 52 (28), 14–17.

IPCC, 2014. *Climate Change 2014: Mitigation of Climate Change Working Group III Contribution to the Fifth Assessment Report of the Intergovernmental Panel on Climate Change* (O. Edenhofer, R. Pichs-Madruga, Y. Sokona, E. Farahani, S. Kadner, K. Seyboth, A. Adler, I. Baum, S. Brunner, P. Eickemeier, B. Kriemann, J. Savolainen, S. Schlömer, C. von Stechow, T. Zwickel, and J.C. Minx, eds.). Cambridge: Cambridge University Press.

Klinsky, S., Roberts, T., Huq, S., Okereke, C., Newell, P., Dauvergne, P., O'Brien, K., Schroeder, H., Tschakert, P., Clapp, J., Keck, M., Biermann, F., Liverman, D., Gupta, J., Rahman, A., Messner, D., Pellow, D., and Bauer, S., 2017. Why Equity Is Fundamental in Climate Change Policy Research. *Global Environmental Change*, 44, 170–173.

Knox, G., 2002. *Religion and Public Policy at the UN: A Religion Counts Report*. Washington, DC: Religion Counts.

Koehrsen, J., and Heuser, A., 2020. Beyond Established Boundaries: Faith Based Organizations in Development Discourses and Practice. *In*: J. Koehrsen and A. Heuser, eds. *Faith Based Organizations in Development Discourses and Practices*. London: Routledge.

Krantz, D., 2022. Climate and Covenant. A Case Study of the Functions, Goals and Tensions of Faith at the 23rd Conference of Parties to the United Nations Framework Convention on Climate Change. *In:* J. Koehrsen, J. Blanc, and F. Huber, eds. *Religious Environmental Activism: Emerging Conflicts and Tensions in Earth Stewardship*. London: Routledge.

Lehmann, K., 2016. *Religious NGOs in International Relations – The Construction of 'the Religious' and 'the Secular'*. London, New York: Routledge.

Lidskog, R., and Sundqvist, G., 2015. When and How Does Science Matter? International Relations Meets Science and Technology Studies. *Global Environmental Politics*, 15 (1), 1–20.

UN Framework Convention on Climate Change, 2017. *Statistics on Observer Organizations in the UNFCCC Process* [online]. Available from: http://unfccc.int/parties_and_observers/observer_organizations/items/9545.php [Accessed 1 July 2019].

Wilson, E.K., 2012. Religion and Climate Change. The Politics of Hope and Fear. *Local Global*, 10, 20–29.

14 Climate and Covenant

A Case Study of the Functions, Goals, and Tensions of Faith at the 23rd Conference of Parties to the United Nations Framework Convention on Climate Change

David Krantz

Bicycle bells accompanied church bells as young and old from all over the world rode bicycles on a carless street in Bonn, Germany, in November 2017, with a few rickshaws for those unable to pedal. Affixed on posters atop the rickshaws and on flags on each of our bicycles was an image of the Earth, held aloft by five hands. The words "Many Faiths One Planet" floated above the image, along with the symbols of Buddhism (a golden Dharma Chakra, aka Wheel of the Law), Christianity (a red Latin cross), Daoism/Taoism (a black-and-white yin and yang), Hinduism (a red Om), Islam (a green star and crescent), Judaism (a blue Magen David, aka Star of David), Shintoism (a vermilion Torii gate), and Sikhism (a black Khanda). After we dismounted, we walked behind a "Many Faiths One Planet" banner with the same imagery to deliver an interfaith statement—titled "Walk on Earth Gently"—to an official at the headquarters of the United Nations Framework Convention on Climate Change (UNFCCC). This was faith in action at the 23rd Conference of Parties (COP) to the UNFCCC.

Introduction

Religion and climate change may represent two of the most impactful anthropogenic forces in society. Their intersection represents a powerful leverage point for change (Meadows 2008), because most people in the world express affiliation with a religion (Hackett and Stonawski 2017), and because faith groups can:

- Use their influence over adherents' worldviews to encourage action;
- Possess an ethical authority that can influence those outside their faiths;
- Control significant economic, institutional, social—and in some cases political—power;
- Reach people at local, national, and international scales and across boundaries (Caniglia et al. 2015).

DOI: 10.4324/9781003017967-18

To date it is unclear to what extent religions have been successful in addressing climate change, although it is clear religion so far is underperforming its potential impact (Caniglia et al. 2015).

Since its first meeting in Berlin in 1995, the Conference of Parties (COP[1]) to the United Nations Framework Convention on Climate Change (UNFCCC) has been the world's largest venue for governments to cooperatively address the climate crisis. The UNFCCC permits nongovernmental organizations (NGOs), including religious NGOs, to apply for accreditation to participate. Thousands of NGOs from around the world do so, and only a small percentage of them are religious in nature (Krantz 2021). But what tensions arise from the functions and goals of religious NGOs at the COP? Serving as an oral-history case study of COP 23, this chapter will discuss the background and existing research of faith at the COP, followed by a description of the methodology, a presentation and analysis of the results, and a conclusion.

Literature Review

There is a small but growing body of literature examining the role of faith at the U.N. Economic and Social Council (Carrette and Miall 2017; Carrette 2013; Petersen 2010; Haynes 2014; Berger 2003; Knox 2002), which operates independently from the UNFCCC, but there has been little research on faith at the COP. Empirical research on the role of faith at the COP is even thinner (Glaab 2017; Glaab et al. 2019).

Katharina Glaab, whose chapter also is in this book, has been the main driver of empirical research on religion at the COP. She, along with co-authors Doris Fuchs and Johannes Friederich, have documented much of the history and motivations of faith at the COP, which began at COP 1 and was enhanced with the 2013 formation of the Interfaith Liaison Committee of COP Interfaith Cooperation on Climate Change (henceforth simply Interfaith Liaison Committee) at COP 19 in Warsaw. Through interviews with faith-based actors, they showed how religious NGOs at the COP promote normative perspectives grounded in their religious ethics and morals, and particularly emphasize justice, at least from Abrahamic faiths (Glaab 2017; Glaab et al. 2019).

Randolph Haluza-DeLay also conducted empirical research on religion at the COP through event ethnography at COP 22 in Marrakech, Morocco, in 2016 (Haluza-DeLay 2018a, 2018b, 2019a, 2019b). In recognizing that one aspect of faith at the COP may be serving a role in mobilizing religious adherents on climate action outside of the COP, Haluza-DeLay sought to find what else religious NGOs bring to the COP, including possible contributions to climate governance; what their goals are; and whether or not the movement of religious NGOs is in concert or is simply "a collection of actors" (Haluza-DeLay u.d.). Haluza-DeLay found six functions and goals of religious NGOs at the COP:

- Networking;
- Providing solidarity for other activists;
- Lobbying negotiators for climate justice;
- Virtue signaling by sending messages to their constituents about the importance of taking climate action;
- Delivering statements on an international stage pronouncing that religious NGOs want governments to take climate action;
- And bearing public witness to the climate crisis through public events. (Haluza-DeLay u.d.)

Haluza-DeLay also concluded that faith-based environmentalism as expressed at the COP merits consideration as a movement, with religious NGOs acting in concert to make transnational linkages among themselves as well as among people of different faiths (Haluza-DeLay u.d.).

The six functions of religious actors found by Haluza-DeLay differ somewhat from the 11 functions of secular non-governmental actors at the COP identified by Mattias Hjerpe and Björn-Ola Linnér. Through a literature review and two surveys of 3,400 participants at 2007's COP 13 in Bali, Indonesia, and 2008's COP 14 in Poznań, Poland, they found that non-state actors attended the COP without participating in negotiations for 11 functions:

- Networking;
- Providing a forum for holistic thinking;
- Coordination and the setting of norms;
- Enhancing institutional legitimacy for their organizations;
- Generating and sharing knowledge;
- Public education and highlighting climate policy;
- Building capacity for their organizations;
- Contributing agenda items to climate policy and negotiations;
- Marketing for their organizations;
- Influencing the implementation of climate rules;
- And piloting and launching new products, projects, and services. (Hjerpe and Linnér 2010)

Lastly, Heather Lovell and Heike Schroeder led a team that conducted 10 interviews and surveyed 117 of 381 UNFCCC-recognized "side events"—talks, workshops, and other NGO activities—organized at COPs 13 and 14 by non-governmental actors and found they had five functions and goals:

- Networking;
- Meetings with peers;

- Learning about quick, effective solutions;
- Debating novel ideas;
- And presenting new research. (Schroeder and Lovell 2012)

My work complements the work of Lovell and Schroeder, Hjerpe and Linnér, Haluza-DeLay, and Glaab through oral histories, adding individualized context to the role of religious NGOs and highlighting tensions between each other and between themselves and society at COP 23.

Methodology

As an observer and participant, I attended formal and informal faith-based activities—including private meetings and public events—at COP 22 in November 2016 and COP 23, hosted by the Republic of Fiji but held in Bonn in November 2017. At the latter, I conducted 17 semi-structured oral-history interviews of key stakeholders among leaders of different faiths, while also participating in interfaith meetings as a representative of Aytzim: Ecological Judaism. In so doing, I recognize that (i) because participants have their own unique histories, bases of knowledge, and implicit and explicit biases, objectivity is an unachievable ideal which cloaks normative perspectives (Wallerstein 2000); (ii) my perspective is one of many (Cayton 2003); and (iii) observing and participating as an insider allowed me to benefit from an intimate knowledge of the players and the faith-based environmental movement.

Interviewees were selected by their willingness to be interviewed, our ability to schedule the interview, through chain-referral (or snowball) sampling, and through their participation in meetings of the Interfaith Liaison Committee—a semi-formal group of faith leaders organized mainly by Valeriane Bernard of the Brahma Kumaris World Spiritual University (henceforth simply Brahma Kumaris), Nigel Crawhall of the U.N. Educational, Scientific and Cultural Organization, and Henrik Grape of the World Council of Churches. Whereas Glaab coded her interviews to find patterns in types of discourse, I approached my interviews individually and as oral history, with each providing a personalized, unique perspective to the COP process. I asked each interviewee similar semi-structured open-ended questions, in the form of a conversation.

The 17 interviewees included: three interfaith leaders; nine representing Christian groups; two representing the Brahma Kumaris; and one each from Buddhism, Hinduism, and Islam[2,3]. The mostly Christian interviewees were reflective of both faith leadership at the COP as well as the outsized role played at the COP by the Global North (Krantz 2021), with Christianity as its primary religion. I have made no distinction between the terms "spiritual movements" (such as the Brahma Kumaris), "faith traditions," and "religions."

Every COP is about two weeks long and I attended for one week each time—common for non-negotiators. Because UNFCCC-accredited NGOs are given a limited number of badges, many split their badges by week, effectively doubling the number of people they can send. Many who only attend one week of a COP never encounter people who only attend the other. While it is possible that the makeup of faith representation was vastly different in the weeks I did not attend, my interviewees told me that my experience at the COP was similar to theirs at past COPs. Nonetheless, this case study is only representative of the work of religious NGOs at COP 23.

There is no formal mechanism for organizing faith leaders at the COP; religion is not among the UNFCCC's nine constituency groups (agricultural, business, environmental, indigenous, local governmental, researcher [higher education], trade union, women and gender, and youth) (United Nations Framework Convention on Climate Change u.d.) that provide input into negotiations. Largely, those leaders who attended solo and did not participate in the Interfaith Liaison Committee fall outside the scope of this case study.

As the vast majority of people in the world belong to a religious group (Hackett and Stonawski 2017), it is plausible that many COP participants also are people of faith. Indeed, religiously motivated people may not be working through religious groups because they see secular groups as the best venue. For this case study I was interested in the perspectives of leaders at the COP who were explicitly representing religious groups. The role of faith among actors representing secular NGOs at the COP calls for future study.

Finally, for this case study I was not interested in religious NGOs that attend for the unrelated purpose of proselytization[4].

Tensions Around the Functions of Faith Actors at COP 23

My case study found that faith actors at COP 23 employed 11 functions and goals, consisting of the six functions observed by Haluza-DeLay along with five more:

- Networking;
- Expressing solidarity with other social movements;
- Lobbying;
- Virtue signaling;
- Publicizing interfaith statements;
- Bearing public witness;
- Piloting projects—using the COP as an organizational launch pad;
- Publicly framing spiritual beliefs as ecologically minded;
- Public education;
- Self-education;
- And promoting plant-based diets.

While I found that the activities of faith actors at COP 23 could be seen as fulfilling all the functions identified by Hjerpe and Linnér—including those not in my list—I excluded those that were neither explicit functions nor goals of faith actors I observed at COP 23. For example, while faith actors at COP 23 provided a forum for holistic thinking and enhanced institutional legitimacy (both identified by Hjerpe and Linnér), such outcomes appeared to be side effects rather than intentional functions or goals of faith actors at COP 23.

Of the 11 functions and goals I identified, six will be discussed in this chapter, one for each of the following six significant forms of tension:

- Solidarity for its in-situ tension that was place-based at the COP;
- Virtue signaling for its intra-religious prophetic tension within faith traditions, where alternative voices within the faith represent neither mainstream leadership nor the flock;
- Lobbying for its intra-religious shepherding tension within faith traditions, where mainstream leadership has embraced ideas that their flock has not yet;
- Interfaith statements for their societal tension between religious groups and society;
- Eco-religious framing for its tension with societal expectations;
- And promoting plant-based diets for its universal tension that includes all of the above.

Overall, I found both the functions and their tensions are not mutually exclusive and some activities may fulfill multiple functions and goals, while some may have secondary tensions. Notably absent was significant inter-religious tension between different faiths. For the most part, aside from minor tensions, I found that religious NGOs at COP 23 often agreed with each other and largely formed a united front on climate change.

Solidarity: An In-Situ Tension

Some actions of religious NGOs at COP 23 were designed to express solidarity with other social movements—such as alleviating poverty and assisting refugees—representing a form of in-situ tension rooted in the unique place-based nature of the COP as an event, which may be the most subtle of the tensions I found at COP 23. For example, a coalition of 21 Protestant and Catholic churches in and around Bonn organized 38 climate-oriented events and art installations—including the display of a refugee's boat, the creation of an ecumenical pilgrim trail and a group walk along it, a discussion on how renewable energy can abate poverty, a mass for climate justice, and a mass for the Earth (Kirchen Und Klima 2017). Donations were collected after the Earth mass—not for the host church, but for Greenpeace.

While some may welcome broadening the discussion at the COP beyond climate change to include issues such as immigration—which could be seen as foundational aspects of climate change and its communication—others may see their inclusion as a distraction (Moser 2016; Boykoff et al. 2009), particularly when the COP may represent the only time when climate change is near the top of the agenda for global governments (Krantz 2021). Even though (or perhaps because) non-state actors may have distinct advantages at advancing various agendas at COPs (Nasiritousi et al. 2016), such broadening may be a factor in the COP's decreased effectiveness at developing international policy to address climate change (Schüssler et al. 2014).

Another kind of solidarity is the way religious NGOs provide solidarity to representatives from other faiths through their very presence. In that regard, while it is not clear to what extent religious communities without NGO representation at COP 23 were disappointed in their faith leadership for their absence, if at all, many faith leaders at COP 23 told me that they noticed the absence of their peers from other (predominately non-Christian) faiths.

Intra-Religious Tension

Although tension between faith groups was largely absent, religious actors at COP 23 still faced tensions between themselves and co-religionists who did not attend—and who may not even believe climate change is happening, let alone that their leadership should address it. This seems to be especially problematic for Christianity (Nche 2020), many of whose members—particularly evangelical Protestants and non-Hispanic Catholics (Hulme 2017; Arbuckle and Konisky 2015; Funk and Alper 2015; Zaleha and Szasz 2015)—hold beliefs that serve as barriers to climate action, such as biblical absolutism and literalism that leads believers to distrust science and/or see it as an opposing belief system (Nche 2020; Simpson and Rios 2019; Rios 2020; Taylor et al. 2016; Arbuckle and Konisky 2015), dominion belief that humanity is supposed to control nature (Nche 2020; White 1967), idolatry belief that caring about environmental issues constitutes "neo-pagan nature worship" (Zaleha and Szasz 2015), eschatological belief that climate change may play a role in the apocalypse and the return of Jesus (Nche 2020; Taylor et al. 2016; Zaleha and Szasz 2015), fatalism belief that the future is predetermined and the outcome inevitable (Nche 2020; Mayer and Smith 2019), and/or political beliefs that glom on to climate change skepticism because of their religious groups' embracement of political conservatives who typically disdain climate action (Veldman 2019).

Virtue Signaling: An Intra-Religious Prophetic Tension

> "I didn't convert to climate change until 2002," Richard Cizik said. "I ended up six years later losing my job for this very reason: Speaking out too much because I became a climate activist."
>
> (Interview: Cizik 11 Nov. 2017)

Virtue signaling—where one's activities or presence is intended to indicate one's values to others—was a common function amongst people of faith at COP 23, and Cizik's presence represented a powerful message for evangelicals. In 2008, he was forced out as vice president for governmental affairs at the National Association of Evangelicals—perhaps the World Council of Churches equivalent for American evangelical Christian denominations—and while it may have been his public support for same-sex civil unions and the administration of President Barack Obama that ultimately cost him his job, he said that his support for taking action on climate change also played a large role in his ouster (Interview: Cizik 11 Nov. 2017).

But in exile Cizik started a different kind of evangelical Christianity that could coexist with Earth stewardship and address climate change—a New Evangelical Partnership for the Common Good—and a significant minority of evangelicals have followed him. He has been attending COPs since 2009's COP 15 in Copenhagen (Interview: Cizik 11 Nov. 2017) and he brings the type of moral speech that is often missing from the secular side of COPs:

> "I would say to my fellow evangelicals and others, 'Wow, you don't think history will judge what we do and dont' do?'" Cizik said before paraphrasing and explaining part of the speech of 24 elders in Revelation 11:18. "'I will destroy those who destroy the Earth.' What does that mean? people ask me. I say, 'Well it may well mean *us*, and we are destroying ourselves by what we allow—and we shan't do that.' And so, you know, some of what is done in the name of development, progress, economy—and all the good things that we in the West seem to associate with destroying the environment—is frankly sinful and will be judged."
>
> (Interview: Cizik 11 Nov. 2017)

Thanks to the prophetic work of Cizik and other prominent evangelicals such as climate scientist Katharine Hayhoe, caring about the environment and climate change is growing among evangelicals in the United States (Warner 2019; Bergman 2018; Goldhill 2019). Those evangelicals who participated in COP 23 may be outliers—making their virtue signaling all the more important—but indications are that environmentalist evangelicals are making strong headway toward shifting the thoughts of their co-religionists (Danielsen 2013; Mills et al. 2015). For example, while this chapter was being peer reviewed, more than 100 evangelical leaders beseeched the U.S. Congress to act on climate change (Evangelical Environmental Network 2021).

Lobbying: An Intra-Religious Shepherding Tension

Lobbying climate negotiators may have been the most impactful activity of religious NGOs at COP 23, but that seems to have been restricted to the largest groups with the most resources, such as the World Council of Churches. Christianity—the largest religion in the world (Hackett and McClendon 2017)—is also the behemoth of religions at the COP (Krantz 2021; Glaab 2017), and the biggest actor among the Christian groups is the World Council of Churches, which claims 350 member church denominations representing more than 500 million Christian congregants in more than 100 countries (World Council of Churches u.d.). The World Council of Churches has been involved in the UNFCCC since COP 1, and its purpose for participation is vastly different from other religious NGOs at the COP in that one of its goals is to affect climate negotiations (Interview: Kurian 8 Nov. 2017; Interview: Grape 10 Nov. 2017).

> "We systematically have teams, groups of people tracking, and lobbying and trying to define what are the issues following these documents, following the discussions, dialogues. And because we have got members from all over the world, we try to influence the different parties, different countries, because you know churches and faith communities are there in every country," said Manoj Kurian, coordinator for the World Council of Churches Ecumenical Advocacy Alliance. "And of course more importantly, after the COP we continue to follow these things during the various negotiations and discussions."
>
> (Interview: Kurian 8 Nov. 2017)

Climate justice is a big issue for the World Council of Churches, added Grape, coordinator of the World Council of Churches climate-change working group.

> "Climate change demonstrates the inequalities on the Earth," said Grape. "Because the more vulnerable to climate change are often the poor ones, and those who have contributed least to climate change. And from a justice perspective, this must be addressed, this must be something we as faith communities, as churches together, must be into that discussion."
>
> (Interview: Grape 10 Nov. 2017)

Faith leaders at COP 23 also met with UNFCCC executives to express the importance of climate issues to people of faith and the importance of the UNFCCC incorporating the perspectives of religion. For example, Bernard and I were part of a small interfaith delegation who met privately with Patricia Espinosa, then-executive secretary of the UNFCCC. The meeting signified that the UNFCCC saw the religious community at COP 23 as one of importance.

However, in the same way that political lobbyists may not discuss whether all their constituents agree with the issues leadership promotes, the unspoken tension in lobbying at COP 23 was that faith leadership may be more willing to promote climate action than their religious base. Yet the talk of faith leadership at the COP may serve as the leading edge of religion's position on climate change with the religious flocks eventually following.

Empirical evidence indicates such a transition is possible (Gehlbach et al. 2019) and already may be happening (Mills et al. 2015)—and that even a "single lecture can significantly alter acceptance of climate science" (Hayhoe et al. 2019).

Societal Tension

Whereas outside of COP 23 the secular are more ardent believers in climate change and climate science than people of faith (Morrison et al. 2015; Funk and Alper 2015; Arbuckle and Konisky 2015; Zaleha and Szasz 2015), religious groups at COP 23 reversed that tension by demonstrating that religion, not society, may be more interested in taking action. And in so doing, faith actors found their biggest tension was not between different faiths, but between a united religious voice supporting climate action and a recalcitrant secular society focused on climate delayism (Krantz 2021).

Framing Religion as Environmental: A Societal-Expectations Tension

Countering society's perception of religion as being incompatible and in conflict with belief in science and climate change (Funk and Alper 2015), religious NGOs at COP 23 used their participation in the COP as an opportunity to frame their faiths as environmentally based. For example, every year since COP 15, the Brahma Kumaris have sent a delegation of about four to a dozen people to the COP (Interview: Bernard 7 Nov. 2017, Interview: Pilz 7 Nov. 2017). At COP 23, Golo Joachim Pilz, the Brahma Kumaris advisor on renewable energy, was in the exhibition hall promoting the work done by the Brahma Kumaris in India to develop solar energy.

> "So we're quite active at these things because we believe that this [spirituality] is a missing dimension and we should share our knowledge, our expertise, how to change the self and how to tackle the problems," said Pilz. "We should share it with the wider audience."
>
> (Interview: Pilz 7 Nov. 2017)

Pilz served on two Brahma Kumaris-organized panels called "Visionary Leadership for a Sustainable Future" and "Transforming Hearts, Lives and Government Policies: 1.5°C without Reliance on Geo-Engineering." The panels framed Brahma Kumaris' work as well as its ethos—personal transformation through meditation and the cultivation of positive thoughts and feelings—as a means to address the root cause of the climate crisis.

The Brahma Kumaris also produced and distributed pamphlets, including one titled, "Renewable Energy: Creating the Future We Want," and one titled, "Creating a Climate for Change." Both pamphlets discuss and depict images of a one-megawatt solar-thermal energy project newly developed in the Indian state of Rajasthan by the Brahma Kumaris' sister organization, the World Renewal Spiritual Trust. And both pamphlets feature small icons of birds, water, trees, homes, the sun, smiley faces, flowers, stars, and people meditating. The messages are consistent with Brahma Kumaris' key belief that the "inner climate impacts the outer climate" (Brahma Kumaris World Spiritual University u.d.)—that personal transformation can affect the natural world.

The presence and active involvement of the Brahma Kumaris and other spiritual and religious groups at the COP tells society that, despite expectations, their faiths care about the Earth.

Interfaith Statements: A Societal-Disconnect Tension

Interfaith statements at COP 23 demonstrated that religions stand together when it comes to climate change and caring for the Earth. "Walk on Earth Gently"—the third in a series of interfaith statements organized by the interfaith nonprofit GreenFaith and released annually at COPs (Interview: Catovic 10 Nov. 2017)—provides an example of what religious NGOs promoted at COP 23. Although leaders of many significant world religions were missing, the signatories represented a greater variety of world faiths than most interfaith statements. It was translated into Arabic, Chinese, Hindi, and Spanish, and was signed by Buddhist, Christian, Hindu, indigenous, Islamic, Jewish,[5] and Sikh leaders (GreenFaith 2017).

While the 2015 Paris Agreement reached at COP 21 calls for global temperature increases to be limited to 2 degrees centigrade, the interfaith statement notes the world remains on track for a 3-plus-degree increase, and calls for dramatic changes to reduce emissions from home energy use; reduce emissions from food by adopting plant-based diets and lessening food waste; and minimize carbon-intensive travel.

The full text speaks from a place of faith without reflecting any specific faith. The absence of the words "justice" and "equity" is also noteworthy because Haluza-DeLay noted religious NGOs emphasized equity and deemphasized overtly religious language in public events at the COP (Haluza-DeLay u.d.). The closest the text gets to "justice" is when it speaks about the most vulnerable suffering "unfairly and unjustly" (GreenFaith 2017). While the ideas of justice and equity are not overtly religious, they nonetheless often code as Abrahamic and Christian in particular (Glaab 2017; Marshall et al. 2016). Glaab observed that justice was an issue on which many religious NGOs at the COP worked—and that the focus on justice claims "can then also be interpreted as a reflection of the dominance of Christian organizations in this setting" (Glaab 2017).

Whenever the text of "Walk on Earth Gently" does mention a specific faith's language, it does so in the context of the language used by other faiths—for example, when referring to the Earth as our cathedral, it also refers to Earth as our mosque, sanctuary, and temple[6]. The text's inclusivity is not an accident—rather, it reflects research GreenFaith conducted around speaking about climate issues to people of different religions (Marshall et al. 2016). And although the statement's title, "Walk on Earth Gently," echoes Abrahamic texts—from the words of the biblical prophet Micah, who said one should "walk humbly with your God" (Micah 6:8), to New Testament apostle Paul, who said to "[l]ook carefully then how you walk, not as unwise but as wise" (Ephesians 5:15, ESV), to the Quran, which says that the servants of Allah walk humbly (or gently) on the Earth (Al-Furqān 25:63)—the idea also reflects Ahiṃsā, aka Ahinsa, a principle shared by Buddhism, Hinduism, and Jainism.

Particularly striking for a faith statement is the absence of overt religiosity. The majority of the text focuses on seemingly secular issues—as Haluza-DeLay observed—but may reflect that faith leaders see matters of the Earth as religious issues. In that sense, caring for the Earth is a sacred act (Durkheim 1915, p. 47)—and that is because most major faiths include religious teachings that constitute Earth stewardship (Hitzhusen and Tucker 2013). Therefore, discussing the number of degrees centigrade by which society needs to limit global warming is religious discourse, as "all acts dictated or inspired by religion—not just those that seem overtly religious, such as attending religious services or observing religious dietary laws—are religious acts using religious methods by virtue of their religious origin" (Krantz 2016). In turn, religious NGOs at COP 23 may have helped provide the UNFCCC and its parties with religious and ethical grounding in pursuit of prudent climate policy.

Promoting Plant-Based Diets: The Universal Tension

Transitioning to a plant-based diet was one of the three planks of the "Walk on Earth Gently" interfaith statement (GreenFaith 2017). Of all the positions taken by religious NGOs at COP 23, the promotion of plant-based diets may be the most contentious, with tensions between and within religions, and between religion and society. But from a climate perspective, it should not be: Meat is among the largest (Hedenus et al. 2014; Ripple et al. 2014; Joyce et al. 2014)—if not the largest (Steinfeld et al. 2006)—contributors to climate change. Yet animal agriculture largely has been ignored by the UNFCCC (Bailey et al. 2014). And some religions, such as Hinduism and Mahayana Buddhism, have been empirically effective at mitigating greenhouse gases through the vegetarianism of their adherents (Tseng 2017; Filippini and Srinivasan 2019). Faith actors at COP 23 assumed a potentially innovative role by bringing the issue of plant-based diets to the table.

COP 23 was the third time that Gopal D. Patel, a GreenFaith staffer and director of the Bhumi Project at the Oxford Centre for Hindu Studies, attended a COP. Patel said promoting vegetarianism is one way religious voices can make a difference. "Hinduism is very clear that the killing of animals and specifically the killing of cows will bring upon the Earth a lot of troubles," Patel said. "The rivers of blood that are flowing across the world from these animals, we're responsible for that" (Interview: Patel 10 Nov. 2017).

Likewise, Saffet Catovic—a New Jersey-based imam and university chaplain who attended COP 23 representing GreenFaith and the Islamic Society of North America—stressed the importance of embracing plant-based diets, or at least reducing Muslims' meat consumption, to address climate change. For him, the first step is initiating a shift in religious attitudes toward meat.

> "Once people adjust their attitudes and their consumption practices, it will become habitual," he said. "People will think climate change when they're shopping in the supermarkets; think climate change when they are turning their lights on; think climate change when they're sitting at their dinner table and having a plate that is rearranged to include much more vegetables and much less meats."
>
> (Interview: Catovic 10 Nov. 2017)

Recognizing the tension between him and his fellow Muslims about eating meat, Catovic suggested that the amount of meat consumed by Muslims could be reduced over time.

> "It's a question of re-sorting the proportions of meat that is placed on our plates, significantly, so that meat begins to disappear," Catovic said. "It will not disappear fully because we have certain holidays connected to meat consumption."
>
> (Interview: Catovic 10 Nov. 2017)

It may be difficult to convince Muslims to reduce meat consumption as religious tensions rise in India, projected to become home to the world's largest population of Muslims in the next 40 years (Diamant 2019). Under renewed Hindu nationalism fanned by Indian Prime Minister Narendra Modi, the government has banned selling cows for slaughter and has shuttered slaughterhouses as well as at least 50,000 stores that provide meat for Muslims (M.N. and Bengali 2017; Gowen 2018). Coupled with Hindu-vigilante attacks on Muslim cattle herders (Human Rights Watch 2019), vegetarianism is being weaponized in India's religious warfare. The corresponding politicization of such an impactful method of reducing carbon emissions bodes poorly for those seeking diet solutions to the climate crisis.

Meat production and consumption is among several major sources of carbon emissions that are not even on the COP negotiating table, and that is reflected on the actual tables at COPs. Much of the food served at the COPs I attended was meat and fish. Others also have noticed how tied COPs are to meat consumption. Three NGOs—the Center for Biological Diversity, Farm Forward, and Brighter Green—analyzed the carbon emissions of food served by the UNFCCC at COP 24 in December 2018 in Katowice, Poland, and found that the heavily meat menu may have been responsible for about 4,500 metric tons of carbon-dioxide equivalent in emissions (Center for Biological Diversity et al. 2018)—or the same as the greenhouse-gas emissions from about 9.7 million miles driven by a 2018 four-wheel-drive Ford F150 pickup truck, which emits 463 grams of greenhouse-gas emissions per mile driven (U.S. Department of Energy u.d.) and was the most popular vehicle sold in the United States in 2018 (Gastelu 2019).

Eleven of the 16 menu items at COP 24 contained meat or fish; 14 of the 16 menu items contained animal products; and only two of the menu items were vegan. All of the menu items responsible for the largest carbon emissions contained meat and fish. Their analysis found that the four worst COP 24 menu items for carbon emissions were beef with smoked bacon and sauerkraut, dumplings with fried pork and beef, cheeseburger, and codfish in butter-wine sauce with mussels. The beef with smoked bacon and sauerkraut alone accounted for 35 times the emissions of the cabbage-and-mushroom dumplings, one of the two plant-based menu options available at the COP (Center for Biological Diversity et al. 2018).

With a much greater emphasis on plant-based options, the menu at COP 26, which was held in Glasgow as this chapter was being revised following peer review, was significantly less carbon intensive than the food served at previous COPs (Adam and Westfall 2021; Vinter 2021). However, because menu decisions are determined by local organizers rather than by UNFCCC policy, future COP organizers are free to return to meat-centric conferences. Further, in spite of the menu changes, as of this writing, COP 26 is projected to have the largest carbon footprint of any COP by far, doubling the emissions generated by COP 25 in Madrid. That may partly be due to a record attendance of nearly 40,000 people—about 10,000 more than at any other COP. On a per-participant basis, however, COP 26 still produced far more carbon emissions than other COPs: an estimated 2.6 tons of CO_2 equivalent per attendee, compared to 1.8 tons of CO_2 equivalent per attendee at COP 25, 1.2 tons of CO_2 equivalent per attendee at COP 21, and 0.8 tons of CO_2 equivalent per attendee at COP 15 (Booth and Stevens 2021; McSweeney 2021).

With so much meat, waste, and unnecessary consumption, the COP—like most environmental conferences—has a mismatch between professed values and practice. That tension is particularly striking given how ineffective the COP has been at addressing climate change (Buxton 2016; Bortscheller 2009; Schenck 2008; Hermwille et al. 2017; Riedy and McGregor 2011; Muller 2010; Kutney 2013; Parker et al. 2012; Bodansky 2010; Cadman et al. 2018)—and given that the public's perception of hypocrisy among scientists and environmentalists offers a convenient excuse for inaction on environmental issues (Attari et al. 2019).

Many acts of hypocrisy at the COP could be avoided. For example, animal products could be greatly reduced or eliminated as a matter of UNFCCC policy. Volunteers could stand by waste bins to help ensure that refuse is placed in its proper bin. Regulations could stem the flow of wasteful uses of paper and tchotchkes. And even though there often is no suitable substitute for long-distance travel by plane—video conferencing can be a poor replacement for relationship building when people do not know each other well (Denstadli et al. 2012)—as a matter of policy the UNFCCC could regularly offset carbon emissions for travel to COPs, as the British government has done for COP 26 (Booth and Stevens 2021).

Conclusion

The active participation of faith-based actors at COP 23—while wearing their religious cloth on their proverbial sleeves—demonstrated that they saw climate change as a religious issue. Working independently and in concert at a range of scales toward a variety of goals for both secular and faith-based audiences, religious NGOs at COP 23 performed a variety of functions. More study is needed, but this work provides an empirical window into the functions, goals, and tensions of religion as an actor at COP 23. Although tensions existed between religions at COP 23, and leaders of different religions engaged with the COP for different strategic reasons, they shared values of Earth stewardship. All expressed an urgent need to act on climate. Instead of fighting along religious lines, the different faith actors at COP 23 were united with each other and with secular actors in fighting for action on climate change. That unity did not solve the tensions, but faith actors at COP 23 showed that tensions can be surmounted by the peaceful mainstream of religious NGOs—meaning that the biggest tension may have been between religious actors promoting moral climate action and arguably immoral societal actors who care more for talk than action.

Even though the ethical call from international religious groups for action on climate change "typically falls short" of its potential impact (Caniglia et al. 2015), and clearly is not being heeded as worldwide carbon emissions continue rising (Boden et al. 2017), people of faith at COP 23 may have succeeded in supporting the expansion of the Overton window (Russell 2006)—the window of discourse or possibility—of what was open for discussion at COP 23. And in the process, it is possible they have contributed to changing the way believers and non-adherents alike see the role of religion in environmental protection, from part of the problem (White 1967; Leopold 1949; Hand and Van Liere 1984; Greeley 1993; Arbuckle and Konisky 2015; Taylor et al. 2016) to part of the solution.

Yet religious representation at the COP remains small (Krantz 2021). Participation from more religious NGOs from more faiths—and following up eco-religious speech with action—is necessary for religion to play a bigger role in surpassing the tipping point (Gladwell 2000) on climate action.

Notes

1 Many U.N. agreements result in COPs; references to the COP in this case study refer exclusively to the UNFCCC's COP, and not to other COPs, such as those for the United Nations Convention to Combat Desertification or the Stockholm Convention on Persistent Organic Pollutants.
2 By comparison, Glaab interviewed 21 faith leaders representing 18 organizations, 15 of which were Christian (Glaab 2017).
3 This chapter includes interviewees whose work represented one of six types of tension.
4 For example, outside COP 23 I saw proselytizers holding signs alluding to a connection between a climate change-induced environmental collapse and a New Testament-predicted apocalypse in which nonbelievers are left to suffer. The role of proselytization at U.N. venues is a subject for future study.
5 I am among the Jewish signatories.
6 Some may note the absence of the term synagogue; however, in many Jewish sects, "temple" is synonymous with synagogue.

Acknowledgement

David Krantz (ORCID 0000-0001-6062-6628) has been supported by an IGERT-SUN (Solar Utilization Network) fellowship funded by the National Science Foundation (Award 1144616). Travel support to COP 22 was provided by the Schusterman Foundation; travel support to COP 23 was provided by the Center for Jewish Studies and the Rob and Melani Walton Sustainability Solutions Initiatives, both at Arizona State University; travel support to meetings of the "Religion and Environmental Change" workshop was provided by the Urban Green Religions project at the University of Basel through funding from the Swiss National Science Foundation. The author is grateful for the advice of Evan Berry and Randolph Haluza-DeLay, the inspiration of Alon Tal, Paul Hirt and Orr Karassin, and the feedback of fellow participants in the "Religion and Environmental Change" workshop. The author thanks Valeriane Bernard, Laurence Blattmer, Saffet Catovic, Richard Cizik, Henrik Grape, Fletcher Harper, Joe Huang, Manoj Kurian, Serafim Kykotis, Rufino Lim, Susanna Mattingly, Junghee Min, Gopal D. Patel, Golo Joachim Pilz, R. B. Sherly Priyanka, Stuart Scott and Edgar Toclo for their participation in the oral-history interview process, as well as for their steadfast efforts to help build a more sustainable world.

References

Adam, K., and Westfall, S., 2021. Hungry in Glasgow? The COP26 Menu Comes Carbon Counted. *The Washington Post* [online]. Available from: washingtonpost.com/world/2021/11/06/cop26-menu-food-carbon-footprint/ [Accessed 7 November 2021].

Arbuckle, M.B., and Konisky, D.M., 2015. The Role of Religion in Environmental Attitudes. *Social Science Quarterly*, 96 (5), 1244–1263.

Attari, S.Z., Krantz, D.H., and Weber, E.U., 2019. Climate Change Communicators' Carbon Footprints Affect Their Audience's Policy Support. *Climatic Change*, 154 (3–4), 529–545.

Bailey, R., Froggatt, A., and Wellesley, L., 2014. *Livestock – Climate Change's Forgotten Sector*. London: Chatham House: The Royal Institute of International Affairs, 1–30.

Berger, J., 2003. Religious Nongovernmental Organizations: An Exploratory Analysis. *Voluntas: International Journal of Voluntary and Nonprofit Organizations*, 14 (1), 15–39.

Bergman, M.M., 2018. What Would Jesus Do? Talking with Evangelicals About Climate Change. *The Guardian* [online]. Available from: theguardian.com/environment/2018/dec/19/talking-with-evangelicals-about-climate-change-south [Accessed 21 February 2020].

Bodansky, D., 2010. The Copenhagen Conference: A Post-Mortem. *American Journal of International Law*, 104 (2), 230–240.

Boden, T.A., Marland, G., and Andres, R.J., 2017. *Global, Regional, and National Fossil-Fuel CO2 Emissions* [online]. Oak Ridge, TN: Carbon Dioxide Information Analysis Center, Oak Ridge National Laboratory, U.S. Department of Energy. Available from: osti.gov/dataexplorer/biblio/1389331-global-regional-national-fossil-fuel-co2-emissions [Accessed 23 April 2019].

Booth, W., and Stevens, H., 2021. COP26 in Glasgow May Have a Record Carbon Footprint, Despite Low-Flush Loos and Veggie Haggis. *The Washington Post* [online]. Available from: washingtonpost.com/world/2021/11/12/cop26-carbon-emissions-offsets/ [Accessed 12 November 2021].

Bortscheller, M.J., 2009. Equitable but Ineffective: How the Principle of Common but Differentiated Responsibilities Hobbles the Global Fight Against Climate Change. *Sustainable Development Law & Policy*, 10, 49.

Boykoff, M.T., Goodman, M.K., and Curtis, I., 2009. Cultural Politics of Climate Change: Interactions in Everyday Spaces. *In:* M.T. Boykoff, ed. *The Politics of Climate Change: A Survey*. London: Routledge, 136–154.

Brahma Kumaris World Spiritual University, u.d. *Caring for the Environment* [online]. Brahma Kumaris World Spiritual University. Available from: brahmakumaris.org/hope/environment/a-different-consciousness [Accessed 21 February 2020].

Buxton, N., 2016. COP 21 Charades: Spin, Lies and Real Hope in Paris. *Globalizations*, 13 (6), 934–937.

Cadman, T. et al., 2018. From Paris to Poland: A Postmortem of the Climate Change Negotiations. *The International Journal of Social Quality*, 8 (2), 27–46.

Caniglia, B.S., Brulle, R.J., and Szasz, A., 2015. Civil Society, Social Movements, and Climate Change. *In:* R.E. Dunlap and R.J. Brulle, eds. *Climate Change and Society: Sociological Perspectives*. Oxford: Oxford University Press, 235–268.

Carrette, J., 2013. The Paradox of Globalisation: Quakers, Religious NGOs and the United Nations. *In:* R.W. Hefner et al., eds. *Religions in Movement: The Local and the Global in Contemporary Faith Traditions*. Routledge, 47–66.

Carrette, J., and Miall, H., eds., 2017. *Religion, NGOs and the United Nations: Visible and Invisible Actors in Power*. London, New York: Bloomsbury Academic.

Cayton, A.R.L., 2003. Insufficient Woe: Sense and Sensibility in Writing Nineteenth-Century History. *Reviews in American History*, 31 (3), 331–341.

Center for Biological Diversity, Farm Forward and Brighter Green, 2018. The Climate Cost of Food at COP24. *Center for Biological Diversity*, 1–2.

Danielsen, S., 2013. Fracturing Over Creation Care? Shifting Environmental Beliefs Among Evangelicals, 1984–2010. *Journal for the Scientific Study of Religion*, 52 (1), 198–215.

Denstadli, J.M., Julsrud, T.E., and Hjorthol, R.J., 2012. Videoconferencing as a Mode of Communication: A Comparative Study of the Use of Videoconferencing and Face-to-Face Meetings. *Journal of Business and Technical Communication*, 26 (1), 65–91.

Diamant, J., 2019. The Countries with the 10 Largest Christian Populations and the 10 Largest Muslim Populations. *Pew Research Center* [online]. Available from: pewresearch.org/fact-tank/2019/04/01/the-countries-with-the-10-largest-christian-populations-and-the-10-largest-muslim-populations/ [Accessed 3 May 2020].

Durkheim, É., 1915. *The Elementary Forms of the Religious Life*. London: George Allen and Unwin Limited.

Evangelical Environmental Network, 2021. Over 100 Evangelical Leaders Call on Congress to Include Climate and Clean Energy Provisions in Final Infrastructure Deal [online]. Available from: creationcare.org/news-blog/overview.html/article/2021/06/07/over-100-evangelical-leaders-call-on-congress-to-include-climate-and-clean-energy-provisions-in-final-infrastructure-deal [Accessed 7 June 2021].

Filippini, M., and Srinivasan, S., 2019. Impact of Religious Participation, Social Interactions and Globalization on Meat Consumption: Evidence from India. *Energy Economics*, 84 (Supplement 1), 1–13.

Funk, C., and Alper, B.A., 2015. *Religion and Science: Highly Religious Americans Are Less Likely Than Others to See Conflict Between Faith and Science*. Washington, DC: Pew Research Center, 1–64.

Gastelu, G., 2019. The 10 Best-Selling Vehicles in the United States in 2018 Were Mostly Trucks and SUVs. *Fox News* [online]. Available from: foxnews.com/auto/the-10-best-selling-vehicles-in-the-united-states-in-2018-were-mostly-trucks-and-suvs [Accessed 21 February 2020].

Gehlbach, H., Robinson, C.D., and Vriesema, C.C., 2019. Leveraging Cognitive Consistency to Nudge Conservative Climate Change Beliefs. *Journal of Environmental Psychology*, 61, 134–137.

Glaab, K., 2017. A Climate for Justice? Faith-Based Advocacy on Climate Change at the United Nations. *Globalizations*, 14 (7), 1110–1124.

Glaab, K., Fuchs, D., and Friederich, J., 2019. Religious NGOs at the UNFCCC: A Specific Contribution to Global Climate Politics? *In:* C. Baumgart-Ochse and K.D. Wolf, eds. *Religious NGOs at the United Nations: Polarizers or Mediators?* Oxon: Routledge, 61–77.

Gladwell, M., 2000. *The Tipping Point: How Little Things Can Make a Big Difference*. New York: Little, Brown.

Goldhill, O., 2019. Evangelical Leaders Are Making Climate Change a Religious Issue. *Quartz* [online]. Available from: qz.com/1709793/evangelical-leaders-are-making-climate-change-a-religious-issue/ [Accessed 21 February 2020].

Gowen, A., 2018. 'Cow Protection' Laws Create Growing Trouble for India's Muslim Cattle Traders. *The Washington Post*, 17 July, A9.

Greeley, A., 1993. Religion and Attitudes Toward the Environment. *Journal for the Scientific Study of Religion*, 32 (1), 19–28.

GreenFaith, 2017. *Interfaith Statement "Walk on Earth Gently"* [online]. Living the Change: GreenFaith. Available from: livingthechange.net/statement [Accessed 1 May 2019].

Hackett, C., and McClendon, D., 2017. Christians Remain World's Largest Religious Group, but They Are Declining in Europe. *Pew Research Center* [online]. Available from: pewresearch.org/fact-tank/2017/04/05/

christians-remain-worlds-largest-religious-group-but-they-are-declining-in-europe/ [Accessed 3 May 2020].

Hackett, C., and Stonawski, M., 2017. *The Changing Global Religious Landscape*. Washington, DC: Pew Research Center.

Haluza-DeLay, R., 2018a. Event-Ethnography of Faith-Based Climate Activism @ COP22. *In: Conference on Ecclesiology & Ethnography*. Winnipeg, Canada.

Haluza-DeLay, R., 2018b. Transnational Faith-Based Engagement on Climate Change Governance. *In: XIX ISA World Congress of Sociology*. Toronto.

Haluza-DeLay, R., 2019a. The Cosmopolitics of Faith-Based Climate Justice Activism at the UNFCCC. *In: The International Society of thr Study of Religion, Nature and Culture Conference: Religion/Water/Climate: Changing Cultures and Landscapes*. Cork, Ireland.

Haluza-DeLay, R., 2019b. Cosmopolitics of Faith-Based Climate Justice Activism at the United Nations Framework Convention on Climate Change. *In: The Elliott Allen Institute for Theology & Ecology*. Toronto.

Haluza-DeLay, R., u.d. The Cosmopolitics of Faith-Based Climate Justice Activism at the UNFCCC. *The King's University*, 1–56.

Hand, C.M., and Van Liere, K.D., 1984. Religion, Mastery-Over-Nature, and Environmental Concern. *Social Forces*, 63 (2), 555–570.

Hayhoe, D., Bloom, M.A., and Webb, B.S., 2019. Changing Evangelical Minds on Climate Change. *Environmental Research Letters*, 14 (2), 1–16.

Haynes, J., 2014. *Faith-Based Organizations at the United Nations*. New York: Palgrave Macmillan.

Hedenus, F., Wirsenius, S., and Johansson, D.J.A., 2014. The Importance of Reduced Meat and Dairy Consumption for Meeting Stringent Climate Change Targets. *Climatic Change*, 124 (1–2), 79–91.

Hermwille, L. et al., 2017. UNFCCC Before and After Paris – What's Necessary for an Effective Climate Regime? *Climate Policy*, 17 (2), 150–170.

Hitzhusen, G.E., and Tucker, M.E., 2013. The Potential of Religion for Earth Stewardship. *Frontiers in Ecology and the Environment*, 11 (7), 368–376.

Hjerpe, M., and Linnér, B.-O., 2010. Functions of COP Side-Events in Climate-Change Governance. *Climate Policy*, 10 (2), 167–180.

Hulme, M., 2017. Climate Change and the Significance of Religion. *Economic & Political Weekly*, 3 (28), 14–17.

Human Rights Watch, 2019. Violent Cow Protection in India: Vigilante Groups Attack Minorities. *Human Rights Watch* [online]. Available from: hrw.org/report/2019/02/18/violent-cow-protection-india/vigilante-groups-attack-minorities [Accessed 26 February 2020].

Joyce, A. et al., 2014. The Impact of Nutritional Choices on Global Warming and Policy Implications: Examining the Link Between Dietary Choices and Greenhouse Gas Emissions. *Energy and Emission Control Technologies*, 2, 33–43.

Kirchen Und Klima, 2017. Churches and Climate: Events During the UN Climate Conference COP 23. *Kirchen Und Klima*, 1–8.

Knox, G., ed., 2002. *Religion and Public Policy at the UN*. Downers Grove, IL: Religion Counts.

Krantz, D., 2016. Shmita Revolution: The Reclamation and Reinvention of the Sabbatical Year. *Religions*, 7 (8), 1–31.

Krantz, D., 2021. COP and the Cloth: Participation of Religious NGOs at the Conference of Parties to the United Nations Framework Convention on Climate Change. *Sci*, 3 (2), 1–27.

Kutney, G., 2013. *Carbon Politics and the Failure of the Kyoto Protocol*. London: Routledge.
Leopold, A., 1949. *A Sand County Almanac, and Sketches Here and There*. New York: Oxford University Press.
Marshall, G. et al., 2016. *Faith & Climate Change: A Guide to Talking with the Five Major Faiths*. Oxford: Climate Outreach.
Mayer, A., and Smith, E.K., 2019. Unstoppable Climate Change? The Influence of Fatalistic Beliefs About Climate Change on Behavioural Change and Willingness to Pay Cross-Nationally. *Climate Policy*, 19 (4), 511–523.
McSweeney, R., 2021. Analysis: How Delegations at COP Climate Summits Have Changed Over Time. *Carbon Brief* [online]. Available from: carbonbrief.org/analysis-how-delegations-at-cop-climate-summits-have-changed-over-time [Accessed 23 November 2021].
Meadows, D., 2008. Leverage Points: Places to Intervene in a System. *Thinking in Systems: A Primer*. White River Junction, VT: Chelsea Green Publishing, 145–165.
Mills, S.B., Rabe, B.G., and Borick, C., 2015. Acceptance of Global Warming Rising for Americans of All Religious Beliefs. *Issues in Energy and Environmental Policy* (26), 1–10.
M.N., P., and Bengali, S., 2017. India Bans Sale of Cows for Slaughter, a Move Designed to Appease Conservative Hindus. *Los Angeles Times*, 27 May, A4.
Morrison, M., Duncan, R., and Parton, K., 2015. Religion Does Matter for Climate Change Attitudes and Behavior. *PloS One*, 10 (8), 1–16.
Moser, S.C., 2016. Reflections on Climate Change Communication Research and Practice in the Second Decade of the 21st Century: What More Is There to Say? *Wiley Interdisciplinary Reviews: Climate Change*, 7 (3), 345–369.
Muller, B., 2010. *Copenhagen 2009: Failure or Final Wake-Up Call for Our Leaders?* Oxford: Oxford Institute for Energy Studies, 1–35.
Nasiritousi, N., Hjerpe, M., and Linnér, B.-O., 2016. The Roles of Non-State Actors in Climate Change Governance: Understanding Agency Through Governance Profiles. *International Environmental Agreements*, 16 (1), 109–126.
Nche, G.C., 2020. The Church Climate Action: Identifying the Barriers and the Bridges. *Transformation*, 37 (3), 222–241.
Parker, C.F. et al., 2012. Fragmented Climate Change Leadership: Making Sense of the Ambiguous Outcome of COP-15. *Environmental Politics*, 21 (2), 268–286.
Petersen, M.J., 2010. International Religious NGOs at the United Nations: A Study of a Group of Religious Organizations. *Journal of Humanitarian Assistance*, 17.
Riedy, C., and McGregor, I., 2011. Climate Governance is Failing Us: We All Need to Respond. *PORTAL: Journal of Multidisciplinary International Studies*, 8 (3), 1–9.
Rios, K., 2020. Examining Christians' Reactions to Reminders of Religion – Science Conflict: Stereotype Threat Versus Disengagement. *Personality and Social Psychology Bulletin*, 47 (3), 441–454.
Ripple, W.J. et al., 2014. Ruminants, Climate Change and Climate Policy. *Nature Climate Change*, 4 (1), 2–5.
Russell, N.J., 2006. An Introduction to the Overton Window of Political Possibilities. *Mackinac Center for Public Policy* [online]. Available from: mackinac.org/article.aspx?ID=7504 [Accessed 23 November 2021].
Schenck, L., 2008. Climate Change "Crisis" – Struggling for Worldwide Collective Action. *Colorado Journal of Imternational Law & Policy*, 19, 319.

Schroeder, H., and Lovell, H., 2012. The Role of Non-Nation-State Actors and Side Events in the International Climate Negotiations. *Climate Policy*, 12 (1), 23–37.

Schüssler, E., Rüling, C.-C., and Wittneben, B.B.F., 2014. On Melting Summits: The Limitations of Field-Configuring Events as Catalysts of Change in Transnational Climate Policy. *Academy of Management Journal*, 57 (1), 140–171.

Simpson, A., and Rios, K., 2019. Is Science for Atheists? Perceived Threat to Religious Cultural Authority Explains U.S. Christians' Distrust in Secularized Science. *Public Understanding of Science*, 28 (7), 740–758.

Steinfeld, H. et al., 2006. *Livestock's Long Shadow: Environmental Issues and Options*. Rome: Food and Agriculture Organization of the United Nations.

Taylor, B., Van Wieren, G., and Zaleha, B.D., 2016. Lynn White Jr. and the Greening-of-Religion Hypothesis. *Conservation Biology*, 30 (5), 1000–1009.

Tseng, A.A., 2017. Reduction of Greenhouse-Gas Emissions by Chinese Buddhists with Vegetarian Diets: A Quantitative Assessment. *Contemporary Buddhism*, 18 (1), 89–107.

United Nations Framework Convention on Climate Change, u.d. *Admitted NGOs: Constituencies* [online]. Available from: unfccc.int/process-and-meetings/parties-non-party-stakeholders/non-party-stakeholders/admitted-ngos#constituencies [Accessed 20 January 2019].

U.S. Department of Energy, u.d. *2018 Ford F150 Pickup 4WD* [online]. U.S. Department of Energy. Available from: fueleconomy.gov/feg/Find.do?action=sbs&id=40222 [Accessed 21 February 2020].

Veldman, R.G., 2019. *The Gospel of Climate Skepticism: Why Evangelical Christians Oppose Action on Climate Change*. Berkeley, CA: University of California Press.

Vinter, R., 2021. Cop26 Menu Focuses on Plant-Based Dishes with 80% Scottish Food. *The Guardian* [online]. Available from: theguardian.com/environment/2021/oct/23/cop26-menu-plant-based-dishes-scottish-food [Accessed 24 October 2021].

Wallerstein, I., 2000. Where Should Sociologists Be Heading? *Contemporary Sociology*, 29 (2), 306–308.

Warner, G., 2019. *Why There's a Divide Between Environmentalists and Evangelicals* [online]. Available from: www.npr.org/2019/11/04/776173312/why-theres-a-divide-between-environmentalists-and-evangelicals?t=1648933608957 [Accessed 4 November 2019].

White, L. Jr., 1967. The Historical Roots of Our Ecologic Crisis. *Science*, 155 (3767), 1203–1207.

World Council of Churches, u.d. *What is the World Council of Churches?* [online]. Available from: oikoumene.org/en/about-us [Accessed 1 May 2019].

Zaleha, B.D., and Szasz, A., 2015. Why Conservative Christians Don't Believe in Climate Change. *Bulletin of the Atomic Scientists*, 71 (5), 19–30.

15 Environmental Action Within Local Faith Communities

Navigating Between High Expectations and Practical Action

Derk Harmannij

Introduction

Human activities are causing major environmental problems across the world. Problems such air pollution and climate change are all well-known and do not only pose a grave danger to the natural world but also to human life itself (Calderón- Garciduеñas et al. 2015; Mora et al. 2018; Haines and Ebi 2019; Diffenbaugh and Burke 2019). Consequently, these and many other environmental problems lead up to an urgent demand to address environmental problems. However, despite the urgency, addressing environmental problem remains a struggle and even people who acknowledge the severity of environmental problems continue give it low priority and are still hesitant to take the necessary steps. Finding solutions to this inaction is difficult and has preoccupied many academics (Gifford 2011; Kollmuss and Agyeman 2002; Amel et al. 2017; Hall et al. 2018). It is in this context that academics, media, and activists alike have responded very positively to the concern that faith leaders have expressed about environment problems.

In recent years, there has been a surge in media outlets, academics, and activists who have portrayed religion, faith-based organizations, and individual believers as very important or even vital participants in addressing environmental problems. From newspapers like *The Guardian* and *The New York Times* to the editorial boards of renowned academic journals like *Science* and *Nature*, they all seem to have high expectations about the way in which religion can help to address environmental problems (The Guardian Editorial 2015; Editorial Board New York Times 2015; Editorial Nature 2015; McNutt 2014). A clear example of these high expectations is a report by the UK Environmental Agency that asked 25 experts to list "the 50 things that will save the planet." On this list "faith groups" came in as second highest, leaving behind, for example, reducing waste and solar energy (Environmental Agency 2007). These high expectations are often based on the idea that with their moral authority, their wealth of teachings, and together with the sheer size and importance of religion in the lives of billions of people, religion can persuade people to act against environmental degradation in a way that science, economics, and politics can't (see, for

DOI: 10.4324/9781003017967-19

example, Rolston III 2006; Donner 2011; Hitzhusen and Tucker 2013). As such, many academics, media, and activists seem certain that the words of faith leaders will persuade many people (including political leaders) to act and that with the help of religion, environmental problems will finally get the priority it needs.

However, such enthusiasm about religion does raise important questions about to what extent "ordinary" believers are listening to the calls for action by their faith leaders, how believers relate their personal faith to environmental problems, and how faith-based environmental concern can be put into practice within local faith communities. There have been some studies that have focused on the ways in which believers and local faith communities relate and engage with environmental issues (Delashmutt 2011; Pfeifer et al. 2014; Bomberg and Hague 2018) but often the interest and enthusiasm about faith-based interest in the environment has only focused on high-ranking faith leaders and their eco-theology (Taylor 2011; Frascaroli and Fjeldsted 2017). Therefore, this chapter will present findings from research on how local faith communities relate and engage with environmental issues. This research was conducted with Christian churches in and around the city of Exeter in the United Kingdom. The goal of this chapter is to move beyond the high expectations of many academics, activists, and media and present a more nuanced picture based on empirical research within local faith communities. The chapter will emphasize the tensions between the high expectations and the struggles that local faith communities face when they attempt to engage with environmental issues.

In the remainder of this chapter there will first be a discussion about the existing research that seeks to investigate how faith communities and faith-based organizations relate and practically engage with environmental issues. This part will focus on Christianity since Christianity is the focus of the conducted research on which this chapter is based. Following this literature review the chapter will briefly explain the used methodology and after that this chapter will present findings from empirical research with local Christian churches. Finally, there will be a discussion and conclusion.

Faith-Based Environmental Concern Beyond Faith Leaders

There is an incredible wealth of Christian theological approaches that seek to propose a faith-based manner to interact with the environment (see Kearns 1996; Delashmutt 2011 for schematic overviews), but how such approaches can help believers to act against environmental problems has been the subject of much debate. Ever since Lynn White published his famous essay about "The historical roots of our ecologic crisis" (White 1967) scholars have been trying to figure out the ways in which religion encourages, hampers, or even discourages the addressing of environmental problems. Many survey-based studies have been published on this topic over the years, but their outcomes have often been conflicting and contradictory (see Taylor et al. 2016 for an overview). However, scholars have also started to conduct studies that focus on the ways in which believers put faith-based environmental concern into practice. But rather than a successful "greening of religion" narrative, these studies often depict entire denominations as well as individual churches as struggling and failing to have a meaningful engagement with the natural world.

Research on denominational responses to environmental problems among Catholics (Agliardo 2014; Danielsen et al. 2021) and Presbyterians (Townsend 2014) in the United States and religious institutions in Canada (Lysack 2014) has shown that even on these high levels it is not easy to engage with environmental issues. Many Catholic dioceses, bishops, and priests have not taken up the issue of climate change in a systematic and vigorous way according to Agliardo (2014, p. 187; see also Wilkins (2020) for a case study of a Catholic diocese). The recent study by Danielsen et al. (2021) even argues that most US bishops were silent about climate change after the publication of Laudato Si' and that those who did speak out often diminished and distanced themselves from Catholic teachings on climate change. Presbyterians were more successful due to a group of social ethicists within the denomination that supported statements about climate change from an early stage, but these statements have had minimal effect "on the ground" according to Townsend (2014). Similarly, Australian research has argued that many denominations have engaged little with environmental issues (Douglas 2009). However, these struggles also persist at the level of local churches. Several studies have argued that the calls for action by faith leaders often fail to reach local churches and that many churchgoers are unaware of their existence (Delashmutt 2011; Douglas 2009; Lawson and Miller 2011). Another important hurdle for many churches is the presence of disinterest, apathy, or even hostility toward environmental issues among churchgoers. Whereas many national faith leaders seem very concerned about the current state of the planet, some local churchgoers do not share their concerns or give them much lower priority (Douglas 2009; Lawson and Miller 2011; Pfeifer et al. 2014; Harmannij 2019). In many faith communities there is a struggle to move the environment from a fringe/activist concern into a church-wide concern. Swedish research by Lundberg (2017) also reveals an ongoing struggle to make the environment a more central concern as he emphasizes how within the Church of Sweden the environment is brought into church life through the interests of individuals rather than through teaching, theology, or church- wide engagements with the environment. How local faith communities and the wider green movement collaborate to save the planet is a topic that scholars haven't researched much yet. But research by Koehrsen (2015) suggests that faith-based organizations only play a very limited role within local energy transition initiatives. A last important barrier for many churches is the ongoing decline of institutionalized Christianity. Many churches are faced with declining memberships, financial struggles, and the forced closure of their buildings and this makes getting involved in addressing environmental problems sometimes almost impossible (Douglas 2009; Lawson and Miller 2011; Lysack 2014; Harmannij 2019). Finally, it is important to remember that churches and faith-based organizations do not only rely on faith-based reasons to engage with environmental issues but also extensively use various scientific, economic, or political arguments when engaging with environmental issues (Vaidyanathan et al. 2018; Freyer et al. 2019).

Many of the aforementioned studies seem to reveal tensions between the high expectations of scholars, activists, and media and the actual practice of faith-based environmental action. However, some scholars have also suggested that churches and faith-based organizations can be successfully involved in environmental issues. According to studies like Kidwell et al. (2018), Bomberg and Hague (2018), and Feldman and Moseley (2003), faith-based groups offer something unique and powerful when it comes to addressing environmental problems. For example, Bomberg and Hague (2018) argue that Scottish churches draw on "spiritual resources that provide a potentially powerful 'transcendent' narrative and strong imperative of climate justice and action" (p. 591) while Kidwell et al. (2018) say that "eco-Congregations can provide a crucial support for incremental, slow change which feeds into wider, longer-term efforts at community building on local, national and international scales" (p. 15). For these studies, the input and aims of faith-based organizations are different from the input and aims of secular groups. Whereas secular groups focus on specific issues and political goals, faith-based groups are more focused on long-term societal change. Feldman and Moseley (2003) argue that "Ultimately, what makes these faith-based efforts unique is their consistent emphasis upon bringing about what we have called personal transformation in order to achieve environmental reform" (p. 246). But this idea of faith-based ideas spurring people into incremental change also remains only "a **potentially** powerful transcendent narrative" as Bomberg and Hague (2018, p. 591) also acknowledge.

Although some scholars see great possibilities for faith-based ideas to stir up the societal change that is needed to address environmental problems, it is also clear that turning the enthusiasm of a group of environmental activists and faith leaders into church-wide interest and action is difficult. No matter how much "potential" researchers say there is, empirical studies reveal the existing tension between the expectations among academics, activists, and media and the actual practice of faith-based environmental action. Making any kind of relevant impact will be a lot harder than many of the enthusiastic writings by academics, media, or activists might suggest.

Methodology

The findings on which this chapter is based have been gathered through focus groups. As the name already suggests, the focus group method is a way to conduct research where the focus is on a selected group of participants. In its most basic form, a focus group is a gathering of selected participants who discuss a selected topic. The goal of focus groups is "to describe and understand meanings and interpretations of a select group of people to gain an understanding of a specific issue from the perspective of the participants of the group" (Liamputtong 2015, p. 4).[1] For many scholars this "group element" is the defining characteristic of focus groups which sets it apart from other qualitative research methods. Or as Morgan (2011)[2] puts it: "the hallmark of focus groups is their explicit use of group interaction to produce data and insights that would be less accessible without the interaction found in a group" (Morgan 2011, p. 2). Therefore, the focus groups in this chapter focus on how the participants see the role of the environment in the church context, rather than their private life. The goal of focus groups is to understand how churches as faith communities understand, relate, and engage with the environment. Churches are communities of believers that share certain (theological) values and beliefs rather than merely being a loose collection of individuals who happen to listen to the same sermon. Churches operate as communities and it is very unlikely that a church will be able to an eco-church if climate change is openly denied by some churchgoers. In such a context the group dynamics that occur when participants talk about the environment are very important. The goal of the focus groups is then not to get some interesting quotes from individual participants to illustrate themes that appear from the analysis but rather to understand the group processes that take place when churches engage with the environment.

In total, five focus groups were conducted with churches in Exeter. The churches that took part in the focus groups were a Methodist Church, a more liberal leaning Anglican Church (Church of England), a charismatic Evangelical church (also Church of England), a Roman Catholic Church, and an independent Evangelical church (this church had been awarded an A Rocha bronze Eco church award). This "eco-church" approach is a scheme developed by the faith-based environment group "A Rocha" and provides churches with a clear set of goals which if met by a participating church are rewarded with a bronze, silver, or gold award. The process to become an eco-church starts out by filling in a survey about what the church is currently doing about the environment.[3] The questions cover all the different aspects through which a church might engage with environment. These are: teaching and worship (does your church pray for creation?), the building (lots of questions about insulation and electricity), the land owned by the church (does the church maintain native wildflowers?), community and global engagement (does your church organize a green fair?), and lastly, lifestyle questions (is your church encouraging the reduction of car usage?). After filling in all the questions a score is awarded and after gaining a certain amount of points an award is given, with gold needing the most points. By rewarding the efforts with an award, A Rocha hopes that churches will be encouraged to improve themselves and not just gain a bronze award but also go further and improve their lifestyle even more, organize more environmental activities and give more space to wildlife and attain a silver and eventually a gold award. The eco-church is this chapter has had a bronze award since 2010 but has not yet moved on to gaining silver or gold[4].

The participating churches were chosen because they reflected different theological views and therefore could very well have different theological approaches to the environment. With these five churches this chapter does not attempt to understand the wider denomination of which these five churches are part (that would probably require something like five churches from each denomination) but the goal of the focus groups is to understand how different churches in Exeter engage with the environment. Somewhat surprisingly, the church where it was not possible to find enough participants was the church that was awarded an eco-church bronze award. After several attempts with the help of a senior lecturer from the Exeter Business School, who also attended the church and who helped the church gain its eco-church status, there was still no success in bringing enough people together. People were too busy according to the senior lecturer. Instead, the researcher had to resort to interviews with a male participant and with the senior lecturer himself. Additionally, there were some informal opportunities to speak with members of the church. The five churches mentioned were the only churches that were approached. There were no churches that refused to participate. The churches were approached through individuals within these five churches who were known to the researcher or to others in the geography department at the University of Exeter and who were willing to help the researcher set up these focus groups. They also arranged the location of the focus group. In the end the focus groups varied in size ranging from 5 participants in the Catholic Church to 12 in the charismatic Evangelical church.

The focus groups were always part of a church gathering. With Methodists it was the theme of a Bible study, with the Charismatic Evangelical church the focus group was part of a walk-in session with lunch which took place on a weekly basis, for the Anglican Church it was the topic of a discussion group, and with the Catholic Church the focus group took place after the Mass had finished. As such, the focus groups were always part of a known church event or happened immediately after it had finished. Most participants usually also took part in the event that the focus group replaced. This might have made some participants feel obligated to participate but participants were always free to refuse participation or leave during the interview. During each focus group someone from the church leadership was present and participated in the focus group. In most cases this was the priest or the minister. This presence of the church leadership might have restrained some participants in their answers, and especially in the Catholic Church the answers from the priest were never opposed by other participants. But this unopposed position of the leadership during the focus groups is not necessarily something that makes the results from the focus groups less valuable. Church leaders are very powerful actors, whose opinions and judgments carry a lot of weight and authority, also during other church activities. Their opinions are also likely to have remained dominant or even unchallenged if they would have discussed environmental issues outside a research setting.

The focus groups focused on three themes. First, participants discussed how they perceived the relationship between faith and the environment. Second, participants discussed what role the environment played within church life. Third, participants discussed how churches can put environmental concerns into action and last, there was a discussion about how churches can join the rest of society to address environmental issues. In the end, the focus groups lasted between 1 hour and 4 minutes and 1 hour and 41 minutes. All the focus groups were recorded with the permission of the participants. Each focus group as well as all the interviews were transcribed. For the analysis, this chapter took a rather critical approach. As already said in this chapter many scholars, media, and activists are very hopeful about religion and argue that "global religion is greening" (Chaplin 2016) or that "religions are entering their ecological phase" (Tucker 2003). However, the researcher of this chapter is more skeptical about whether such a faith-based engagement with environmental problems beyond a small group of faith leaders and faith-based environmental activists is really taking off. Therefore, during the focus groups and the interviews there was much focus on how churches and individual churchgoers were not only theologically relating to the environment but also how they were seeking to put their ideas and concerns into practice and how their ideas and practices compared to the hopeful expectations of academics, media, and activists. Also, the analysis of the focus groups and interviews focused on how churches were practically engaging with the environmental problems beyond expressing theologically grounded environmental concerns.

Within this chapter the participants will be quoted by using their denomination, gender, and an assigned number. This simple way of describing the participants was chosen because with focus groups the focus is on the group interaction rather than in-depth personal insights and second because the participants were often rather homogenous (mostly white British, retired/close to retirement, active churchgoer, and long-term/lifelong residents of Exeter) and as such putting in lots of extra details wouldn't have added much extra insight.

Five focus groups might seem like a very low number in order to understand how churches in Exeter engage with the environment, but research has shown that data saturation comes quickly with focus groups. As little as two to three focus groups already reveal more than 80% of discoverable themes (Guest et al. 2017). Also, the aim for the focus group is rather modest as this chapter is more of an explorative study about the role of the environment with a limited set of churches within the specific context of Exeter.

Outline Empirical Sections

The following sections of this chapter will present and discuss the results from the focus groups. The first section will discuss the different theological approaches to the environment that were present within the participating churches. The second section will discuss how the participating churches seek to put these approaches into practice but how regardless of their different theological approach all participating churches did not to move beyond individual actions and thereby failed to meet the high expectations that many academics, activists, and media have. The third and final section will link this failure to the very narrow roles that participants envisioned for churches in the addressing of environmental problems.

Relating Faith to the Environment

All the churches that participated in the focus groups expressed a desire to care for and protect the natural environment. Within all churches, except for the Liberal Anglican church, this desire to care for and protect the environment was embedded within stewardship language which emphasized that caring for the environment was a God given commandment.

> The whole thing is created by God. . .. we are trustees or stewards of what we have been given.
>
> (Catholic priest)

> God has given us the world and made us custodians and has given us responsibility to look after it.
>
> (Male member, Eco-church)

However, this description of humans as stewards of creation also remained a rather distant concept for participants as it did not reflect their concerns for people who are affected by climate change, flooding, or drought. For example, a male Methodist argued that his motivation to address environmental problems came from a desire to be "loving your neighbors as yourself" (Male Methodist 5) because environmental problems always affect the poorest the worst. Another female Methodist argued that addressing environmental problems is like "the where were you when I was hungry or naked" (Female Methodist 3). Other churches also shared this view. For example, the interviewed male member of the Evangelical eco-church said:

> Looking after our shared environment is an important part of loving our neighbor because we see an increasing frequency of disasters around the world.

Similarly, a Catholic participant argued that we should address environmental problems "by focusing on the care for others" (Female Catholic 1). Many participants felt that the connection between environmental problems and "loving your neighbor" was much more relevant for them because it linked environmental concerns with long-standing interests in development work and social justice issues. As such, for many participants, relating environmental problems to justice issues was a logical step that linked the environment up within wider concerns and interests within their church. However, the charismatic Evangelical participants remained "loyal" to stewardship but embedded it within much more explicit "Evangelical" language while the liberal Anglican participants abandoned anthropocentric thinking altogether.

For Charismatic Evangelical participants, the central idea of stewardship wasn't about the fact that humans were standing above the rest of nature as stewards but rather that both humans and the rest of nature are subordinate to God. Not humans but God is in control. However, according to the Evangelical participants, humans are selfish and don't want to submit to God and instead want to be God. This desire to become like God is where everything goes wrong. A male Evangelical participant explains:

> What we're trying to do is try to restore the correct order in creation which puts God at the top, in charge of everything and we're kind of below God, which is a bit of a shock to a lot of people, they think they are God. Now the whole of creation goes wrong and people put themselves first and they think that it's really important to have as many SUVs as possible and to have diamond collars for their dog and things like that.
>
> (Evangelical Male 2)

To statements such as those previously quoted, other respondents replied with phrases like "I couldn't agree more." or "Absolutely smashed it." (Evangelical Females 1 & 2). The idea that environmental problems were caused by spiritual problems was widely supported by participants. People are willingly polluting the environment or abusing natural resources because they lack Jesus in their life. Again, the same male participant said:

> People who are Christians, really disciples of Jesus, are trying to become more like him. Now He didn't go for flashy jewellery, he didn't even have a house, he didn't have a car. We should be more like him.
>
> (Evangelical male 2)

As such, for the Evangelical participants the solution to environmental problems is to be more like Jesus and follow his (eco-friendly) example. By weaving the concept of stewardship so deeply into wider Evangelical theology and by linking environmental degradation with a lack of "proper" Christian faith, the participants made stewardship into a distinctly Evangelical concept. As such, rather than being an anthropocentric concept, stewardship was theocentric, God in the centre.

For the participating liberal Anglicans, the incentive to care for and protect the environment didn't come from a Bible passage but rather from an acknowledgment that the whole of the planet is interrelated and interconnected through evolutionary processes. A male Anglican said the following:

> We are physically interconnected . . . all our atoms were created in the stars . . . all things are interconnected and you can't simply chop off one bit and say this bit is more important than that bit.
>
> (Male Anglican 2)

A female participant argued in very similar terms:

> It's all connected really . . . there is that sense of wholeness that we are all here sort of together.
>
> (Female Anglican 1)

This connection was especially fostered through an emphasis on evolution and how everything has a shared origin "in the stars." For the participants, this shared origin created a sense of wholeness and togetherness with the rest of creation.

Despite the different theological underpinnings of their environmental concerns, all the participating churches also argued that the roots of environmental problems were linked to injustice and inequality. For example, participants from the charismatic Evangelical church blamed "The injustice of the corporate" (Evangelical Female 1), the liberal Anglican priest stated that "we've got a very, very, very crazy system" while another liberal Anglican participant argued that "It's the invested interests that are holding back" (Male Anglican 3), and Catholic participants blamed "our consumer society" (Female Catholic 1). It is also important to emphasize that although this chapter focuses on faith-based environmental concern many participants from all churches also had many "secular" reasons as to why they wanted to address environmental problems. For example, the interviewed male member of the eco-church expressed much concern for the planet and argued that "you don't need to be a scientist to notice that we had a string of the hottest years that we have ever had" while various members from across the participating churches stated that they accept climate science or that they want to give their grandchildren a less polluted world.

Putting Faith Into Action

However, when asked how to put faith-based environmental concern into practice and how to challenge, for example, "our consumer society," regardless of their theology the answers of all participating churches, apart from the Liberal Anglican church, focused narrowly on things like recycling, reducing air miles, and making more use of public transport.

> If you think about the choices that we have as consumers, if you can choose between a holiday which involves flying or going on a boat . . . we have to keep our carbon emissions as low as possible. We have lots of things in which we have choices as consumers.
>
> (Male Methodist 5)

> I don't know what lightbulbs are used, what energy we buy? Could we have solar panels on our roof? There are loads of things that we can look at.
>
> (Female Methodist 4)

> Actually, I think we do care about it because we're recycling here, we care about who we get are electricity from for our bills here so I do think it's a big deal actually. We don't like waste generally. . .. So, I don't think we're ignoring it. As a church we take it seriously.
>
> (Group leader Anglican Evangelicals)

> I'm very conscious of my water usage every time I shower . . .. I turn the tap off when I brush my teeth. I try to do that as much as I can.
>
> (Catholic Priest)

As the previous quotes show, during the focus groups almost all discussions about addressing environmental problems focused on improving personal behavior. Regardless of whether participants followed stewardship of more justice-oriented reasoning they all strongly focused on individual behavior. Participants had extensive debates about recycling old furniture, car sharing, and buying locally grown vegetables, but beyond such individual actions there was very little. Even after being explicitly asked by the researcher to think beyond individual actions, participants remained focused on it. For example, after a long discussion about greening your personal lifestyle one female Methodist participant remarked:

> The difficult thing about it is . . .. It needs a collective response, it is something that as one individual we can't make that much of a difference.
>
> (Female Methodist 4)

In order to make certain that participants would discuss this comment, the researcher commented: "*A church would be a good place to respond as a collective. . . . ?*" To which she replied: "Yeah, it would." (Female Methodist 4). However, instead of talking about church as a community or what actions they could do together, the participants started talking about the church building, solar panels, recycle bins, and mandatory environmental audits.

The Liberal Anglican participants and their more evolutionary oriented thinking stood apart from the other participating churches in the sense that all the participants in the focus group saw a clear need to go beyond greening individual behavior. Changing individual behavior is "not a bad idea but it won't solve all the problems" as the Anglican priest put it. Others agreed with her assessment and said that individual action "doesn't make much difference unless we know that we are working as part of a larger corporate network and that's really where, we got to get busy, I think" (Anglican Male 1). However, although acknowledging the importance of collective action they also found it difficult to come up with ways to become active as a church community. They felt that it was hard for churches to instigate any practical collective action.

> A Catholic Church can act corporately in the sense that the Pope can get up and say something. . .. For many other churches it is actually extraordinarily difficult, for other religions too.
>
> (Anglican Male 2)

Even though the Liberal Anglican participants said that more collective action needed to happen, they didn't know how to do that.

All churches that took part in the focus groups did not have activities or services about environment. This wasn't done deliberately but it was rather a case of "there should be more attention to all sorts of things" (Evangelical Male 2). Many participants were already involved with other church-related activities and especially the liberal Anglicans, Methodists, and Catholics were feeling the pressure on their resources from the ongoing decline of institutionalized Christianity. For example, Methodist participants argued that there wasn't much attention to the environment within their church because:

> The Methodist Church is a struggling church and when you're struggling for your own survival as a church and as a community that will take higher priority.
>
> (Male Methodist 1)

Similarly, the Liberal Anglican priest mentioned how due to financial hardship "we haven't got resources for solar panels anymore" (Anglican priest, female).

From the interviews and informal conversations that the researcher had with participants at the eco-church, it seems that the "eco-church" approach does manage to bring the environment much more prominently into church life. The eco-church devotes substantial amounts of time to creation care during Sunday services and has an active "eco-action team." For example, the church invites speakers to give talks about faith and the environment and the leader of "eco- action team" occasionally gives sermons about faith and the environment during Sunday services. The church leader also told that apart from some initial hesitation and doubt most church members were supportive of the idea of being an eco-church. Other activities that the church had organized were prayer breakfasts on the role of supermarkets, the making of their own communion bread, and farm visits. Children and youth were also educated about the environment during Sunday school. The other churches had no such resources or activities and did not invite speakers. However, also the eco-church seemed to have the same problem as the other participating churches. Namely, a struggle to go beyond stimulating individual behavioral change. At the eco-church people were encouraged to recycle and reduce their emissions but going beyond individual actions was also problematic for the eco-church. The church leader openly admitted that "most of the practical action is through the lives of the people who worship with us." Similarly, the interviewed male member of the eco-church said: "It makes me happy that I am a member of an eco-congregation but I couldn't tell you what that actually means in practice" (Male member, Eco-church). As such, also within the eco-church there were few opportunities to engage with the environment as a community.

This struggle to move beyond individual behavioral actions does reveal the tension that exists between the high expectations of academics, activists, and media and the actual practice of the churches that participated in the focus groups. These high expectations assume that believers engage with the environment as a collective through their church or environmental faith-based organization rather than as individuals (see, for example, Chaplin 2016; McLeod and Palmer 2015; Dasgupta and Ramanathan 2014). Those favoring the inclusion of faith-based groups in the addressing of environmental problems are not interested in believers who buy more organic or who recycle more of their waste. In fact, many academics have criticized the strong focus on "greening personal lifestyles" as failing to address the underlying societal roots of environmental problems (Slocum 2004; Johnson 2008; Corner and Randell 2011; Moloney and Strengers 2014). The reason why so many academics, media, and activists are so enthusiastic about churches and faith-based organizations is because they are seen as excellent places to challenge the status quo, mobilize people, move beyond individual behavioral action, and start collective action to save the environment. But the participating churches in this study do not seem to be meeting these expectations. Instead they engage with environmental problems as individuals rather than faith communities. As such, this chapter reveals the tensions that exist between the high expectations of scholars, activists, and media and the struggles and difficulties that local faith communities face.

The Narrow Role of Churches

However, the struggle of engaging with the environment as a church community goes deeper than just a struggle to move beyond individual actions. Many participants from all focus groups held a narrow focus on the way that churches could engage with environmental problems. The interviewed male member of the eco-church summarized the difficulty to find a role for faith communities when he commented the actions which he thought needed to happen were on "a very big scale by big organizations or governments" or they needed to be done on an "individual scale where we need to make our own decisions." (Male member, eco-church). From these quotes it is clear that the participant struggled to find a role for his church. The participant perceived that there was a large gap between individual actions like recycling and (inter)national political decision-making processes with little in between. Issues like homelessness or foodbanks fit churches well because they are local, practical, and can be done together. But for environmental problems to be solved a mix of individual actions and (inter)national agreements is needed according to the participant. Therefore, he could not really find a role for church communities as the actions need to be taken at either the personal or (inter)national scale. As such he stated that:

> The best way that the church could be part of the solution . . . helping people to more understand their moral obligations.
>
> (Male member, eco-church)

The role that the participant envisioned for the churches is one of encouraging and stimulating churchgoers to act more sustainable. For him, the thing that churches can do is helping people to understand their moral obligations and making them see the need to change their lifestyle. As such he seems to place much emphasis on teaching believers about the importance of the environment and its relevance to their faith. However, it also confines churches to individual lifestyle improvements that the other participating churches without an eco-status have also emphasized. Although this participant from the eco-church was the only participant who explicitly mentioned this issue during the various focus groups it was very clear that other churches also struggled to find ways to engage with environmental problems as faith communities. For example, the liberal Anglican participants wanted to move away from individual actions but struggled to name actions that were not personal or aimed at political action on a national or international level. Also, the Catholic participants wanted to engage as a faith community with the environment but struggled to move beyond recycling and solar panels or the need for action by politicians. Similarly, the Methodists talked about recycle bins and solar panels but struggled with the role of the church as a faith community. All participating churches struggled with finding a role for church as a community. The only suggestion for a more community-oriented activity came from the Charismatic Evangelical church. They were in the process of buying their own church building and when the group leader mentioned that the church building had a back garden where they perhaps could start growing their own vegetables other participants replied very enthusiastically and started to mention all the benefits of growing your own vegetables. This vegetable garden was the only mentioned practical action during all the focus groups that did not involve individual behavior. During the research and the writing of this chapter the Charismatic Evangelical Church did not have its own church building. However, they recently purchased their own church building and gained an eco-church bronze award in 2021. Since then they have been working hard on making the environment an important part of their church life. For example, they held a Climate Sunday and switched energy supplier according to their church website. Unfortunately, this happened after the research had already been concluded. But it is a very positive development that this church has been able to make such progress and might possibly even have overcome some of the struggles mentioned in this chapter.

Discussion and Conclusion

The findings from the empirical research reveal a picture which churches are struggling to engage with the environment. The participating churches do care about the environment, but engagement beyond individual actions is difficult. Participants do want their church to do more, but an individualized approach combined with a narrow focus on what roles faith communities can take up makes more collective involvement difficult. For the participating churches, the main role that local faith communities can play in the addressing of environmental problems is convincing and encouraging believers to change their behavior. Such a role seems very inward looking and far removed from the enthusiastic calls for the inclusion of faith-based groups.

However, it is important to realize that this failure to meet these high expectations is not caused by the Christian faith. The strong focus on individual behavior is found throughout society and often an official governmental policy (Jones et al. 2010, 2011). Approaches like social marketing (Kotler and Zaltman 1971; Peattie and Peattie 2009) and nudging (Thaler and Sunstein 2008; Sunstein and Thaler 2003) have been very popular policies to stimulate people to "green" their behavior. As such, the dominance of individual actions is a much wider spread phenomena than just churches. It is also important to realize that religion alone will not make any churchgoer "green." Rather religious ideas about the environment are always intertwined with "secular" arguments about, for example, water quality, anthropogenic climate change, biodiversity, and the health and well-being of future generations. Religion is not a standalone argument that can make its followers green. If faith leaders and the wider green movement want believers to practically engage with environmental problems it will require local faith communities to also sincerely engage with things like climate science, environmental politics, and animal welfare and incorporate these arguments into their faith-based ideas. Therefore, improving the engagement of churches with environmental problems also means improving the engagement with topics like science and politics. As such, to improve the engagement of faith communities with environmental issues not only an engagement with eco-theology is necessary but also an engagement with science and the political and economic debates surrounding the addressing of environmental issues is needed.

This chapter has shown that churches struggle to engage with the environment as communities within the specific context of the city Exeter (UK). As such, this chapter did not explore how local faith communities in small villages or rural areas relate and engage with the environment nor did it study how faith communities in highly urbanized areas or immigrant churches relate and engage with the environment. However, beyond conducting more empirical research with local faith communities, the interest in the participation of churches and faith-based organizations in the addressing of environmental problems also raises important questions about the role of religion in society because giving more space to religion within the context of environmental problems will inherently lead to a more visible and more politically engaged form of religion that needs to be accommodated by the wider society. Therefore, the wider green movement will need to be open and reflective and willing to learn from them religious insights, but also faith-based organizations and individual believers need to be willing to learn, listen, and collaborate with others and not just stay within their "religious bubble." As such, the interest in faith-based environmentalism presents an interesting new angle for questions about the role of religion within society, the relationship between science and religion, and how faith-inspired ideas can become beneficiary to the wider society.

Notes

1 Liamputtong (2015) is a book but for the purpose of this chapter an electronic copy was downloaded from the website http://methods.sagepub.com/ Page numbering will be different from hard copies.
2 Morgan (2011) is a book but for the purpose of this chapter an electronic copy was downloaded from the website http://methods.sagepub.com/ Page numbering will be different from hard copies.
3 The survey can be found here: https://ecochurch.arocha.org.uk/wp-content/uploads/2018/01/Eco-Church-Survey-v2-January-2018.pdf But it can also be found on website of A Rocha (https://ecochurch.arocha.org.uk/)
4 The first ideas of this chapter were written in 2019/2020. However, since finishing this chapter the eco-church very recently did gain a silver award (more than 11 years after getting a bronze award). However, since the chapter was already

finished and the researcher already had passed his PhD defence and left Exeter by then it was not possible to conduct further research into this new development.

References

Agliardo, M., 2014. The US Catholic Response to Climate Change. *In*: G. Veldman et al., eds. *How the World's Religions Are Responding to Climate Change. Social Scientific Investigations*. London: Routledge, 174–192.

Amel, E., Manning, C., Scott, B., and Koger, S., 2017. Beyond the Roots of Human Inaction: Fostering Collective Effort Toward Ecosystem Conservation. *Science*, 356, 275–279.

Bomberg, E., and Hague, A., 2018. Faith-Based Climate Action in Christian Congregations: Mobilisation and Spiritual Resources. *Local Environment*, 23 (5), 582–596.

Calderón-Garcidueñas, L., Calderón-Garcidueñas, A., Torres-Jardón, R., Avila-Ramírez, J., Kulesza, R., and Angiulli, A., 2015. Air Pollution and Your Brain: What Do You Need to Know Right Now. *Primary Health Care Research & Development*, 16 (4), 329–345.

Chaplin, C., 2016. The Global Greening of Religion. *Palgrave Communications*, 2, 1–5.

Corner, A., and Randell, A., 2011. Selling Climate Change? The Limitations of Social Marketing as a Strategy for Climate Change Public Engagement. *Global Environmental Change*, 21 (3), 1005–1014.

Danielsen, S., DiLeo, D., and Burke, E., 2021. U.S. Catholic Bishops' Silence and Denialism on Climate Change. *Environmental Research Letters*, 16 (11), 1–11.

Dasgupta, P., and Ramanathan, V., 2014. Pursuit of the Common Good. Religious Institutions May Mobilize Public Opinion and Action. *Science*, 345, 1457–1458.

Delashmutt, M., 2011. Church and Climate Change: An Examination of Attitudes and Practices of Cornish Anglican Churches Regarding the Environment. *Journal for the Study of Religion, Nature and Culture*, 5 (1), 61–81.

Diffenbaugh, N., and Burke, M., 2019. Global Warming Has Increased Global Economic Inequality. *Proceedings of the National Academy of Sciences*, 16 (20), 9808–9813.

Donner, S., 2011. Making the Climate a Part of the Human World. *Bulletin of the American Meteorological Society*, 92 (10), 1297–1302.

Douglas, S., 2009. Religious Environmentalism in the West II: Impediments to the Praxis of Christian Environmentalism in Australia. *Religion Compass*, 3 (4), 738–751.

Editorial Board New York Times, 2015. The Pope and Climate Change. *The New York Times* [online]. Available from: www.nytimes.com/2015/06/19/opinion/the-pope-and-climate-change.html?_r=1 [Accessed 25 April 2020].

Editorial Nature, 2015. Hope from the Pope. *Nature*, 522, 391.

Environmental Agency, 2007. The 50 Things That Will Save the Planet: Results of a Poll of 25 Experts. *Your Environment Extra*, 17, 4–18.

Feldman, D., and Moseley, L., 2003. Faith- Based Environmental Initiatives in Appalachia: Connecting Faith, Environmental Concern and Reform. *Worldviews: Global Religions, Culture, and Ecology*, 7 (3), 227–252.

Frascaroli, F., and Fjeldsted, T., 2017. From Abstractions to Actions: Re-Embodying the Religion and Conservation Nexus. *Journal for the Study of Religion, Nature & Culture*, 11 (4), 511–534.

Freyer, B., Aversano-Dearborn, V., Winkler, G., Leipold, S., Haidl, H., Brand, K., Rosenberger, M., and Wallnig, T., 2019. Is there a Relation Between Ecological Practices and Spirituality? The Case of Benedictine Monasteries: Ecological Practices in Benedictine Monasteries. *Journal of Agricultural and Environmental Ethics*, 31, 559–582.

Gifford, R., 2011. The Dragons of Inaction. Psychological Barriers That Limit Climate Change Mitigation and Adaptation. *American Psychologist*, 66 (4), 290–302.

The Guardian Editorial, 2015. The Guardian View on Pope Francis: An Unlikely Voice for the Environment. *The Guardian* [online]. Available from: www.theguardian.com/commentisfree/2016/sep/01/the-guardian-view-on-pope-francis-an-unlikely-voice-for-the-environment [Accessed 16 March 2020].

Guest, G., Namey, E., and McKenna, K., 2017. How Many Focus Groups Are Enough? Building an Evidence Base for Nonprobability Sample Sizes. *Field Methods*, 29 (1), 3–22.

Haines, A., and Ebi, K., 2019. The Imperative for Climate Action to Protect Health. *The New England Journal of Medicine*, 380 (3), 263–273.

Hall, M., Lewis Jr, N., and Ellsworth, P., 2018. Believing in Climate Change, but Not Behaving Sustainably: Evidence from a One-Year Longitudinal Study. *Journal of Environmental Psychology*, 56, 55–62.

Harmannij, D., 2019. Is It Possible to Give Environmental Issues a More Prominent Role in Church Life? *In*: W. Leal Filho and A. Consorte McCrea, eds. *Sustainability and the Humanities*. Cham: Springer, 97–114.

Hitzhusen, G., and Tucker, M., 2013. The Potential of Religion for Earth Stewardship. *Frontiers in Ecology and the Environment*, 11, 368–376.

Johnson, J., 2008. The Citizen-Consumer Hybrid: Ideological Tensions and the Case of Whole Foods Market. *Theory and Society*, 37 (3), 229–270.

Jones, R., Pykett, J., and Whitehead, M., 2010. Governing Temptation: Changing Behaviour in an Age of Libertarian Paternalism. *Progress in human geography*, 35 (4), 483–501.

Jones, R., Pykett, J., and Whitehead, M., 2011. The Geographies of Soft Paternalism in the UK: The Rise of the Avuncular State and Changing Behaviour After Neoliberalism. *Geography Compass*, 5 (1), 50–62.

Kearns, L., 1996. 'Saving the Creation' Christian Environmentalism in the United States. *Sociology of Religion*, 57 (1), 55–70.

Kidwell, J., Ginn, G., Northcott, M., Bomberg, E., and Hague, A., 2018. Christian Climate Care: Slow Change, Modesty and Eco-Theo-Citizenship. *Geo: Geography and Environment*, 5 (2), 1–18.

Koehrsen, J., 2015. Does Religion Promote Environmental Sustainability? – Exploring the Role of Religion in Local Energy Transitions. *Social Compass*, 62 (3), 296–310.

Kollmuss, A., and Agyeman, J., 2002. Mind the Gap: Why Do People Act Environmentally and What Are the Barriers to Pro-Environmental Behavior? *Environmental Education Research*, 8 (3), 239–260.

Kotler, P., and Zaltman, G., 1971. Social Marketing: An Approach to Planned Social Change. *Journal of Marketing*, 35 (3), 3–12.

Lawson, J., and Miller, K., 2011. Green Revelations in a Country of Drought, Flood and Fire: A Case Study of Abrahamic Faith Communities and Sustainability. *International Journal of Environmental Studies*, 68 (6), 965–979.

Liamputtong, P., 2015. *Focus Group Methodology: Principles and Practice*. London: SAGE Publications Ltd.

Lundberg, A., 2017. Constructing the Environment as a Faith Issue: Environmentalists at Work in the Lutheran Church of Sweden. *Worldviews: Global Religions, Culture, and Ecology*, 21 (2), 152–174.
Lysack, M., 2014. Stepping Up to the Plate. Climate Change, Faith Communities and Effective Environmental Advocacy in Canada. *In*: G. Veldman et al., eds. *How the World's Religions Are Responding to Climate Change. Social Scientific Investigations*. London: Routledge, 157–173.
Mcleod, E., and Palmer, M., 2015. Why Conservation Needs Religion. *Coastal Management*, 43 (3), 238–252.
McNutt, M., 2014. The Pope Tackles Sustainability. *Science*, 345, 1429.
Moloney, S., and Strengers, Y., 2014. Going Green? The Limitations of Behaviour Change Programmes as a Policy Response to Escalating Resource Consumption. *Environmental Policy and Governance*, 24 (2), 94–107.
Mora, C. et al., 2018. Broad Threat to Humanity from Cumulative Climate Hazards Intensified by Greenhouse Gas Emissions. *Nature Climate Change*, 8 (12), 1062–1071.
Morgan, D., 2011. *Focus Groups as Qualitative Research*. Thousand Oaks: SAGE Publications.
Peattie, K., and Peattie, S., 2009. Social Marketing: A Pathway to Consumption Reduction? *Journal of Business Research*, 62 (2), 260–268.
Pfeifer, J., Ecklund, E., and Fullerton, C., 2014. How Evangelicals from Two Churches in the American Southwest Frame Their Relationship with the Environment. *Review of Religious Research*, 56 (3), 373–397.
Rolston III, H., 2006. Caring for Nature: What Science and Economics Can't Teach Us but Religion Can. *Environmental Values*, 15 (3), 307–313.
Slocum, R., 2004. Consumer Citizens and the Cities for Climate Protection Campaign. *Environment and Planning A*, 36 (5), 763–782.
Sunstein, C., and Thaler, R., 2003. Libertarian Paternalism Is Not an Oxymoron. *The University of Chicago Law Review*, 70 (4), 1159–1202.
Taylor, B., 2011. Editor's Introduction: Toward a Robust Scientific Investigation of the 'Religion' Variable in the Quest for Sustainability. *Journal of the Scientific Study of Religion, Nature and Culture*, 5 (3), 253–262.
Taylor, B., Van Wieren, G., and Zaleha, B., 2016. Lynn White Jr. and the Greening-of-Religion Hypothesis. *Conservation Biology*, 30 (5), 1000–1009.
Thaler, R., and Sunstein, C., 2008. *Nudge: Improving Decisions About Health, Wealth and Happiness*. New Haven: Yale University Press.
Townsend, P., 2014. How Many Presbyterians Does It Take to Change a Light Bulb? Confronting Global Climate Change in the Presbyterian Church, U.S.A. *In*: G. Veldman et al., eds. *How the World's Religions Are Responding to Climate Change. Social Scientific Investigations*. London: Routledge, 193–208.
Tucker, M., 2003. *Worldly Wonder: Religions Enter Their Ecological Phase*. Chicago: Open Court.
Vaidyanathan, B., Khalsa, S., and Eklund, E., 2018. Naturally Ambivalent: Religion's Role in Shaping Environmental Action. *Sociology of Religion*, 79 (4), 472–494.
White, L., 1967. The Historical Roots of Our Ecological Crisis. *Science*, 155, 1203–1207.
Wilkins, D., 2020 Catholic Clerical Responses to Climate Change and Pope Francis's Laudato Si'. *Environment and Planning E: Nature and Space*, 5 (1), 146–168.

Index

Note: Page numbers in **bold** indicate a table on the corresponding page.